COSMIC HERETICS

On the cover: Isodensitometer tracings of comet
Morehouse 1908 III,
in J. Rahe *et al, Atlas of Cometary Forms*
(Washington: NASA, 1962), 63-4.

Alfred de Grazia

COSMIC HERETICS

A Personal History of Attempts
to Establish and Resist Theories
of Quantavolution and Catastrophe
in the Natural and Human Sciences,
1963 to 1983

Metron Publications

ISBN: 978-1-60377-084-2
Library of Congress Catalogue Number: 2012921083
Copyright © Alfred de Grazia 1984, 2012
Second Edition
All rights reserved
Metron Publications, P.O. Box 1213
Princeton, N.J. 08540-1213 USA

This book is dedicated
to whoever figures in it,
whether or not by name.

The most elementary books of science betrayed the inadequacy of old implements of thought. Chapter after chapter closed with phrases such as one never met in older literature: "The cause of this phenomenon is not understood;" "science no longer ventures to explain causes;" "the first step towards a causal explanation still remains to be taken;" "opinions are very much divided;" "in spite of the contradictions involved;""science gets on only by adopting different theories, sometimes contradictory." Evidently the new American would need to think in contradictions, and instead of Kant's famous four antinomies, the new universe would know no law that could not be proved by its anti-law.

To educate — one's self to begin with — had been the effort of one's life for sixty years; and the difficulties of education had gone on doubling with the coal-output, until the prospect of waiting another ten years, in order to face a seventh doubling of complexities, allured one's imagination but slightly.

Henry Adams

From: The Education of Henry Adams: An Autobiography. Privately published in 1906, in 100 copies, and sent to interested persons for comment. General publication ensued in 1918. In 1975 republished by Berg: Dunwoody, Georgia.

TABLE OF CONTENTS

In Search of Times Past — 13

Part One

1. Royal Incest — 17
2. The Prodigal Archive — 22
3. Cheers and Hisses — 36
4. A Proper Respect for Authority — 64
5. The British Connection — 79

Part Two

6. Holocaust and Amnesia — 99
7. From Venus with Love — 115
8. Homo Schizo Meets God — 124

Part Three

9. New Fashions in Catastrophism — 148
10. ABC's of Astrophysics — 178
11. Clockwork — 200

THE COSMIC HERETICS

Part Four

12	The Third World of Science	213
13	The Empire Strikes Back	257
14	The Foibles of Heretics	292

Part Five

15	The Knowledge Industry	313
16	Precursors of Quantavolution	347
17	The Advancement of Science	371
Epilogue		399
Index		402

IN SEARCH OF TIMES PAST

I did not obtain Alfred de Grazia's materials for this book without remonstrance and persiflage. I had thought that he would be pleased to have someone writing about his activities, especially someone like myself who could be counted upon for sympathy. Rather not! He said that he was quite capable of writing his own autobiography, and indeed intended to do so, in several volumes, no less. Strange, for Immanuel Velikovsky had responded to me in the same way!

When I muttered something about reminiscence and the consolations of old age, he was primed for the retort, and I learned that Leonard Woolf had written *his* autobiography in his eighties, in five volumes, and Woolf was then old enough to be his father, and Bertrand Russell at the same age in three volumes. And I had better read them.

Furthermore, said he, I have a lot to recount, think of it, a boyhood spent sniffing the stench of the Chicago stockyards, shivering in the icy blasts off the prairies, a small critter's glance up the skirts of the Roaring Twenties. Then the University of Chicago in the heyday of Robert Maynard Hutchins. And more, seven campaigns of World War II, and still more, an island of the Aegean Sea, an experimental college in the Swiss Alps, intelligent women, singular, even beautiful, women, even beautiful men, for that matter. No, I can't let you take it away, there's too much to say.

Let me try, I said, there'll be no conflict of interest. I'll hew to the line of the Cosmic Heretics as they tried to break into the halls of science. It's got to be dull. It'll save you doing the chore. I can't take in your *enfants terribles* or your politiking, your love affairs or your friends who escaped your involvement in cosmic heresies. Or your poetry or attempts at educational revolution. No Naxos, not the beautiful ideas by half. No grueling trips, failures, pains, unless they're cosmical. No Vietnam, no University life.

Then Deg began to reproach me for taking a person's life out of its context, arguing that you have to talk about everything to say the truth about anything, whereupon I argued that no field of science could exist if most of everything weren't left out of the investigation of a single thing.

Well certainly, he granted, you'll have a better chance of excising the insignificant details of life. Yes, exactly, I said, but I thought there's the problem and the genius of biography, fixing upon the detail which may be the fulcrum of a change of life, precisely the sort of thing that is often lost in sociology and history.

Where will it start, where will it end, he wondered. I'll start, I said, at the time when you met Immanuel Velikovsky, the beginning of 1963, and carry it down to the publication of your *Quantavolution Series,* that is, the beginning of 1984. Not in chronological order of course. The story will lurch from side to side and pitch and roll.

Using your iconoclastic word "quantavolution" will help to define the *dramatis personae.* If a person's been observed by you amidst the mêlée provoked by the claim that nature and mankind have been fashioned by disaster, then that person belongs to the cast of characters.

Deg told me that the cosmic heretics were many, and their number would grow with the acceptance of the heresy. But, he warned me, if the heresy were to fail, I would be guilty of slandering decent citizens by inclusion. In either event, he said, history will be rewritten; it always is.

To whom will you dedicate your book, he asked, which was tantamount to giving his blessing to the project. To the Cosmic Heretics, naturally, I answered. O.K.; anyhow I've already taken care of Velikovsky with a dedication of my first book in the field. V. died four years ago, seventeen years after we met, and before we met had done almost all of his writing. For my own part, previously I had done a lot in political behavior and methodology, but nothing that might be called quantavolution. It was a sociological problem that brought us together in the first instance — the "reception system of science," I called it afterwards. Although I might have known better, I almost immediately entered into the substantive theory of catastrophe; I couldn't resist the challenge.

And I am just about finished now. (I grinned, and so did he.) I'm beginning to repeat myself, too, so it's not a bad time to end your book. By the way, have you read everything that I've ever written? Yes, of course. Just wondering, he mused, because V. tried never to talk to a person about his works who hadn't read the pertinent volumes. It makes sense and saved his time.

I don't feel strongly about it: my books are children who have gone off somewhere, on their own responsibility. I don't possess them, though I ask that they not be mistreated — the same as I would for other people's children. Who is entirely read, anyhow, he asked of me almost angrily, as if I had raised the subject.

I said I didn't know. Once I had met a psychologist who had read the 24 volumes of Freud's collected works. Still, commented Deg, some of his pieces escaped the Hogarth Press. William Yeats dedicated his autobiography "to those few people mainly personal friends who have read all that I have written," but probably noone qualified. It's good that nobody has read everything of anybody. It might abet the idea that where the pen stops the person vanishes. Rather, although the powers of expression tower above life, life rampages uncontrollably below.

PART ONE

CHAPTER ONE

ROYAL INCEST

Alfred de Grazia was entering his forty-fourth year when he met a self-styled cosmic heretic, Immanuel Velikovsky, who was already sixty-seven, and for the next twenty years a wide band of life's spectrum was colored by their relationship. As with a love affair, all that happened in the beginning presaged what would happen later, stretched out on the scale of time, themes doubling back upon themselves, attractions and reservations never to be erased, continuing accumulations.

The men changed, the world of science changed, too, and also the political world, yet this latter less; for, after all, one man died and the other grew old, whereas science and politics, those statistical behemoths of collective behavior, go on forever, compounded of many millions of individuals whose average age hardly varies, exhibiting trends whose progress, if it could be called such, is hardly discernible and might indeed have constituted a regression. At least so it seemed to these two men who were trying to affect the science and politics of their time.

Velikovsky died a heretic, with scattered generally unfavorable press, while his friend de Grazia moved on with a spirit that could be called existential, convinced as before that politics (and he insisted upon regarding science, too, as politics and often included politics in psychopathology) — that politics, although probably irredeemable, was the elemental hydrogen of human behavior, no matter how compounded into life styles.

As the winter days of 1962 became 1963 in Princeton, New Jersey, 08540 U.S.A., families and friends gathered into clusters like the last of the leaves, so that half-consciously and driven by eddies of customs and calendar, de Grazia saw more of his friends like Livio Catullus Stecchini and of his brother Sebastian. He did not know Velikovsky, and if he had been asked about him, he would have replied that he had never heard of him.

This may appear strange, considering that Deg was to be numbered, by whatever scales a social psychologist might invent to distinguish the "informed and involved" from the "ignorant and apathetic," as a high-scorer on information and involvement. He had enough children in the Princeton school system, a half-dozen, to catch the sound of names from all quarters. He spent part of each week in New York City and at Greenwich Village where, of all places, the name of Velikovsky might have been bruited about. He had since 1957 published and edited a magazine, the *American Behavioral Scientist,* which pretended to cover those matters that were or should be the concern of social scientists. He personally scanned a hundred-and-fifty magazines in the social sciences and current affairs each month. He had many students, several of them close friends. His parents and the families of two brothers were living most of the time at Princeton.

He was not socially pretentious, nor a prideful man, not a University snob, and had had to pawn his professional reputation several times on behalf of scholarly and political iconoclasm. Withal, when it came down to it, he claimed that he had never heard of a man about whom a million or more Americans could have delivered him a rancorous account. One feature that makes mass society a horror-show is the actual anonymity of the famous. (However, the mass scatoma of social realities may be a worse feature.)

This he confessed when Livio Stecchini, as they walked along Nassau Street on that cold day, brought up the matter, disjointedly, as happens with men walking down the street to no end, intellectuals with minds chock-full of oddly related and far-off affairs, old friends whose thoughts needed no introduction nor conclusion. Knowing the two men, 1 imagine that their conversation would have gone something like this:

"There is a man in Princeton with good material on the scientific establishment... Cosmogonist ... They suppressed his books."
"What do you mean, suppressed his books?"
"They smeared him."
"Like Reich? Like Semmelweis?"
"Yes."
"What does he do?"
"He lives here. He writes."
"About what?"
"Mythology, astronomy, the Bible, ancient catastrophes."
"What does he live on?"
"His books. They are very well sold."
"That's not our topic."
"No. The ABS could take up the sociological side. It's rich."

Deg was skeptical. Although his *American Behavioral Scientist* would stop at nothing, every scientist had his one or two little scandals of defamation, every professor his Dean's secret crime, his edgy paranoia, and you had to take his word for it. It was the same in politics, dirty tricks everywhere and defamation as a matter of course. As for the juggernaut of science, it rolled along smashing unconscionably the god's celebrants who crowded in upon it from all sides with fresh ideas and reputations.

"His materials are rich." Again that remark.
"Really?"
"I can introduce you. We can go to his house. He lives on Hartley Avenue."
"Down near the Lake."
"To take a look at his stuff."
"Maybe... What's his name?"
"Velikovsky"
"Never heard of him."

* * *

A few days later Stecchini received a phone call from Deg. Deg had been to dinner at Sebastian's home. There was the usual babble and movement afterwards. He circled around the front room with its piles of papers and open bookshelves, pausing at the one where books of high mobility and heterogeneity sunned themselves for a few days. He picked out a forcefully jacketed book, *Oedipus and Akhnaton,* the author: Velikovsky. First the large photograph of the author, then the flyleaf, then the table of contents, then the index — he is grasping now for the thesis: the ill-fated incestuous Oedipus was none other than the Egyptian monotheistic pharaoh Akhnaton — more riffling of pages — the small definite sparking of the book browser.
"What's this?" He poked the book at Sebastian. "Any good?"
Sebastian was non-committal; probably he had not read it.
"Mind if I borrow it?"

He began to read it that evening. It was "True Detective," connecting two eminent figures never before joined. He finished it the next day.

How did he find the time to read it so promptly? A man who attends to a wife, a passel of kids, a dog, a cat, a station wagon, a large house with many doors and windows to mind, fireplaces to dampen, a busy telephone, a fat folder marked "action now", with half a dozen jobs, including a professorship and an editorship, with a propensity to daydream, and in that American society which tries in a hundred ways to pry into one's time and makes life tough for readers, and needing seven hours of sleep — how does he read a book? They say, "When you want something done, go to a busy man." His urges are compelling.

This act of devouring the book was typical of Deg. He would seize things out of his life-stream like a bear grabbing fish and do something with them, a compulsion to undertake and a compulsion to complete, not unlike Velikovsky, and the tie between the two men had something to do with V's recognition of this similarity, and perhaps with his growing problem of completion after the compulsion to take on matters lingered; but both men too sometimes had to drop affairs that needed completion or stuck to them beyond their point of pay-off, beyond hope also, so I would not stress the trait, and I even think that it may be so common as to be undistinguished. Velikovsky had made wide turns in his life, too, architecture, medical practice, psychoanalysis, politics, and now all this catastrophism which had something of everything.

Outwardly, they differed most apparently, Deg of medium height and compact build, V. tall and spare, the one with a midwestern background and accent, the other with a heavy Russian accent, Jewish above all. To V. outrage was a simple, direct emotion; Deg had the youngness of Americans that comes from promiscuous outrage and wide dispersal of feelings inimical to authorities. Pablo Picasso used to tell Gertrude Stein: "They are not men; they are not women; they are Americans." So how could Deg become outraged at the enemies of V.? Living was parcelled among sporadic outrages; indignation cropped out all over the American landscape.

While I am at it, I might say something, too, about Deg's attitude to his own writing because this also explains how he might view V.'s troubles. It is also about Gertrude Stein: "In those days she never asked anyone what they thought of her work, but were they interested enough to read it. Now she says if they bring themselves to read it they will be interested."

Victim of the Rule of Three, Deg added a first phrase: at first he thought what he wrote was interesting and everyone should be required to read it. Then, after he had passed most of his life in Gertrude Stein's second stage, he postulated a final stage, a nirvana where what he wrote was objectively of interest but neither he nor anyone else should be interested to read it.

This is too early to be analyzing character, but I cannot refrain from another comparison, a fatal difference. Whatever V. completed, he fiercely possessed; whatever Deg completed he relinquished. This made their cash flows, you might say, very different. And their advice to each other very different. Deg was saying to V.: "Give it away. Let it go!" and V. to Deg, baffled; "Why didn't you hold on to that?" Moreover V. overvalued whatever he gave, and undervalued what he received.

* * *

Halfway through the book — before Akhnaton had espoused his own mother, Queen Tiy, Deg was committed to V., the author. A literary *tour de force* of the rarest kind, it succeeds in making a single person out of two of the most famous heroes of antiquity. Nor are they of the so numerous type of military heroes. They are the active substances of the raging intellect, flourishing amongst squirmy snakes of psychology and religion.

Should the temporal sequence be right, then the book would be valid, that Moses preceded Akhnaton and Akhnaton came before Oedipus. The legendary, historical, psychological and archaeological evidence marched in brilliant composition and concordance on behalf of V.'s thesis. That Moses had come first follows from V.'s book, *Ages in Chaos,* already a decade old, which was to be read and to convince Deg in a matter of weeks. That the Oedipus legend developed after the history of Akhnaton was established in the book itself to Deg's satisfaction, and he confirmed it once again when it came time to write *The Disastrous Love Affair of Moon and Mars,* years later.

By then he was convinced of V.'s theory that the Greek Dark Ages were in fact several centuries that had never existed, and then, within a couple of years, the masterful work of young Eddie Schorr effectively closed up the gap in two articles on Mycenae, Pylos, Troy, Gordion, and other sites. Velikovsky himself here speculated that Nikmed of Ugarit became Cadmus the founder of Thebes and carried the Oedipus legend from the East to the North. V.'s reconstructed chronology closed the centuries like a vise, to where Akhnaton could readily reach to Nikmed and Nikmed to Cadmus and out of it all came the Oedipus Rex of Thebes, the fabled character who gave name to the most popular concept of Sigmund Freud, and it was Freud who had brought on all of this work by his psychoanalytic disciple, but had himself missed both the precession of Moses and the identity of Oedipus as Akhnaton, although he had written directly about all three figures.

The book was the best produced of V.'s works, which were ordinarily drab. *Oedipus and Akhnaton* carried many fine illustrations, a superior jacket, and excellent typeface and good printing paper. Still, it did not sell as well as any of a dozen detective novels of the day, and, vibrant and valid, was marked by its publisher for abandonment in 1984.

Deg could be sure that practically none of his hundreds of friends and colleagues, students and acquaintances had yet read the book or would ever do so... But then he too had written books of which none but the textbooks had sold over a thousand copies. And he could recite the names of many distinguished scholars whose books had sold less. The dream of best-selling great books nevertheless carries on, a myth, deadly to most and profitable to a very few.

CHAPTER TWO

THE PRODIGAL ARCHIVE

The other book, that which won Velikovsky fame, income, and scientific disgrace, was a happy accident of publishing. It could hardly have become a best-seller on its merits; very few books do, and this one was not easy to read or flamboyant. *Worlds in Collision* was reluctantly published, deceptively publicized, and foolishly attacked. It was written in the 1940's, after *Ages in Chaos* had been completed and had been circulating among publishers and collecting one rejection after another. Evidently the later work had the better chance, because of its larger, more explosive message.

But *Worlds in Collision,* too, was rejected time after time, this all during a period of high prosperity when publishing company shares boomed on the stock market and practically anything might be brought out. Velikovsky was desperate. One evening he walked the Upper West Side of Manhattan with Elisheva, telling her of how he would buy a typesetting machine and they would compose the book at home and he would sell it himself. He would have done so.

All of his publications before then — there were not many — had been in some sense subsidized, the articles appearing in psychoanalytic journals, supported by small intellectual circles, the pamphlets appearing under the shadowy imprint of the Hebrew University of Jerusalem when this was only a few dedicated utopians enjoying an impetus from Simon Velikovsky's purse. V. knew something about publishing, as he did about many things.

V. would never have been "himself", a revered image to countless readers and a buffoon to scientists and scholars had he not fallen into the crazy typical pattern of a popular author. He was able to catch the attention of John J. O'Neill, Science Editor of the *New York Herald Tribune,* who was thrilled by the manuscript and wrote about it in an article of August 11, 1946. James Putnam, an Editor of Macmillan Company took it up, praised it among his acquaintances, processed it through several readers and achieved a favorable vote. A chapter of the book was sold to the *Reader's Digest* and other selections to *Collier's Magazine*. *Collier's,* struggling for circulation, took a large ad in the *Herald Tribune,* headlining that modern science had now proved the Bible correct, while the *Reader's Digest* carried the story of the Sun's standing still at Beth-Horon by the command of Joshua, so as to let the Israelites finish off their enemies.

Both stories and the publicity attendant upon them played directly to a large audience of bemused Jews and "Old Testament" Christians, including what would be called creationists and millennialists. Then, even before its readers could discover that it was not quite what they had expected, the wrath of scientists descended upon the book. Velikovsky's figure, until then only that of a minor personage in psychoanalytic reading circles, was elevated to a pyre of fame and burned to the ground. Macmillan hastily sold its rights to Doubleday Publishers.

Of all this that occurred between 1950 and 1962, Deg learned upon his first meetings with V. "I want you to read everything," he said and handed over to him two monumental manuscripts entitled *Stargazers and Gravediggers.* "Everything" meant also *Worlds in Collision* and *Ages in Chaos.* Deg complimented him upon the Oedipus book and wondered at the documentation piled upon the living floor for examination.

Velikovsky wondered, too, for noone came to him as innocently as his new acquaintance. He was thankful but also dismayed at this walking effect of the suppression of his books. (It hardly occurred to him that his book might have sold under a thousand copies if it had been published by a university press without the publicity that he himself found rather obnoxious, in which case practically everyone might have been expected to be ignorant of it, but the ilk of Deg might have known it).

V.'s correspondance was still heavy after a dozen years. His readers sent him every scrap of publicity that they found and he kept it all and tried to reply, far more so than any other author of Deg's acquaintance. A large public was out there somewhere, a heterogeneous network of bright students, people suspicious of the scientific and academic establishments, and Bible believers.

Mrs. V. was present; she tried always to be on hand when visitors came and to Deg at least, hers was always a welcome presence. V. kept nothing from Elisheva that he was not also keeping from his visitors. Sheva's grand piano stood in the next room, between a desk loaded with papers and a great cabinet stuffed with books. In the front room were couches and chairs, none too comfortable, and a large coffee table accommodating the tea, crackers and cheese, cakes and dry Israeli white wine that would be brought forth. There were ashtrays, too, for then many were smokers, not V., for he had quit years before after he had suffered a stomach cancer, whose removal had forced a lightened diet as well. Oriental rugs stretched across the floors.

The ponderous front porch let in little light nor did the rooms have much place for an elegant style; or perhaps they reflected an empiricist, not a philosopher. Their charm depended upon the objects in themselves: Sheva's piano and the music resting on it, her strong marble sculptures, several handsome and less useful books on art and archaeology that had entered lately, like those at Sebastian's from which Deg had plucked *Oedipus and Akhnaton.*

From the porch, one penetrated into the sitting room through heavy grey stone walls in five stages: first up the flagstone walk through thick bushes, then up the stairs, then through the first heavy door into a tiny hall, then another heavy door, then an anteroom with a mail-cluttered table and clothes-closet, and finally into the front room.

Elisheva, like her husband, had a strong character and great energy. She had large hands and a solid body, maintained a direct and friendly stare through thick glasses, and was perhaps of his age. She had mastered the arts of music and sculpture. Perhaps all the laborious functionalism of its occupants gave the rooms a lack-luster belying the considerable value of their contents. Poor cooks have dazzling automated kitchens; disemployed people have smart interiors. Much later on, when he finally released his books to Dell Publishers for publication in paperback and received a hundred thousand dollars, V. went into a fit of remodelling, building a garage and new airy light-struck rooms, redistributing books and papers for greater efficiency, buying flashy cars for himself and his grandchildren, reminding Deg of Parkinson's "Law," that, as an Empire enters upon its finale, it builds extravagantly.

Deg had often to consider, when he taught courses on leadership and creativity, whether a person's appearance correlated with his mind and effectiveness. The stereotype is, of course, "Yes, it does." A great general has a martial air, a scholar looks like a parsnip, an athlete is musclebound, and so on.

Deg had arrived at the all-answering concept of sociology — the mutual interaction of physique and role. Little Napoleon looked more imperial than tall de Gaulle, who was obstinate and oafish. But de Gaulle thought of himself as a Great Leader and worthy husband to La Belle France, and played the part and *became* a great leader. ("France is a widow," Pompidou orated when he died.)

"The Russian Jews are the handsomest of all," Stephanie Neuman told Deg, and he, looking at her, had of course to agree. The best explanation of the phenomenon comes in a note by V. himself, published posthumously. The "lost Tribes of Israel" had been moved North, and passed through the Caucasus between the Black and Caspian Seas into the lower Volga River Basin. There they mingled genetically with the ever-changing population, with always at least a critical fraction maintaining the Judaic culture-core. Deg had won a piece of the action; his wife's family, with its cluster of Teutonic cognomens — Oppenheims, Lauterbachs, Weinsteins,

Fleishackers, etc — had managed some handsome blonde alternatives in the aftermath of the Diaspora.

"But see here..." to use a common interjection of V., Velikovsky stretched his large spare frame a full two meters, his face with all its big bones and high forehead was cleanshaven and forceful, his large brown eyes open and direct behind his reading glasses, his movements from his favorite low chair, up and down, across the room, untiring and easy, not graceful but neither awkward. His voice was sure, slow, deep, his words marvellously well-chosen, uttered in the language that he knew least well of Russian, Hebrew, and German, while Arabic and French came after. He couldn't match Stecchini, who had these, plus Italian, Latin, Greek and Arabic, plus the dead languages of Babylon and Egypt, while Deg with his modest portions of French and Italian and smattering of German, Latin, and Spanish was in a pitiable state.

V's English was formal, never Americanized; his dignity forbade slang or the vernacular, though it amused him to have the vernacular explained. Deg was fond of H.L. Mencken and played loose with the language when let off the field of science. "Sand-bag them," he remarked when V. was expostulating over the attempts of a panel of the American Association for the Advancement of Science to get hold of his finalized paper without revealing to him their final replies to it. "What does 'sand-bag' mean?" V. asked. "It's what thugs use to hit people with from behind. Let them have the paper; let them rewrite their papers; then withdraw your paper." Then he explained how in some impolite poker games, if you have a good hand, you sometimes pass on it, enticing the other players to bet on their own hands, then double their bets. That's sand-bagging, too.

V. wrote well, better than Deg, I think, although he denied it and had to make liberal use of copy-editors. For he explained his every step carefully and was rarely abstract or harsh, whereas Deg usually wrote condensedly, abstractly, and stridently.

Looking at V. in these first meetings in a more analytic way, Deg questioned whether a person so physically modelled to the ideal expectation of a heroic figure could nevertheless be a genius and not an actor, an honest victim and not a charlatan. Of what could V. complain; he was famous; his books sold by the tens of thousands; his messages had carried throughout the English-speaking world, into several language-areas of the western world besides.

Deg flipped through the loose-leaf volumes as they talked. He could read fast and V. was alternately suspicious and admiring of this facility. "I am a slow reader," he announced on occasion. "Yes, but I don't have your memory," grumbled Deg. V. had a superb memory for details. Deg gulped down batches of material, retained their forms, and excreted the details. This is what happened when he read: the stuff was gobbled up by pre-existing forms.

Every detail of the volumes before them was remembered by V. though he could hardly have seen most of it for some years. Every few pages contained another foolish review, comment or letter by a scientist or historian or archaeologist. Just to be preserved and collected, side by side, they damned themselves and each other as envious, illogical, irrelevant, ignorant, narrow, and incompetent.

Why haven't you published this, it's great? he asked V. V. had strung together a large and complicated story with only rare descriptions and without editorial comment; it was not vainglorious or egocentric; the documents marched along by themselves, calling out their message in turn. V. blew hot and cold on the idea of their publication. Mainly he feared legal action were he to reprint letters several of which had come to him deviously. Of these Deg could not feel sure, but he argued that persons in a public controversy in which their reputations were at stake might publish private correspondence. A menacing letter from Professor Fred Whipple to the Macmillan Company might be published because it injured and defamed the author and was associated with letters of the same type from other academicians. His publishers, Doubleday, were unsure, said V.

In fact the volumes were not published until after his death. By then the whole Macmillan archive of those years had been given to the New York Public Library and Warner Sizemore, who knew the case as well as anyone alive, located them there, with all the papers that had been so guarded for a few years. When Leroy Ellenberger reviewed them in 1983, he noted especially Brett's account of the final interview with Velikovsky when the President of Macmillan informed Velikovsky that *Worlds in Collision* could no longer be tolerated on the Macmillan list, but had to be transferred out, and luckily Doubleday was ready to assume the risk. When asked how the two versions of the meeting compared, Velikovsky's and Brett's, Ellenberger, who was by then most sensitive to contradictions in the Velikovsky story, granted that substantially they agreed, save that V. had understandably portrayed himself as less shaken and more in command of the situation than Brett had viewed him to be.

The materials that V. showed Deg were a sociologist's wishful dream. Deg decided immediately to publish in the *American Behavioral Scientist* the story of science *vs.* scientism, as he put it. He carried home the manuscripts and *Worlds in Collision,* which Velikovsky carefully autographed, a little touch that Deg was unused to: books were books: he was never into first editions or autographed copies, and in those days had to be reminded by his publishers that a page was reserved for a dedication if he wished to use it.

The journalistic papers he hurried through and put aside. They would give an example here and another there. Some readers no doubt would be astonished at the behavior of their sacred scientists, but the case was mere basic social psychology. The scientists and their coterie of publicists were behaving very much as might be expected in the face of disturbing theories, like politicians, like administrators, bishops, and all other elites of organized networks.

He decided to take upon himself the most difficult task, the theoretical analysis of the system that exuded injustice normally. The historical section would go to Stecchini and deal with scientific precedents to V.'s catastrophism, an approach quite new to the discussions of a decade earlier, and one which Stecchini, using the principle of contradictions, executed beautifully, calling up Whiston, Boulanger, La Place and Kugler as unexpected witnesses on behalf of the defendant. The straight history of the affair went to Ralph Juergens, who had been introduced to Deg by V. as a mechanical engineer, much interested in electrical theory, who had moved his family down from Ohio in order to be near to where V. was working; he was now a scientific editor working in New York for McGraw Hill.

Juergens had published nothing; he knew the facts, however; he was a careful worker, Deg was quick to note; he worked very hard; he held V.'s confidence (not easy to achieve) and won Deg's sympathy and respect. Noone else could have done the job without a year's study; even then it would have had to be a historian of science, who would risk his career if he accepted the challenge of the facts, or a publicist, such as Eric Larrabee, who would have produced a recital much like Ralph's but probably too late for publication. As a matter of fact, his name came up and V. reported that he had been under contract for years with Doubleday to do a book on the controversy. No sooner had Deg's *ABS* decided to publish the story than V. got in touch with Larrabee and prevailed upon him to sell the idea of an article to *Harper's Magazine,* which Larrabee did, by virtue of an old connection there, and so wrote a piece that actually appeared several weeks before the special issue of the *ABS.*

After examining the files on the case, Deg turned to reading *Worlds in Collision,* telling himself that it might be wrong, harmful, mythical, distorted, and incompetent; still his intuition was prompted by all that he had learned thus far: V. could not do a bad job on anything. So he found the book was none of these things, and was not surprised. Then he worried and never ceased to worry that his taking up the cause of V. came about because he thought V. to be correct in his theories rather than because his rights were violated.

***

Worlds in Collision is a book in two parts, one on the Venus catastrophes, the second on the Mars catastrophes. These conform to two sets of events that are claimed to have befallen the world in the years around 1450 and 700 B.C., about seven hundred years apart. The planet Venus, argued Velikovsky, began its career as a comet that probably exploded from the giant planet Jupiter sometime, whether a few years or thousands of years before its disastrous encounters with Earth. (V. never used B.C. preferring BCE, "Before the Common Era" or a simple negative [as -1450], begrudging the calendar of world history to the Christians, which Deg agreed to in principle but thought was only quibbling, given the huge contortions history has suffered. Better he thought to settle on the year 2000 as the present, use B.P. back from this date, thus to give us some standardization for a generation or so, or perhaps to settle upon 1919, the year when the first association of the nations of all the world was formed, the League of Nations).

Flaming Venus passed with its huge cometary tail close by the Earth occasioning general disaster by flood, fire, pestilence, electric shock, and fallouts of various materials, and incited a horrendus fear that affected all areas of culture everywhere down to the present day. Mankind lived virtually in a Venusian world for seven centuries, for other near passes occurred at 52 year intervals, until the comet disturbed Mars, sent Mars to molest the Earth and Moon, and brought a Martian period that endured for rather less than a century. All of this had severe and prolonged after-affects geologically, biologically, and culturally.

V. endeavored to be exact, allowing the series of Mars incidents to occur between the years -776 and -687 on the basis of legends and historical-archaeological evidence from around the Mediterranean and wherever else in the world it cropped up. For example, an incident of the year -776 would be the founding of the Olympic Games, those sacred manifestations of aggressive competitive sport that brought the Greek communities together and were said to have been founded by Hercules, who has been identified by several scholars with the god Mars or Ares; an instance of the year -687 would be the destruction by natural disaster of the army of the Assyrian emperor Sennacherib while besieging Jerusalem.

Thus the bare plot. Its importance derives from the shock it gave to conventional natural science and history, its extension of the use of legendary materials to reconstruct history, and the excitement it caused among many people eager to escape the toils of modern science.

The most disturbing claim of *Worlds in Collision* was that the planet Venus as a comet approached and devastated Earth. Several excellent writers, as I shall explain later, had claimed that comets had devastated the Earth, and mathematical exercises on the putative effects of comets in passages and collisions with Earth are conventionally acceptable. Not so planets, that are believed to be fully and nicely bound to their present orbits.

The sequence of thoughts occurred to V.: first, the Egyptian accepted chronology is wrong and Moses preceded Akhnaton; next, at the time of Exodus, there was heavy natural turbulence; third, the turbulence was incited from the skies, and took numerous forms well recounted in legend and sacred scriptures; finally, evidence came in rapidly from all parts of the world to support the idea that the planet Venus was involved as prime cause. A mosaic of legends from the Near East, Greece, Italy, China, and the Americas could be fashioned, and enough geological evidence would be assembled to tolerate the suppositions of the legends.

V. was not as rooted in Newtonian and Darwinian prejudices as the typical Anglo-American scholar. He could also contemplate ancient evidence without contempt. (A psychiatrist might recall, "Ah, yes, he loved and respected his father Simon who worked long for the revival of Israel.") V. also knew that natural laws must rest upon evidence, not dogma; if evidence contradicts the laws, the laws must change. The immensity of the topic; the difficulties in finding and handling the data; the roundabout way in which the books were published; and many other intervening and confusing variables concealed the essentially proper progression of V. 's mind, which behaved in ways both psychologically understandable and logically proper. (Often, private motives lead men scientifically astray; here, as sometimes happens, V.'s private motives led him along the path to significant scientific theses and discoveries.)

To Deg's view, from the beginning, the ethical duty of science was clear. Confronted with V.'s claims, the scientist should weigh the evidence, first, for the chronology, second for the Exodus disasters, third for the exoterrestrial involvement, and finally for the identity of the forces. In each case, there is, then, a probability, low or high, of validity. Actually the only policy problem for science here is how much additional scientific energies should be directed at the intriguing hypotheses. This implies the possibility of proving (disproving) them; and the effort required to raise the probabilities of valid answers to a respectable level.

In American politics and law, case after case had imprinted upon all concerned the notion of a right to due process of law and to certain basic freedoms as distinct from the desirability or correctness of a position. There is a religious right, when forbidden by one's religion, to not salute the national flag; there is a right to not confess to a criminal act. And so on.

Scientific behavior is not so clearly mannered. It is not governed by the coercive physical force that gives more distinct form to the organs of the state. Also a general belief in individualism among scientists, amounting to a kind of philosophical anarchism, makes each scientist judge and executor of his beliefs.

Deg was enough of a philosopher and practitioner of science to recognize a widespread belief, that a truth exists upon a subject and that no consideration needs be given untruth or antitruth. There was, on the other hand, the reputable principle that all scientific positions are basically hypothetical; nothing is proven now and forever. And there was even the principle, espoused by many contemporaries, that there are as many scientific truths as may be useful in solving a practical problem: in other words, never mind the principle: perform the operation and the principle, if the operation is successful, will come trailing after.

But the vulgar and predominant belief is a belief in truth and antitruth, especially when dealing with outsiders, and V., by this view, deserved no more than he received, there being numbers of established truths violated by his assertions. He should have banked his receipts and joined the outcaste company of the von Danikens.

However, according to the other views, all of which merge in this regard, nothing that V. could possibly say should deprive him of a hearing, save that he should present his views in a format suitable for passing judgement. Deg had to make up his mind whether the basic offering was appropriate for judgement and whether a hearing was provided. Still he could not but feel that the organization of science would fall apart if no advantage were given to the accepted "truth" just as the state would become defenseless if everyone refused to serve in the armed forces on constitutional grounds.

What happens ordinarily, he observed often, is that the more "obviously untrue" a proposition and its proof appear, the less due process of law is used and need be used in dealing with it. We have to reconcile ourselves to the "miscarriage of justice", at least in science and probably in every area of conflict, the "Bill of Rights" notwithstanding. If for no other reason, the burden of treating every statement with all the respect due and owing to the best and most correct-seeming statements would be impossible for the economy of science to bear.

In return, Deg told himself, we can ask for some minimal formatting of a case prior to processing it through the reception system of science. This, it appeared to him, V. had done, and much more, and some scientists had nevertheless pilloried him and ruined his chances of obtaining scientific respectability — not affirmative agreement, but just simple honest respect for a remarkable job.

V. had approached the altars of science with the assiduous ritual of Aaron before the Holies of Holies. And, when, like the drunken sons of Aaron, his books were struck by the Lord's Fire, he was stunned. "What sacrilege have I committed?" he asked himself repeatedly. And the answer, from all sides, if not from heaven, was "None." It is true that he had won literary fame and supported his family meanwhile, a rare success among non-academic writers in America. So what? Have the rich no right to complain? Who else can send the steak back to the kitchen?

The scene was familiar and the opportunity presented: the establishments of academia had offended a man who was a fighter and had his evidence in hand. Something rare and good in the history of science might be achieved. With the contaminants of politics and religion absent from the mixture, and the publishers acting as catalysts, it was as clean a case of pure science in action as one might ever hope to come upon.

***

The work on the special Velikovsky issue of the *American Behavioral Scientist* had been mostly done when Deg addressed a letter to his Advisory Board explaining Velikovsky's position and justifying a special issue in support of him.

March 8, 1963

To: ABS Advisory Board
Subject: Notes on several current matters

I. We plan to devote a major portion of our June issue [actually it came out in September] to a topic called: "The Politics of Science: The Velikovsky Case." Immanuel Velikovsky, as you probably know, is a highly controversial figure whose book *Worlds in Collision* incited the wrath of a number of astronomers and geologists twelve years ago. Several other works dealt with similar themes of prehistoric catastrophe, social upheavals, and the origins of myth. Another book, somewhat distinct, is *Oedipus and Aknahton*. I believe him to be a brilliant theorist and am not persuaded that his criticisms of various astronomical principles are as wrong as Shapley and others have made them out to be. The recent Venus probe has brought some surprising information in accord with his views, for example. However, our main interest in the topic lies in its relation to numbers 3, 7, 8, 9, 12, 13, 14, and 16 of the ABS Program. A basic question is *the canons which science uses to appraise work that is offered.* As we move into the Velikovsky case, we observe that both the normal and the peculiar features of the criticism of this work throw much light on the workings of the scientific establishment. Additionally the evidence of boycott of a publisher in the case leads one into the question of the relation of scientists to freedom of the press. The proposed table of contents would include first a history of the Velikovsky case, a comparison of the case with various episodes in the history of science by Stecchini, a content analysis of the reviews of Velikovsky's book, an article by Velikovsky reciting ten important instances in which his theorizing led him to correct or at least now respectable statements about natural events (this one to give a flavor of the substance of the case), and an appraisal of the operations of the scientific establishment. We have abundant material. We lack funds, as usual, for the kind of content analysis and investigation that should be engaged in. If any of you can find a few dollars to lend to this enterprise, it will be helpful in improving the product (especially in the reliability of coding the book reviews, and increasing the number sampled from 100 up to 500)....

The "good will and advice" was there; as for the money, the Board knew Deg was bluffing: the magazine would continue, one way or another.

Also, to attack frontally an array of scientists, Deg thought to assemble a special committee of notables that would protect his flanks. He sent the manuscript of the *ABS* issue to his friends Harold D. Lasswell, Hadley Cantril, and Luther Evans, all three well-known, distinguished and innovative social scientists. He also contacted, at Velikovsky's suggestion, Salvador de Madariaga, Moses Hadas, Horace Kallen, Harold Latham, R.H. Hillenkoetter, and Philip Wittenberg. Madariaga and Hillenkoetter admired V's work; Hadas respected the learning evidenced in it; Kallen was a grand liberal educator who had run interference for V. when V. was trying to obtain a reading from Harlow Shapley; Latham had shepherded *Worlds in Collision* through Macmillan; and Wittenberg was an expert on libel law. Deg also invited Harry H. Hess, Chairman of the Geology Department at Princeton, who had given V. a forum, and was helpful on several later occasions; V. counted him as *a* friend; Deg had met him and found him *simpatico* and every inch what an Admiral in the U.S. Navy (Reserve) should be. He was a top leader in the wartime and post-war revolution in oceanography. Hess replied by hand:

June 4, 1963 Washington, DC.

Dear Editor de Grazia:
    The manuscripts you sent me reached me at a particularly bad time: PhD exams, department budget construction, a request to appear before a committee of Congress and finally orders to two weeks of active duty in the Navy starting yesterday. I have spent two days reading the material and trying to analyze my own thoughts.
    I can't urge you to publish it. Velikovsky is a friend of mine. You will reopen old wounds and create more antagonism against him, though at the same time you will support his position and bring out the injustices. I am not sure that this is a net gain.
    Why were scientists outraged by Velikovsky's books? This is the question I have been asking myself because I too felt a sense of outrage even though I have a kindly feeling towards him as a friend. The reasons given by Stecchini are plausible and perhaps true with respect to some scientists. The real reason is something much more fundamental — at least the reason why I rebel is, and I am a fairly good guinea pig example of an ordinary scientist.
    I haven't time to write the essay that might be written to explain the phenomenon correctly. Velikovsky is partly to blame because of the way he handles his data. This is no excuse for most of those who criticize him. Nor is it an excuse for the manner in which they have treated him.
    Thank you for sending me the manuscripts. I wish I could do more for you than I have.

    Sincerely,
    H. H. Hess

Deg was not surprised nor did he feel Hess's refusal at all unworthy.

Hess was not the Admiral Nelson to violate Admiralty orders and take his fleet into battle; still, as Deg remarked to me, we already had an admiral (referring to Admiral Hillenkoetter), we certainly could have used a geologist on the team. Years later, Deg was able to persuade Hess to join the Board of Trustees of a foundation for studies of catastrophe.

A problem of concern to me was that, in the years following, there was no evident opposition to V., whether as to his treatment or his ideas, carried in the *ABS* files and the later book, *The Velikovsky Affair,* and I badgered Deg on this point repeatedly. He puts up a kind of general defense that has some merit: "Under the circumstances, we did what we could to excite an opposition. We had no money to conduct research. Everyone was unpaid and working at other things for a living. The issue on V. was itself only one of ten issues to appear that year, each on different topics. Mainly the expressions of disagreement were directed at the substance of V.'s theories, which were, strictly speaking, irrelevant to the discussion. Juergens went farther in explaining these and defending them than I would have gone. It was like pulling teeth to get a scientist to enter upon the politics and sociology or even the methodology of the case. One received simply arguments on the stability of the solar system and the unreliability of legends and ancient history."

Deg talked on, as the tape spun on its roll:

> I wrote Otto Neugebauer, a hostile critic of V. and renowned expert on Babylonian astronomy, but he did not reply for a long time, for years, in fact. I met with Harold Lasswell, who was a psychologist, political scientist and professor of Law at Yale; he was favorable to the issue, which he read, but concerned that the bridge he perceived as abuilding between the natural and human scientists might be damaged. (This was then the well-publicized thesis of C.P. Snow, physicist and novelist, who decried the existence of these two uncommunicative worlds.) I visited Freeman Dyson, the mathematician, who was at the Institute for Advanced Studies and had been President of the Federation of American Scientists, of which I was a member, and which was agitating against the "Cold War." Dyson was lukewarm about the matter; he had been approached by V. some time before, and had no desire to enter the lists; furthermore he found the scenario of V.'s work unacceptable. There was noone, it seemed, on the first call for debate, and very few ever, who were ready to defend what had happened, as there was noone ready to defend V.'s substantive views on exoterrestrially-produced disasters. Worse, there was hardly a notable scientist of the Establishment of physics, geology, astronomy who was willing publicly to acknowledge the legitimacy of the discussion. I approached Tom Kuhn, a neighbor, who was beginning to win fame as a historian of science. He shied away.
>
> I will say more. You have been presenting my analogy of this case with cases in the law and courts. Actually, this is only one side of the coin. Just

as the law and courts are utterly inadequate to their tasks when a society is failing, so too in science the reception system is inadequate when the institutions and politics of science are failing to begin with. That is, unless you have a liberal, open-minded republic of science, you'll have too many cases of injustice in the reception system. I spent some time developing the problem of the institutions that are needed in science as in politics to back up a proper reception system, but noone of competence has come around to discuss this subject, which is as critical today as it was then. Criminality in science, if I may use the word, or misbehavior, is common throughout the sciences and ultimately its origins dissolve into the background of an illiberal, non-pragmatic, materialistically competitive, and philosophically ignorant environment where scientists are bred.

I felt that Deg's tone was becoming strident. I still doubted that he had exhausted the possibilities of a debate, and later on I will tell of other forensic episodes. He might have talked to Dr. Norman Newell, of the New York Natural History Museum; Ted McNulty, one of his aides and squash-playing friends had learned that Newell had something to say; he might at least have tried to speak to the king-pin Harlow Shapley, who was old but still feisty; he might have approached George Brett, President of Macmillan, to corroborate that he had "dumped" V. and explain why. Further, Deg might well have been more rigid, and might have excluded all substantive comment of V's theories, admittedly to the point of losing some of the excitement of his story. It is true, however, that copies of the issue were sent to potential opponents among natural scientists, inviting and expecting comment. There were none. Nor did the thousands of normal readers produce from among their number calls or letters of protest.

Nor, with one or two exceptions, did any evidence appear for decades that would affect the statements made on the affair by the three authors. In May of 1983, Leroy Ellenberger told me that he had found at least one bit of evidence in the Macmillan files giving scientists reason to attack Macmillan for advertising the book as a work in science. A regular catalogue of Macmillan books in science carried *Worlds in Collision* as a possible supplementary reading in general courses. This was a trifle, to be sure, but a red cloth is no trifle to a goaded bull.

Still the annoying question once more arises: why should not the book have been advertised as a contribution to science, even if it were ultimately to go into oblivion with most other books that tried to make contributions to science? So again I prodded Deg on the matter and this time got what amounted to a lecture.

> Formal law has the strongest means to avoid consideration of the merits of a case in judging whether the case properly belongs in a certain court and has been properly heard in that court. It insists that the accused be given his day in court, with a defense lawyer, an unprejudiced jury in most cases, and a

full account of the testimony against him and the right to confront his accusers. Formal law of course often falls short of its expectations.

Formal science has roughly similar rules for judging every work coming before it. The book is the defendant, you might say. It should be penalized, that is, dismissed, reproached, vilified, sentenced to non-reading and non-propagation only after it has had its day in court. And, it should come up for a parole hearing almost on demand. This, too, often does not happen. Anybody but V. would have taken his lumps — I would — and cry all the way to the bank.

When the law or science does not live up to its rules, then one appeals to a higher court or authority that created the institution in the first place. In the matter of a book, intelligent readers form themselves into a kind of court of consensus on the matter. That is actually what happened in the Velikovsky Affair, but still the court refused to remand the case for trial to the numerous special fields. The closest thing to this was the AAAS Panel a decade after my book and two decades after the events.

Now when the court or scientific establishment finds the defendant 'crazy' or 'delinquent' or 'fraudulent' or 'concealing the truth' or 'non-cooperative,' but there is still evidence that the court or science is wrong, then the higher court — that is, those institutions sponsoring the establishment, including the reading public, may call the lower court to order, reprimand it, force the remand for a re-hearing, or transfer the case to another jurisdiction.

In order to face down the court or science, the higher court or critics must look as far as necessary into the facts of the case to determine whether the defendant is indeed frivolous, delinquent, fraudulent, concealing the truth or non-cooperative. For these purposes, some degree of substantive worthiness of the defendant must be present to justify the intervention. This was indeed the situation here; the content and presentation of his theories were therefore legitimately at issue and part of the presentation of his full legal case. We therefore had to judge the defendant in a sense on his merits and let him speak briefly on his own behalf.

Scientists are understandably annoyed by ungovernable antics and criticism, noone more than us political scientists, who must suffer the most abusive, crazy and unscientific ideas and behavior every day in the newspapers, in legislative halls, and in political meetings, indeed wherever politics and public opinion generate, even at the dinner-table. They still must operate a clean shop, a decent court, which in the end serves best themselves...

He had more to say, but this is more than enough for now.

CHAPTER THREE

CHEERS AND HISSES

Deg found himself losing status in the eyes of his children, who had through their earlier years seen and heard much of important personages, partly because all of them went through rebellious adolescence during years when he was respectful, helpful, and obviously orienting his thoughts toward V., so that they found a weakness in their father — his rare complaisance — and could, through being critical and slightly disdainful of V., get at him twice, directly in himself and indirectly through rejecting V. It was not, as it had been put from time to time at home, that he gave too much of his crowded time to his venerable friend; indeed, the children could have done well in their troubled group life at school by carrying the banner of Velikowsky (and their father) for V. could easily be fitted (noone knowing his character) into the mold of anti-authoritarian ideas and leadership exceedingly popular among those in that era, town, and age-group.

On a summer day in 1963 Deg ushered his family of eight persons aboard the U.S. ocean liner "Atlantic" bound for Lisbon, Naples and Genoa. The boat was a slow last effort of the collapsing merchant marine but, he thought, just as several years earlier they had crossed the American continent on a railroad train from California to Chicago, they ought to have the experience of an ocean voyage. He then returned to Princeton and moved the family's possessions and his office from Queenston Place to Linden Lane, from a large old house to a small old house, aided by daughter Jessica's lovesick young boyfriend. His magazine was left in the custody of Ted Gurr. Then he flew to Lisbon, joined his family on the boat, and all sailed for Italy.

Deg made final corrections to the *ABS* Velikovsky issue at Marjorie Ferguson's villa in Marina di Massa, fuming at his four boys on the beach across the street who, instead of swimming out to sea like little Shelleys, had transferred with insouciance from the pinball machines of Princeton to soccer machines in Italy. "Dear Ted," he wrote,

> You will be pleased to note that I have incorporated most of the suggested changes... I could not accept the idea that the political network paragraphs were irrelevant and unnecessary. [This referred to intimations that the furious attacks against Velikovsky were prompted in part by frustrations of Shapley and other scientists at being attacked for "red" affiliations by Joe McCarthy and his ilk during these years.]
> I felt forced to deal with them and did all I could to make them objective. What is 'innuendo,' after all, is a question of motive. There is no innuendo here therefore. If a trace of poison is found in a deceased's blood, do you ban its reporting on grounds that it constitutes an innuendo? Every generalization of science implies a stereotype, to take another case. Must we then never generalize? [Later, Norman Storer and others picked up the theme, which social psychologists might best appreciate, most historians of science being too narrowly educated for such subtleties, or too constrained to deal with them.] By the way, Lucca Cavazzo [an Italian supporter of the ABS and wife had a baby. He was dining with me just before it happened. He calls his Federico Julio, two emperors yet! [Ted had begun his family.]

Now the special issue of September 1963 appeared and before long was reprinted. The response was strong, but within the ABS orbit was almost entirely of social scientists and humanists. Prompted by free copies and alerted by word of mouth, natural scientists nevertheless played deaf and dumb, and so did those dependent upon them directly.

In the files of Deg no new voice from a natural scientist comes forth amidst the many letters of a type to warm the cockles of an editor's heart. The scientists simply stooped low to avoid the flying bullets and returned the silent message, "Science is truth; truth is one; who defies the truth is no scientist; whatever happens to him he deserves." A few ducked because they had no recourse and feared the collective or public opinion of science, perhaps retaliation. It was a small step, which the sociologically untrained scientific mind can easily take, from witnessing a fellow supporting the case of Velikovsky to disdaining him erroneously for supporting his theories. Some would have been just normally lazy. Dr. Robert Jastrow, Director of the Institute for Space Studies, wrote Deg on October 20, 1980: "I had, of course, read your earlier very fine pieces on Velikovsky and his theories and had drawn on them in preparing my own article." But maybe this was later.

The *New York Times* ignored the *American Behavioral Scientist* and did not review the book when it later appeared. A brave letter came from an editor of the *Christian Science Monitor* (This newspaper, you may appreciate, is one of the world's finest, and has a disproportionate scientific audience.) "May I say," wrote G. Wiley Mitchell to Deg, on December 12, 1966, "that I have read your book through, consider it a real contribution and am very regretful that neither my efforts, nor those of some of my colleagues who agree with me, have been successful in getting my paper to publish a review. The Velikovsky smearers have been effective! (Mind you, I am not at all sure I endorse his theories in toto. But I think his method is sound and his theories are certainly no weaker than others that gain a hearing simply because they come with the right 'credentials.')"

An attorney at NASA (and I must point out that he was Dan, the son of David Arons, a Gimbel Bros, executive and an acquaintance of V.) wrote happily to his father that he had "received a call from Dr. Newell [head of the National Aeronautics and Space Administration] this morning bright and early who told him that

> .. he had read the articles in the American Behavioral Scientist which I sent him and was 'aghast at the inquisition' to which the Velikovsky books have been submitted.
>
> He said he had noted some of the comments made back in the '50's but these articles place them all in a pattern. He particularly noted a remark of Fred Whipple to the effect that scientists ought to send back the postage paid postcards to publishers who use them to advertise such books as Velikovsky's. Dr. Newell thought this was very 'vindictive' and 'uncalled-for.' While Velikovsky 'might be wrong' he is entitled to 'dispassionate review and criticism.'
>
> Dr. Newell said that he had already discussed this matter with some of the 'leading lights' at NASA including Arnold Frutkin, Director of International Programs. He requested that he be permitted to keep the copy he has and be provided with additional copies.
>
> I wouldn't be surprised if someone here makes a statement on Velikovsky in the near future...

But of course, there were no actions taken. Involve NASA in such a demonstration? Impossible!

There was another case, which V. pinned his hopes upon for a time, pathetically, a President of the grand University of Southern California, Murphy by name, who had indirectly voiced sympathy for the Velikovsky problem and V. had barged in to suggest that he appoint a commission of inquiry. The response: polite, and routinely cordial; but no interest, the matter being out of bounds. No University was going to dirty its hands with the nitty-gritty of scientific conflicts. If V. had been more of a sociologist, he could draw the appropriate parallels with the Catholic Church at the time of Galileo, reluctantly drawn in to support his enemies, a case V. knew well — up to a point.

There came Peter Tompkins to Princeton and Jill and Deg had him to lunch, along with their neighbor, Thomas Kuhn. Peter had published the story of his wartime escapade in German-occupied Rome a feat which Deg, a few miles away at the time, thought to do but had not done, and Peter had written *The Eunuch and the Virgin,* which Stecchini had shown to V. and which he had rejected, even though Tompkins could throw light on two points of importance: the sexual derivations from cosmic disaster (which V. had recognized) and the descent of great bureaucratic institutions from the same obsessional terror (which Deg but not V. was attending to). His *Secrets of the Great Pyramid* was ultimately to achieve fame.

Tom Kuhn's book on scientific revolutions was beginning to gather *kudos* for him as a historian of science. Deg had footnoted it in his study of the reception system, for old time's sake, since the book hadn't come to hand until the manuscript was ready to print, and Deg had wondered why so little attention was paid to the materials of politics and sociology on revolutions. When the *ABS* was publishing its Velikovsky Issue, Kuhn was publishing an essay on the function of dogma in scientific research in a book edited by A.C. Crombie; there he argued that science is and must be dogmatic and the present balance between dogmatism and openmindedness appeared to be a healthy one.

Kuhn and Tompkins got into a bristling argument over parascience. They were such formidable-looking men, especially at the moment. Deg felt embarrassed, as their host. Neither had the energy to spare for Dr. V. Tompkins was rebuffed because of V's heavy anxiety over associating with the scientific fringe, especially if sex reared its head. Tom volunteered no support, not then, not later. The presence of the great Velikovsky archive went unnoticed by him, too. Deg thought, well, Kuhn is in the grip of the Princetonian academia and is an historian of science, a field of nitpickers, excepting a few like Kuhn, ignorant of the springs of human ingenuity, clumsy handmaidens of the technical scientists.

Deg could see continually in science the ghosts of politics concealed by their shrouds. One of his old-time acquaintances was Don Price, an epiphenomenal career man of the public service, who launched from the pioneering Public Administration Clearing House alongside the University of Chicago to Washington, to the headship of the John F. Kennedy Center at Harvard, to the Presidency of the American Association for the Advancement of Science. Deg wrote him concerning the Velikovsky affair, seeking moral support. The answer: bland, perfectly unobjectionable, priceless.

Not having gotten his support for the report of 1963, Deg wrote Price again in 1966 asking him to intervene to get a communication of V. into *Science.* He repeated the pledge and passed the buck. Thus, on December 22, 1966, with "a very Merry Christmas and a Happy New Year" Price writes:

> I am glad of course to have the opportunity to read it and will forward it immediately to the Editor of *Science.* It is the general policy of the Officers and Board of Directors of AAAS not to interfere with the editorial judgment of the Editor and his editorial advisers. Since I believe that the Editor should be aware of your opinion, and that of Mr. Wigner, I am sending a

copy of your letter as well as the note itself on to Dr. Abelson, and I am sure that they will be useful to him.

For many years, Deg had preached that science could be regarded as a branch of administration and administration, the huge corpus of civilized routines, as the outward expression of human habits, largely unconscious, and therefore excusably termed obsessions.

Journal, undated, Spring 1963

Science, and all that goes by that name in discourse and actions is almost entirely a process of administering deductions in the name of an ideology. [Actually, this is a paraphrase of what Deg had written for the *Administrative Science Quarterly* a decade earlier. I am trying to exclude from this book whatever he has printed elsewhere, as I promised him, but I am like the oaf who quit his job grading potatoes because all the choices between big and little made his head hurt: at times I find such distinctions imperceptible.]

On December 9, 1966, not long after the publication of the *Velikovsky Affair* in book form, Dr. Douglas Shanklin delivered an address on childbed fever at the College of Medicine, University of Florida, applying Deg's model of the reception system to J.P. Semmelweis and Oliver Wendell Holmes. They had independently proposed infection as the source of the often fatal puerperal fever, and are famous therefore. But Charles White of Manchester, England, had insisted upon absolute cleanliness in the lying-in hospital in 1773 and Alexander Gordon of Aberdeen, Scotland, stated the theory of infection in 1795. Holmes was an illustrious poet before he published in 1843 his theory of infection as the source of the fever that killed so many women in the hospitals of the nineteenth century; he did not hold an academic position at the time but later became Professor of Anatomy and Physiology at the Harvard Medical School. The dogmatic opposition persisted until the science of bacteriology of the next generation overwhelmed it. Holmes died at 85, highly regarded.

Semmelweis was a Hungarian practicing medicine at the Maternity Department of the Vienna General Hospital when, in 1847, he introduced the practice of washing hands with chlorinated water before examining women in labor. Although the results were a fivefold decrease in the mortality rate, he was attacked and forced out of his position, and took a new post in his native Hungary. There he published a massive book on the etiology, concept, and prophylaxis of childbed fever (1861). Four years later he cut himself during a post-mortem examination, became infected, was mentally deranged, and died soon after at 47 years.

Holmes' essay was well-written and without first-hand experience.

Semmelweis' work was intimidating, ponderously written and he was fully experienced. Holmes republished his own essay a dozen years after its first publication in a medical journal, declaring: "When, by the permission of providence, I held up to the professional public the damnable facts connected with the conveyance of poison from one young mother's chamber to another, for doing which humble office I desire to be thankful that I have lived, though nothing else should ever come to my life, I had to hear the sneers of those whose position I had assailed, and, as I believe have at last demolished, so that nothing but the ghosts of dead women stir among the ruins."

Semmelweis was persecuted for his heresy. Shanklin writes of Semmelweis' tragedy:

> A few people acted with bold imagination and foresight, accepting the data at its face value and effectively saving many lives...the overwhelming majority dealt either from a power base or a dogmatic base, steeped in the irrational. The net effect for an interval was described in the indeterminacy model. Truth was accepted here and rejected there and by gradual exchange assimilation was finally achieved. Additional proofs with the evolution of a new technique wrote the final chapter of the saga of Semmelweis.

It took about a century from White's obsessive insistence upon cleanliness in Manchester's lying-in wards to consensus about a matter that should have been simple enough to grasp, if one recalled that peasants used salt, alcohol, and herbs on wounds and they isolated persons associated with plague by the most cruel means. That the use of hospitals for parturition increased and that the doctors and their students increased their postmortem dissections in this environment escalated the puerperal fever mortality rate. These two "advances" confused the issue, just as "advances" in agriculture, particularly in the U.S.A., have caused devastation of the soil, water resource depletion, and new chemical diseases. In the middle of advances, regressions are minimized or even denied scornfully. Obviously the scientific process is largely understandable by sociological and psychological analysis.

Deg did not enjoy any illusion that there would be a direct rational line from publicizing V.'s poor reception in the sciences to the acceptance of his views and their incorporation into science. For one thing, he felt certain that if V.'s ideas, or anyone else's including his own, would succeed, they had to be first disassembled, torn to shreds, and then reassembled by thousands of people from the nearly unrecognizable shreds. Only much later might some historians recognize the many truths and even the valid general theories in their work.

Nonetheless, the exposition of such large ideas and the controversy over them would perform the first major task of any revolution, namely the refocussing of attention and the conditioning of the minds of scientists and teachers to the new frame of thought. In these very days of the 1960's, the leaders of the movement for women's liberation were stressing "consciousness-raising;" many blacks were doing the same by stressing *"negritude;"* (as the French blacks called it) and accusing pro-black liberal whites, "their best friends," of necessarily being racially prejudiced; radical students caught on also to the effectiveness of "irrational," often destructive, behavior as a way of getting the attention of the civil and educational authorities.

Adverse publicity is a shock to the generally sheltered scientists and effectively alters their perceptions. The demoralization of a supreme power such as the scientific establishment with its credo and foci can occur by the exposure of weaknesses among a few leaders and heroes and proceed with the underlying economic forces that limit rewards and positions; demoralization then moves to the rank-and-file individuals who pay less respect, work less hard, ask more money and benefits, and pay attention to supernatural or heretical interests.

In a democracy, the withdrawal of any substantial amount of public support for the ideas and position of any institution, including science, results in some demoralization. A perfectly normal remark, if publicized, can invite latent opposition to take form. When the renowned astronomer and public scientist *par excellence,* Harlow Shapley, declared "If Dr. Velikovsky is right, the rest of us are crazy," what would appear to be a humorous truism set up when publicized a rallying point for all who were even slightly concerned about this or that fallacy of science; what many scientists believed to be only an absurd contrast gave to many a premonition that, yes, all scientists are crazy.

Although Deg believed that he had substantially accounted for the scientific behavior witnessed in the Velikovsky case, one of the most common questions asked of him in discussions and at lectures over the following years was "Why did the scientists make such a fuss?" It did not seem to matter that often the people assembled had come because they already knew the answer. There would, of course, always be on hand for analysis new cases of idiotic name-calling and denigration of V., but the causes agitating the scientists remained essentially the same: dogmatism (fueled by the need for respect), expressions of power (agitated by personal ambitions and feelings of insufficient influence), indeterminacy (the frustrated wish to know, and the denial of confusion and uncertainty) and rationalism (narrowly defined, and therefore inadequate against ideas of quantavolution, which seem so easy to refute and dismiss but turn out to be remarkably rich and resilient).

Exposing the mental and social operations of science produced an effect almost entirely favorable. Some addressed Deg for bringing justice to V. Others praised him for introducing the issue of justice into the scientific process. Some others commented upon the novelty of the approach. Mentions of unusual courage were frequent. Social scientists recognized the phenomena of establishment defensiveness and crowd behavior; they expressed little surprise. The letters of surprise came from persons who had undergone a conversion experience; they professed humiliation and disenchantment because of scientific conduct. Several urged that Deg turn his attention to cases which they believed to be similar.

Deg objected, when I thought to print some of the encomia that his magazine (1963) and book (1966) evoked, saying that rehearing old praise can be bittersweet, to editors as to stage and screen agees. To most it is a bore, old or new. Blurbs are the medium of exchange between producer, salesman, and customer. If it is necessary, if it's never been printed, OK, let it be brief.

So this is brief— but it's important, because it shows that the message was intelligible, and got through in the larger intellectual world. A comparison may be pertinent: it was widely believed that scientists took up their pens *en masse* to castigate Macmillan Company when it published *Worlds in Collision*. In 1983, when Leroy Ellenberger delved into the appropriate files he found only twenty-one of such letters.

The favorable correspondence received by Deg and the *ABS* in 1963 and 1966 exceeded the unfavorable mail received by Macmillan Company in what the Company regarded as a massive assault upon its integrity and its ability to do business with scientists. The gutless behavior of well-intentioned institutions is proverbial; Senator Joe McCarthy and a few assistants reduced the mammoth State Department and other agencies of the Federal Government to terrorized submission around the same time.

Some figures in the forefront of scientific method in the social sciences, then or later, responded to the issue forcibly, a "most interesting" from Herbert Simon; "used to very good teaching purposes" from Bernard Barber; "both fascinating...and important... a splendid account," from Hadley Cantril; "beautifully makes the point about the psychology of scientists... grateful" from James C. Davies, a "signal service" from Arthur S. Miller; "a superb example of the sociology of knowledge," from Wendell Bell; "sobering and helpful," from Renato Tagiuri; "an outstanding contribution on so vital an issue...not only the matter of methodology but also one of political toleration and scientific craftmanship" from Ralph M. Goldman; "fascinating...excellent..." from Wayne A. R. Leys; "splendid...outstanding...personal congratulations," from William Anderson (U. of Minnesota); "congratulations" from George A. Lundberg; and a grumpy reassessment by Stuart Chase, "I can see your point."

Sociologist George Lundberg's letter to Deg pointed to a different type of reception system problem in science, one in which he had once been personally involved:

> The question has a great many aspects. In the first place, there is the problem all editors face in discriminating between work of a crackpot and the work of a genius. As has often been pointed out, they are hard to

distinguish, especially on the more advanced levels. A very different problem (not involved in the Velikovsky case) faces the conscientious editor when he gets a paper the validity of which he does not question, but which, if published, will in the editor's opinion give aid and comfort to a group hostile to a viewpoint which the editor personally shares, on grounds reflecting the most creditable public spirit.

Lundberg also noted, "It appears that Velikovsky's ideas have been widely circulated in spite of the hostility of the Establishment... Is it possible that the enormous growth in communication technology has made it practically *impossible* to suppress new ideas for long?"

Stuart Dodd wrote from the University of Washington:

> I think you have done a magnificent job of l'affaire Velikovsky in the September *ABS*. The care with which you worked up and presented the complete case in the three articles, with excellent refereeing throughout, was a historic achievement in challenging and improving methodology in the Behavioral Sciences. I particularly admire the way you did not go into the controversy of the correctness of Velikovsky's theories, leaving that to the specialists concerned. Your editorial statement of the issues involving the mores of both the physical scientists and the social scientists as scientists in accepting and sifting new scientific work is a skillfully done job.

On the humanities side Mose Hadas, Horace Kallen, William T. Couch, Jacques Barzun, William Sloane and August Heckscher wrote Deg supportively. Medicine, social work, psychiatry, and law were among the fields of applied science reporting interest and conveying congratulations. Several *ABS* readers arranged meetings for Dr. V. at their campuses. Articles based on the *ABS* issue originated in Italy, England, Australia, and elsewhere during the 1960's. Reviews of the book when it appeared two years later were favorable; however, no scientific journal dealing with the natural sciences reviewed it. Ultimately, the book was republished in England, and translated and published by Bertelsman-Goldmann in Germany.

Deg introduced the second, English Edition of the *Velikovsky Affair* in 1977. Brian Moore, the librarian of Hartlepool and a cosmic heretic, reviewed the work in the *Society for Interdisciplinary Studies Review,* III: 2 (1978), 38. Crediting the book "a 'classic' in its field" with "the renaissance of scholarly interest in Velikovsky" he quoted its preface:

> We dedicate this book to people who are concerned about the ways in which scientists behave and how science develops. It deals especially with the freedoms that scientists grant or withhold from one another. The book is also for people who are interested in new theories of cosmogony — the causes of the skies, the earth, and humankind as we see them. It is, finally, a book for people who are fascinated by human conflict, in this case a

struggle among some of the most educated, elevated, and civilized characters of our times.

The area to which the *ABS* addressed itself was apparently much in need of attention. Sociologist Lundberg thought "that the AAAS, not to mention individual scientists and groups, must now prepare a detailed answer," and he added, as did others, various matters for investigation in the reception system of science. David Wallace wrote happily, "I hope you get sued."

The *American Political Science Review,* which had carried negative reviews of, or ignored, Deg's iconoclastic or deceptively simple works in political science sprang to attention with the *Velikovsky Affair.* John Orbell opined that "it represents a most significant contribution to the sociology of science." He applauded Deg's "most valuable chapter" on the scientific reception system and concluded: "Behavioral scientists might be expected this time to have been on the side of the angels; they were, after all, nearly alone among scientists in not having some fundamental notions challenged by Velikovsky."

Stecchini wrote to Deg, then in Italy, on Oct. 2, 1963: "There has just appeared a manifesto by [Robert Maynard] Hutchins and others of his coterie on *Science, Scientists, and Politics.* It says in general what the *ABS* has said, but it does not give any evidence. Hutchins begins by saying that in his experience the scientists are the most unscrupulous and power-motivated members of the academic community. The concluding paper by Lynn White, Jr. [historian of science] declares that scientists do not understand philosophical issues and often have philosophical prejudices."

One sponsor of this manifesto was Harrison Brown, a renowned scientist whose reviews of V.'s books were madly mediocre, which goes to say something of the significance of works of the Hutchins kind that do not name names, and makes recommendations that are not specific. Deg liked and admired Hutchins, even when strongly critical of him, ever since he had attended a seminar of that handsome, brave, relatively intellectual, self-contained, and slightly phony cavalier, then President of the University of Chicago.

There came shortly afterwards to Deg another letter from Albert Schenkman, Publisher of Cambridge, Mass., breaking a lance against the *ABS.* Ted Gurr, minding the *ABS,* wished to publish it and Deg replied "Dear Ted: It is cruel of you to hound me across the Big Pond with Mr. Schenkman's letter with a request that I reply. He is in a state of awful confusion. Print it if you will, with or without my comments," and he suggested that Gurr put the comments alongside the appropriate paragraphs of the letter. Gurr did not print them.

Philip Converse, who at this writing is President of the American Political Science Association, on Oct. 9, 1963 congratulated Deg on "a superb document." Unlike most, he had followed the case from its inception in the early 1950's. Unlike most, too, he directed his thoughts to measures of policy and control.

> ...In accordance with the principle of open public challenge and rebuttal, why not publicly invite those of the principals on the other side (certainly Shapley, Gaposhkin, Harrison Brown, perhaps Abelson, etc.) who are still active to respond to this issue in an ensuing number? I assume they would be willing actually to read the whole issue before writing rejoinders.
>
> I trust such an invitation could be handled without devolving into a Counter-Inquisition. That is, the profound ignorance in some sectors of the hard sciences with respect to what you loosely call the sociology of knowledge has been apparent for a long time. This ignorance, coupled with the arrogance of success, has had material consequences for the development of the behavioral sciences, and I am sure leaves many social scientists in a counter-inquisitional frame of mind. On the other hand, it is we who purport to understand the psychology of the inquisition, and we contend among other things that they are unlikely to. I think it is fair game to make the basic points and make them vigorously, while a classic case is still fresh. Yet if our claimed perspective on such matters has any merit at all, it should both permit us and require us to handle the matter with some *noblesse oblige,* out of respect for the gross differences between the two camps in comprehended information concerning these social and psychological processes. This is true not only because of the negative consequences of the unfettered inquisition spirit, but also because of our beliefs that the problems are principally system-level ones, not good-guys and bad-guys, and ones moreover that social scientists have not to date resolved operationally themselves. So a personal vote for increased discussion and allocation of resources toward remedy, but not the pillory or the witch hunt.

* * *

Deg at Florence was sent a copy of the *New York Times* of August 16, 1963 about "the first definitive list of books assembled for the White House Library," John F. Kennedy being President and Jacqueline, his wife, being interested in such matters as the White House decor and French poetry. Professor James Babb, Librarian of Yale University, directed the task. "Those on the arduous project included the best brains of the Library of Congress, the editors of the Adams and Jefferson papers, members of the White House Fine Arts Advisory Committee and a host of distinguished scholars, librarians, publishers and experts in many fields throughout the nation." Deg's book, *Public and Republic,* was on the list, his father said, and in response to a plea from the allegedly poverty-stricken White House for donations, his father had sent in the autographed copy Deg had given him years before.

Deg examined the list and wrote a brief essay about it. In his usual way, he managed to scold everybody, the pretentiousness of the scheme, the great works left out, the silly books entered, the illiteracy of Presidents, and the antiquated view of the methodology of politics and history evidenced by the list. Most pertinent here are his remarks on the treatment of science in this super-list:

> Nor do we understand why the natural sciences are excluded. Certainly there is room for some principal articles and books. If readability is the criterion, they are as likely to be read as several hundred other works in the collection. Besides the originals, there should be present at least Sarton, Conant, Whitehead, and Santillana. It is as important that the mythical President who reads should read science as that he should read "Little Women."
>
> This is probably another aspect of the escapism which shuns the future. The immense and fertile American planning community is scarcely heeded. The best predictions and estimates of what can be done in the natural sciences in the next century are absent. The best proposals for the control of war are not available. If indeed the President were to read randomly in this collection, we should fear for the nation.
>
> The tools with which an active presidential mind might work are not dominant here.

The incident displays Deg as something of a *misanthrope,* but what meaning has this word — a hater of one's fellow humans or, like *Le Misanthrope* of Moliere's drama, an idealist and severe critic of others? It is clear that he was the latter, he had the two tell-tale signs of this *Misanthrope:* he was a harsh judge of himself, subjecting himself to daily Augustinian interrogations of his activities, his use of time, his ideas, his conduct towards others, his intellectual and logical rigor, and his failures. Second, he had an inflated hope for others: for educating the uneducable, giving to the undeserving, organizing the unorganizable, loving the unlovable, bringing peace to the world; worse, he could see good in everyone — his opponents, madmen, silly women, gangsters, wicked politicians. Even at the moment of judging harshly, he was sympathizing secretly. One reason why he was attracted to V. was V.'s simple unidimensional moral quality: there were enemies and friends; the friend of your enemy is your enemy; the enemy of your enemy is your friend; the friend of your friend is your friend. The fourth category — the enemy of your friend is your enemy — was not so well accepted by V., or to most others who went so far as to accept the first three propositions. So it is not all simple, but nothing is, and all generalizations are false to a degree.

* * *

Let us move to Deg's journal.

Princeton, April 7, 1966

I was abruptly pulled out of the relaxation of homecoming when I visited Velikovsky. He was haranguing me about Livio's misspelling of the Pharaoh's name and I was sipping tea and listening respectfully but comfortably and even amusedly when the telephone rang and he answered it. I could hear him asking who it was and then "jail," and "marijuana," and "most regrettable," and "I am in full agreement", but then "I am not the man for you. I have here with me Professor de Grazia, Professor Alfred de Grazia," and "Let me have him speak with you... He is better qualified to deal with this subject."

He lumbered in and explained that a gentleman on the phone wished to have a Dr. Timothy Leary introduced. This Dr. Leary had been sentenced to thirty years in prison for possessing marijuana. He was a psychologist... I began to recall Leary... Harvard... experiments with LSD... and reluctantly but with some interest I picked up the receiver and received an invitation to come to Town Hall on Tuesday (this was Monday) at 8 pm and introduce Dr. Leary to the audience. The caller, Mr. Bogart, stated that under the circumstances of the sentencing, it would be helpful if Dr. Leary were not to go 'cold' onstage but be preceded by some supportive words. I replied that I might do so but wished to look into the matter and call him back the same afternoon. I hung up and V. said "You should do it, Alfred, it is a very good and useful thing to do." I felt that I should probably do it but did not finally decide until I had read a little of the background of the case and an article of alarmist nature in *Life* magazine regarding LSD.

Sizemore joined us at V.'s and we examined some of the long-sought-for Macmillan correspondance on V.'s case. Miraculously, after it had appeared first that Macmillan would never let us see what they had in their files from the days of the crisis over the publication of *Worlds in Collision* and then later they said that they had destroyed the files, Sizemore learned that the files had actually gone with many other files over to the New York Public Library for some future literary historian. Well, history had already begun. Sizemore requested the materials and they were brought up for him. He was not supposed to remove them, but he did so temporarily, reproduced them by xerox, and returned them immediately. So now we might read the full texts of the letters of the scientists Shapley, McLaughlin and the rest to Macmillan, the notes of Mr. Brett of Macmillan agitating the question of whether or not to ditch V.'s book, and related letters and papers. We were now in position to back up what some people regarded as exaggerated statements concerning the dispute with actual quotations corroborating our charges.

The matter of introducing Leary bothered me a bit. V. and Jill both spoke of my acceptance as an act of courage. So did Eddie [Deg's brother] when I called him that evening for information. So also several others in the next day or two. I feel uneasy when people say I am generous, kind, understanding or courageous. Partly I doubt that I am any of these things. Or if I think I am, it is upon occasions when nobody in the world notices; but then when I act normally and naturally, it seems to me, as in the case of Dr. Leary, I am explicitly informed of my virtues. I have long been convinced intellectually of the absolute lack of coordination between good deeds and rewards but their lack of coincidence in practice never ceases to bother me and unsettle me. I don't know how to put it: it seems that I do praiseworthy things in quiet, boldly, but when a public approves my conduct, far from plunging forward even more enthusiastically, I tend to pull up a bit and examine my conduct: am I being rash; what am I doing that is extraordinary. I almost never find that I am fully in accord with the applause.

Eddie told me on the telephone from Washington that Leary's case had several legal possibilities, that it was worth trying in court. He urged me to talk to Allen Ginsberg about Leary, since he recalled Ginsberg having an interest in the matter. He then spoke with A. G., I believe, the next morning, for G. phoned me at my office, speaking unexpectedly in a smooth, organized way, and we arranged to meet at the Faculty Club at 3:45 that afternoon for the first time.

At the appointed time, having speedily dispatched a batch of phone calls, letters, papers, and other miscellany from the piles of homecoming mail, I was at the Faculty Club and Ginsberg came in soon thereafter. The apparition is nothing to dismiss, especially if it occurs in the framework of the old Federal architecture and furnishings of Washington Square North. He was more completely uncouth than I thought possible. Full grown hair and beard flying in every direction, dishevelled attire of ditch, barn, and beach. He said Peter was parking the car and would be in, so we began to talk while we waited and after twenty minutes Peter came in with his tarn, long red braids, and grimy gym suit and tennis shoes, bringing along also his brother. By then Allen and I had come to terms and he could introduce Peter's brother nonchalantly as "Julius, Peter's brother. We've taken him out of the insane asylum where he's been for thirteen years. He's become our ward." Peter said, "Sit here, Julius!" and Julius staring far far out of this world, sat straight and mechanical on a chair and said nothing nor scarcely moved a muscle for the hour or more that we talked thereafter.

The trio was spectacularly disgusting. Several professors and the manager poked their heads inquiringly our way and I gave them a polite "hello!" My own feeling was of warmth and fondness. They were completely reversed characters. All the evil in them was in their appearance, while inwardly they revealed a beauty and kindness that was holy. They are in the great tradition of the blessed spirits — the hermits who live in caves and on poles, the beggars of St. Francis. Ginsberg is a man of surpassing intelligence, aside from all else, and Peter a kind of saintly inquirer. They are not mere celibates, or even better-than-ordinary men. They stand on the other side of Evil, having passed through it or flown over it.

I invited them to the bar downstairs for a drink, but they took me instead to their party, where they were tardy. Present when we arrived was the hostess, Miss Beach, daughter of the first publisher of Joyce, a Frenchman who has just translated Ferlinghetti, a Solomon who had just been freed from nine years in a mental hospital (this must be Allen's great early friend) and a pretty young man who looks like Edgar Allen Poe and publishes *Fuck you: a Magazine of the Arts*.

I stayed for a while, then left despite their invitation to dinner, because I had to put down some words for my Introduction, I signed into the Stanford hotel for the night, scribbled hastily for half an hour and then walked to Town Hall (taking a cab the last couple of blocks, since I turned E rather than W) and arrived a little late to spend time with Leary before the address. It was as well for he was busy with the press and TV until the moment he had to appear. He welcomed me and we went onstage to a house three-fourths filled. A young crowd, I observed. My introduction went off well, and Leary's small strange eyes lit up warmly when I finished and he shook my hand cordially. He rambled on nicely for over an hour under painful white lights. They bothered me more than him but he had indicated he wished me to sit on stage alongside the rostrum and I complied. (Now I must see what mode of exploitation there will be of the films that were made. If I am on display I shall want to be sure of the context and qualifications.)

Leary's message was simple and harmless. He spoke of the levels of consciousness and asserted that the deepest was provoked by LSD. He argued that the knowledge one gained thereby was to the good (automatically, I suppose, as the naturalist fallacy has it that all fact and truth is good and wreaks good, no matter the context or the controls). It wasn't much. Leary has been the patient *amicus adolescencis* of boys and girls seeking self-awareness and thrills of sensation, and is adulated for this and for his troubles and for his pursuit of a vague set of psychological and theological ideas that hover in the experiences of drug-taking.

I bid him goodnight afterwards, ate a poor solitary meal at a late diner, and slept well.

Princeton, October 6, 1966

Bad headache. Hot flashes, apparent heart palpitations after lunch. Query: alcohol? Alcohol plus fine crop of my garden mushrooms "coprinus" for dinner last evening? barometric pressures possibly related to hurricane Inez? something more functionally severe? Poor mood, anyhow. Louise S____ our house guest again. A beautiful woman, so well turned out, and 52 years old. She had a torrid affair with a young Greek and spent weeks with him on a primitive island in the Aegean this summer.

Walked with Franny (their shepherd dog) along the streets in the balmy night air. Stopped by Velikovsky to give him an article on "Magnetic Pressures" that describes the newest successes in building up tremendous magnetic charges. What artifice can do, nature may have done and may do. Hence V.'s theories about the possible role of electromagnetic charges in cosmic events and catastrophes may be supported or considered in new light.

He insisted I stay and despite my headache, we talked for nearly two hours. He had me read his latest correspondance and advise him on letters to Sullivan of the NYT and others. We spoke of his archives and I repeated my thoughts about a foundation to take over his home and archives. He is very anxious about his many remaining tasks. Fifteen they were, he said. I said "I have fifteen not counting you as a project." He joked about the peasant pushing the old ass and saying, in response to a remark of a bystander: "Between us we are 100 years old."

Deg's Journal, Princeton, October 9, 1966

It is as difficult to make a little change as a big change in politics. Or is it? I sometimes think the former and usually act upon it. But I am a radical by temper and I resent being involved in little changes when bigger ones are needed.

I wonder: can it be that in the measurement *NOT* of the difficulty of change, but whether the changes brought are big or little, that the conservatism of a society should be determined?

Deg's Journal, Princeton, October 8, 1966, 11pm

At 9 am Edward de G. calls and we discuss his problems in finishing "Congressional Liaison." At 10 V. calls and tells me we should publish his Brown University speech and the accompanying talks of his critics, together with the Neugebauer reviews and correspondance, as a book. I agree, but he takes a half-hour to unload his early morning thoughts upon me. I should charge the old psychoanalyst a psychiatrist's fee (professional discount, of course). At the end he says "I feel better now. We have this straightened out. Now I will go back to the miserable German translation of my book." I feel compassionate. At every turn of the road, a further obstacle to communicating one's ideas arises — when nothing else, there will always be the damnable errors of a typist, a translator, or an editor.

Deg's Journal, Princeton, 1967

The afternoon of Sunday, December 17, Jill and I bicycled down the hill to the Velikovsky house for a tea party, with Francesca, our German Shepherd dog, toping along nicely beside us. When we arrived she insisted upon coming in, or rather, behaved in such a confused fashion that we finally brought her in with us, and she finally discovered her place under the grand piano, where she had lain on prior occasions. Present were the Ralph Juergens, Dr. Kogan, Velikovsky's son-in-law and a Professor and Research Scientist from Israel, with whom I had met on his previous trips to the United States. So were the Bigelows, he from the Institute for Advanced Study and she a psychologist. I had not met them before although Velikovsky spoke of Bigelow from time to time. He is one of the few natural scientists who has lent sympathy to Velikovsky in recent years. A newly met acquaintance of Velikovsky, Spelman Waxman, was in the company with his wife. He is retired now from the Center for Antibiotics Research, that he had established at Rutgers University on the basis of the returns from his discovery of certain antibiotics, especially streptomyocin, for which he had received the Nobel Prize some years ago. The Waxmans had scarcely heard of Velikovsky. I had only vaguely recollected *them* as well. The Juergens didn't know the others. The Bigelows did not either, so all in all, except for Velikovsky, who has a great memory for everybody and everything, it was a typical gathering of specialized intellectuals who had heard little or nothing of one another despite the feeling that some of those present had that they might have met or that they were worthy of being known to others. Jill later told me that Mrs. Waxman seemed offended when Jill did not recognize her name, and of course Mrs. Waxman and Dr. Waxman were probably surprised when I asked him how he spelled it later on when he was asking me to send him a copy of "The Velikovsky Affair" which I of course felt that he should have known about, and I am far too aware of the networks of acquaintanceship in The Great Society to expect anybody to know me before meeting, unless they come from certain circles the existence of which I am well aware of. Under the circumstances, it is easy to see why there is so much trouble in gathering together a public opinion among scientists except at the most superficial level of the top associations and those who agitate among them and in the mass media, denoted by prizes and the like.

I learned about Kogan's work in desalinization of sea water. He is now constructing a model in Israel that is supposed to be a great improvement over existing distillation types that require much expensive copper alloy tubing. His method is a kind of open channel way that cuts down the considerable proportion of cost of the installation that come from tubing. He has also worked in physics and astronomy. He is a large man, wall-eyed, pleasant and highly intelligent, persuaded, I believe, of the validity of Velikovsky's general theory. We discussed the force fields that could have been operative during the encounter of Venus and Earth about 1500 B.C. He explained in answer to my questioning that it might be possible to set up a model to duplicate the forces involved, but it would be a very costly affair. Natural forces are not easy to set up in a natural state. He felt that the force of electromagnetism exerted presently among the planetary bodies and the sun might be enormously modified because its cube principle follows gravitational force very quickly and provides a very different relationship between the two bodies. Hence, one cannot say that the force between Earth and Venus would be negligible at all. Furthermore, we could venture a number of different positions, charges, currents, axial coordinates and the like that would determine a very wide range of possible forces between Earth and Venus during the period in question. And of course the present slow retrograde motion of Venus does not at all indicate what might have been the position and rotation of Venus at the time of the encounter. Unless someone comes up with a brilliant scheme, it will be difficult to reconstruct the historical incident with details more specific than those rather general ones provided already by Velikovsky. (However, I feel that there is some possibility that we might be able to use a more intensive and exhaustive scrutiny of ancient documents to discover somewhat more detail about the motions of the heavenly bodies during the encounter period.)

Dr. Waxman is an old Russian Jew of about the same age as Velikovsky, and they were able to recall passing by one another at different points in their early wandering lives. Dr. Waxman began to recollect his experiences in the years following his discovery of antibiotics and his naming of the field. I asked especially "How long would you say it was from the time you made your discovery until the time you finally had a full research institute set up and operative with the people you wanted?" He replied, after much clarification of the question, partly because he, like other natural scientists, do not think in sociological process terms, that ten years was the period from the time that he made his discovery until the pharmaceutical industry purchased rights to use them, to the payment of royalties back to the University, to the voting by the Trustees of a new Center for Antibiotic Research at Rutgers to be set up by Dr. Waxman, to the construction of the building and then the hiring of a first group of deliberately temporary people who were space occupiers to prevent other ill-housed faculty of the University from taking over Waxman's facilities before he had a chance to bring in the permanent first-rate men that he was seeking. Finally, at the end of ten years the cycle concluded. I commented that this was a very short cycle of this type. It had to do with the nature of the discovery, of the fact that a market was present, and a few unique factors, including, of course, the shrewdness of Dr. Waxman himself throughout the total operation. A much more thorough study of this experience would be very worthwhile from the standpoint of the history of science and the sociology of science, as well as comparable studies of other experiences.

The tea itself was only a small part of a rather elaborate Russian type of menu that Elisheva Velikovsky provided — sweet pickled herring, cheeses, hams, several kinds of cake, and the company enjoyed itself at table, Franny having lodged herself below the table and under the feet of everyone, somewhat to the embarrassment of Jill who was never really embarrassed about this sort of thing but thought that poor Elisheva had enough to do without concerning herself with the physical presence of a large bitch. Numerous stories were recounted. Velikovsky told of the legend of Solomon in which was apparently involved a bit of radium that had been picked up somewhere and was carried in a lead box and was used from time to time for performing miracles, and finally after generations was exhausted. I thought the story showed very well the terrific power of Velikovsky's mind in looking at stories and seeing beyond the simple words facts at an entirely different level. He is unquestionably a great detective.

Juergens caught me aside as we were leaving the table and the dining room to show me a long letter he had just received from John Lear, the Science Editor of the *Saturday Review.* In this letter, Lear was defending himself against Juergens' assertion in his essay on the history of the Velikovsky controversy that Lear and Stuart McClintock of *Collier's magazine* had attempted to go beyond Velikovsky's wishes in jazzing up and popularizing *Worlds in Collision,* something that we have felt contributed to the original hostility to the Velikovsky book on the part of the scientists. Nothing in my experience would make me surprised at a popular magazine's handling of a scientific issue. It is almost impossible, given the rules of journalism, to do justice by science. Among many other reasons, the journals themselves are unequipped to handle distinctions between fact statements and scandalous exaggerations. However, in this letter, Lear again said that he had a most difficult time in working with Velikovsky; he disputes that there was ever any intention of serializing the book itself instead of condensing it (something that Velikovsky himself later confirmed and said that he had misremembered this fact when he looked up his agreement), and went on at great length quoting copiously from a letter written by McClintock to him a few months before McClintock's death last year, in which McClintock gave the most harrowing account of an evening spent at Velikovsky's home when he and Lear and later he alone, after Lear went out to wait for him, had tried to escape the wrath of Velikovsky and to appease him and at the same time to try to present an article that they thought would be printed by the magazine. In fact, McClintock accused Velikovsky at one point in his ranting and raving of bringing out a gun from the cabinet, putting it on the table and saying "Let this settle the matter right now." McClintock wrote, if Lear is correct in having such a letter, that he (McClintock) left the place shaking and with an eruption of the ulcers that he had thought once cured and for a year felt poorly as a result of the meeting. I laughed rather grimly when I heard the story. Of course one would have to check the reliability of both Lear and McClintock in respect to the incident at which Mrs. Velikovsky was supposed to be present. But again I would not put it past Velikovsky. I could see that a man coming out of a dozen years of every day in the stacks all day long and with his whole life work and magnificent set of theories at stake and with all the driving power, and determination that was required for that effort, being confronted by what had to be a shallow, glancing misrepresentation of what he was trying to say, and considering also the enormous domineering quality of Velikovsky and of how he wants to control every single thing that has to do with himself, he would be most intemperate, disagreeable and could even have pulled out the pistol. Juergens wondered whether he should show the letter to Velikovsky or Mrs. Velikovsky. I said hold it another day or two until I could look at it more thoroughly, and then we went into further conversation with the group, the Waxmans having departed and Jill having gone onto the subject of forming a foundation for the study of some of the theories in which Velikovsky was interested. He would like me to organize it. I am thinking strongly of it but I would like a much more clear definition of our respective roles.

I arranged to see Juergens several days later and did on Thursday afternoon. Then I read through the letter again, we joked about it some more, and I said to Juergens that I saw no reason why it should not be shown to Velikovsky. I believed it worked out all right because the next day Velikovsky called me on another pretext and raised the subject again just to hear my response. He didn't mind my treating it in a jocular way. And he certainly did not express the right amount of indignation, I thought, at the fact that I appeared to believe the story. But he denied it and said that he had never owned a pistol since he had one many years ago in Russia or was it Israel. He weakened my belief in the letter a little, but it would seem hard for McClintock to make up the story completely, so specific was it. He also claimed that Lear was not there at all during the meeting.

Juergens and I then discussed the foundation, and he agreed completely with me that prior to the establishment of the foundation it should be determined that it would carry a full range of objective studies of the many types of problems in numerous disciplines that we had come upon in the course of the Velikovsky experience. Furthermore, he agreed that we should ask for the rights to almost all of the Velikovsky archive because it is from his voluminous notes and the total collection of commentary that we could fashion many a first-rate hypothesis for our colleagues to research, both in the history of science and the substantive areas of concern. I am now drafting such a letter to Velikovsky explaining the conditions under which we would have to work. It is impossible to be in any dependent position with respect to Velikovsky and get out any kind of regular journal, or series of publications, or systematic argument in opposition to his theories. I could not work otherwise; I would find, as would everyone else concerned with the foundation and its publications, that he would gobble up all of our time whether it was necessary or not in the affairs of the foundation and we would be able to do nothing with our lives otherwise. The pretext I referred to above that Velikovsky called me about had to do with Professor Neugebauer. Neugebauer had apparently accused me of "dishonesty" in some letter to Delaplaine, a science writer, because I did not print or acknowledge a letter that he had written me (the *ABS*) in 1963. But I don't recall having received such a letter until 1965, at which time, O.N., probably feeling threatened by an imminent visit of Velikovsky to Brown University, N.'s own school, sent me an explanation of why he had distributed "only one hundred" copies of his review of Velikovsky's book containing a serious error that would make Velikovsky appear foolish or treacherous with facts.

* * *

Every month of the decades of 60's and 70's there would be an alarm raised to rally to V.'s cause, and the volunteer firemen would rush to the scene. For persistent devotion to duty over the whole period Warner Sizemore gets the prize. He was out of Georgia originally, became a Presbyterian minister, studied for his doctorate at Temple University. He never completed his dissertation, which he might have written ten times over if he had not given so much time to Velikovsky. Sizemore was an artist as well, a modest painter who would not stretch himself to create. He devised, too, a method of reproducing in wood a painting, whether classical or banal, and sold his productions at fairs, in shopping centers and fairgrounds.

I must not give the impression that V. would not help his supporters. When it was safe to do so, and would not compromise himself, he would write letters; since almost always the cosmic heretics needed letters that would recommend them to academic foes of V. and cover up their friendliness to V., there were not many of such letters. In Sizemore's case, V. guaranteed a mortgage on a house in Trenton, so that Sizemore and his family might settle down. They did and found their life-paths successfully.

The interventions of Sizemore on V.'s behalf were to be numbered in the hundreds. A minister of the many, he became a minister of the one. Hardly a week would go by without some assistance. He gave counsel, wrote letters to the media, made phone calls, solicited support, attended every related public assembly, taped miles of discussions and lectures, gave his own funds to publish the magazine *Kronos,* kept hostilities to a minimum, and maintained a good-natured concern through thick and thin and down the years. He became Professor of Philosophy and Theology at Glassboro State College and persuaded the authorities to authorize a Velikovsky Center, which began to collect items of interest and which served as a background screen for *Kronos* magazine. There was little gain here except the prestige of an academic address. V. never did consign a copy of his archive to the "Center."

Friends like Sizemore come mostly in fairy tales and epic poetry. V. took him for granted, as indeed he took everyone for granted who did not hold some prestigious place or manage a power center. He bequeathed Sizemore nothing — nor anything to anyone else except his wife, and then by descent through her to his family. It is continuously remarkable how gratitude in life, where it exists, is typically decapitated in the performance of a last testament. It was disgraceful, after having taken up so much time over decades talking about making his archives available and helping others carry on his work, that V. did nothing to that effect nor did his wife and daughters, and in fact his books and materials and funds were held more tightly than ever after his death. I have already said that V. undervalued what he received from others and overvalued what he gave them. Lewis Greenberg, to take another case, had for a decade edited *Kronos* without compensation (unless his profligate telephoning were to be counted as such) and could only wrench a few articles out of V. and his heiresses. Very late, Jan Sammer, the family's assistant, helped to pry loose some pieces. As we shall see, *Mankind in Amnesia* is not much as a book, but would have appeared gracefully and appropriately as articles in *Kronos.*

Meanwhile *Kronos* was weakened by its top-heavy reliance upon Velikovsky's case. When the magazine was very young, Deg had proposed, in a fateful meeting of several cosmic heretics in a Chinese restaurant of Philadelphia, that the magazine "go public." It should define its mission in general terms and seek a wider audience. Greenberg, whose paranoiac outlook he was the first to confess, felt threatened and drew back. Deg, who should have pursued his aim more gently and privately, let it drop, and hardly had personal contact with Greenberg in the years that followed.

But this is true, that V. would have been outraged if any of his circle, and certainly *Kronos,* would have essayed to count him as only a leading figure among cosmic heretics, rather than as their *raison d'être*. Those who thought such "evils" were evicted, like the Talbotts, or dropped out, like Stecchini and Bill Mullen. Only Deg, I must say, pushed over the years for an opening up to the world, and only once did what seemed like an awful break occur, which lasted for a couple of days. Then the British began to skirmish, and opened up frontally with the Glasgow revisionism; Deg began circulating his own manuscripts and coining doubly heretical terms like "revolutionary primevalogy;" and ultimately *Kronos* began to carry non-Velikovskian material and theory.

Withal Deg could note with interest how in published articles of *Kronos* and the British *Review* and wherever else a piece might appear, the writer would be sure to interject a mention or quotation from V. in the first paragraphs, as over the years, in American political science journals, one felt he must refer to the latest book of the "hit parade," one year being the year to cite V.O. Key on political parties, next year David Truman on political processes, then Robert Dahl on democratic theory, and so on, or, in a more stable setting, the communist scientific writers who seem hardly able to put a pen to paper without promptly keying in a reference to Marx or Engels, no matter what the subject and "the state of the art;" and the Chinese for a while with Mao, and so on. The issue was not "giving credit where credit is due" but of political-social game-playing. When a man writes much, he must ultimately mention everything from sex to the weather, and every phrase can become Biblical in its marvellous "perceptiveness" and "prophecy."

Deg was not of course alone in detecting this in-gathering effect of fame, as I discerned in reading the Journal of Andre Gide for 4 February 1922:

> Freud, Freudianism... For the last ten years, or fifteen, I have been indulging in it without knowing it. Many an idea of mine, taken singly and set forth or developed at length in a thick book, would have made a great hit — if only it were the only child of my brain. I cannot supply the initial outlay and the upkeep for each one of them nor even for any one in particular.
>
> "Here is something that, I fear, will bring grist to your mill," Riviere said to me the other day, speaking of Freud's little book on sexual development. I should say!

It would be impossible to carry in any interesting manner an account of Deg's interventions on V.'s behalf, just as it would be to list Sizemore's multitude of favors. Instances would include: setting up with John Bell a meeting for V. to address at New York University (Mar. 1, 1968); offering to the President of the Franklin Institute of Philadelphia (Feb. 20, 1967) to take the platform with V., if it was the presentation of "another side" that was truly wanted; dealing with publishers (Dell, Feb 27, 1968, Simon and Schuster, *et al.*) to publish more of V.'s rebuttals of the "establishment;" writing letters to the Editor of *Newsweek* (May 29, 1968) and to other media directors; appearing on radio discussions; helping to arrange television programs; addressing a "Social Order in Science Study Group" at the George Washington University (Jan. 18, 1965) meanwhile conducting general research in the field and carrying on another complicated life.

On occasion (rare because his obduracy was known) intimates remonstrated with Deg for spending too much energy upon V.'s problems. His attitude was typical: give me a better cause in the intellectual world, a more worthwhile victim; a better archive; most victims are dull, or psychotic, or trivial..."Think of your own interests," they would say. But that only confused Deg. He didn't feel actually that he was giving V. so much. His "own interests" were for affection, good food, good company, sex, beauty, travel, and there seemed a good supply of all these to be had. As for "other people's interests," he would gladly save the world and did make a couple of literary stabs in that direction, nor was there any world movement worthwhile; he tried to save higher education by starting a school. He jumped into the Vietnam vortex but could do little. He took initiatives to advance his field of learning by inventing a computerized information retrieval system. Other things as well, such as a stint to help erase anti-semitic elements in the Catholic rite, offers to reorganize his department, etc. It was not so easy, I conclude, for him to have found a better cause. Recall it was the "richness" of V."s materials that that attracted Deg, and allowed the science of sociology and the history of science to progress.

Let me dip into his journal to see what was up otherwise. On March 8, 1968 is an entry that combines food, presidential politics, Vietnam, economic development, the arts, and religion:

> Lunched 1-3 pm with Rod Rockefeller at "Pireaus, My Love," rolled lamb and stuffed flounder in a second floor saloon lined with portholes. Decided:
> 1) We might set up a company to study possibilities of large-scale condominium conversions of slum properties. I'll form a committee.
> 2) It would be well to set up a committee of ten for Nelson R. for President among scholars and from that I might send a larger mailing to the 15,000 political scientists of the country, and then all the other fields.
> 3) IBEC would be interested in VN if United Fruit could come along and develop the economic output of a new city. [Deg was pushing to create a new city in Vietnam.] We'll see what Julian Turner [U.S. Army Colonel, formerly logistics chief in Vietnam] has to say next week when he comes from Fort Lewis.
> 4) The fine arts corporation and antique properties holding corporation can be gotten to whenever the means and times are right.
> 5) We'll try to get the National Council of Churches to do a practical and strong job of handling its 3-year program on the social responsibilities of corporations.

I scarcely need say that none of this succeeded, but perhaps it goes to show how Greek cuisine can help to vent hopeful dreams. Every now and then the two men would lunch together and concoct schemes that didn't seem to go far beyond the lunchtable. Deg stopped seeing Rod without saying anything because when the big crunch descended with the School in Switzerland, Rod gave a mere $100 to the cause. They were used to dividing their lunch bills, but this Swiss fare was too exotic for Rod to share.

The same night, he was writing a poem on the train:

> How many Fridays we thanked for
> not being Mondays,
> wish we life away so.
> Draw back all those weeks, dear breath,
> into the fresh lungs of youth and
> fill them with the best of life,
> skimmed of complications,
> Humpty Dumpty splatted where he fell

and tra la la la for him.

Just a dog lying in the sun
Waters creeping up a beach
A long walk to nowhere
An enthusiastic argument
A book on the wide harmless world
No shocks and jolts of rioting
but sweet time, soft time
fall stilly, pass gently
around our retracements
drink long and cool
wet and stretch these cords
from Monday to Friday.
Will the little god to rest
and give the big one a chance to work.

Some of the life he was leading in these years is reflected in the following letter from Naxos to Dr. Zvi Rix of Jerusalem, dated July 19, 1976:

Dear Dr. Rix:
    Greetings! I hope my letter finds you well — and not too impatient with your friends and colleagues of the field of revolutionary primevalogy. I have settled down in Naxos for a few weeks (until August 15), after visits in London, Amsterdam, Delft, Dusseldorf, Dornach (the Rudolf Steiner Center), Athens, and Thera Santorini. On the 15th of August, I go to Athens, the Dordogne (to spend two weeks around the caves and digs), Nice for the IX International Congress of the Union of Pre- and Proto-Historical Sciences, and then probably straight back to NYC and Princeton. I have been carrying your letter of April 2 (terrible!) with me for months. Let me "respond" to it.
1) As I have said, you only need a) to be able to come and b) to find out whether I am here, to come to Naxos as my guest *any time.*
2) If you ask him, Sizemore will probably duplicate for you a set of the Glassboro papers, which I see are beginning to appear in *Kronos.*
3) Did I send you the "Jupiter and Saturn" piece? No! I have searched my folders here and, alas, I must have given the copy I had carried with me for you to somebody in the English group (I became generous and present-oriented under the influence of good company and whiskey). I will send it to you when I return; it is only a brief piece with a well-phrased hypothetical formula.
4) Did your piece not appear or is it not promised for publication in *Kronos* (I have no copy of the Birthday Symposium myself.)
5) Your "psycho-politics" was gratefully received and read by my seminar at NYU.

6) I wish it were as easy (*cf.* your compliment *re* my article on Michel son's Moonshine) to set up our own elaborated time frame and scheme for myth analysis as it is to knock down those set up by others.

7) The model for the new Holocene that I set up views it as an age of the "Unsettling of Heaven and Birth of Man," the age of catastrophes, using Greco-Roman terminology: Urania, 14,000-11,500 (BP 2000 AD); Lunia, 11,500-8000; Saturnia, 8000-5700; Jovea, 5700-4400; Mercuria 4400-3450; Venusia, 3450-2750; Martia, 2750-1600; Solaria, 1600-0. The greatest catastrophes occurred with the birth of the Moon from the Pacific Ocean *ca* 11500 for much crust was lost as the larger element of outer planets (Uranus-Neptune, etc. possibly) passed closely and the water canopies fell cataclysmically. The scheme appears too radical at first sight, but in hundreds of pages of working back and forth logically and with the scraps of available evidence, it seems to hold together. I propose it in order that we may begin to fit in all of the scattered pieces of myth, evolution, paleontology, behavior. Whenever the exposition is ready I shall send it to you.

7a) as for the dynamics of the birth of *Homo Sapiens Schizotypicalis*, I have at least a pamphlet nearing reproduction on the subject and will send you that too. I shall try to find H. Gunkel's book; thank you.

8) I do have access to the sourcebooks that Corliss is publishing on ancient riddles and reports. I agree with you that St. Brendan-Quetzalcoatl follows a universal pattern; the ultimate problem is to fix the first age (Urania?)of the practice of these rites and to show how they emerged from the brain (double-brain?) of the new *homo sapiens schizotypicalis cum* geo-celestial terrors.

In the sourcebooks that you mention (Corliss'), did you remark upon the vitrified Scottish forts? I am going into this matter now. This seems to be lightning, and on a grand scale, i.e. the protracted withdrawal or rush of charge from the Earth via the most convenient modes of exit towards an accumulated and approaching extraterrestrial charge (opposite). Hypothesis: at a certain point in time (Mercuria?), thousands of points on Earth were mobilized to discharge electricity (*cf.* my article on Troy IIg, which might be synchronized with the vitrification found in many places). Query: does the Tower of Babel case belong here? Did the languages of man disperse in shocked amnesiac behavior? Do the ziggurats and pyramids evidence vitrification or an intent to facilitate *(ex post facto)* future current-flows? (Troy IIg is in pyramid-building times.) Note Mercurial qualities? When did Hermes flourish as a god? (under overall aegis of Zeus, perhaps). If people on an eminence feel current starting to flow, they get out before the heavy scorching from the heavier flow occurs. Are there vitrified eminences and walls, mid-3rd millennium, in the ruins of your area? Perhaps, and even probably, this phenomenon, like quakes, flood, fire, whirlwinds, occurs whenever a major extra-terrestrial approach or major planet disruption occurs.

A young Dutch geologist, Poul Andriessen, is here in Naxos drawing samples for 40K-40A tests, that he performs himself. We've spent many hours discussing the valididty of the technique. There are serious questions that he admits, although he defends the results of his other radiochronometries. It is all so difficult, a seemingly endless set of important problems concerning which one must make up his mind.

But enough for now. The sea is too rough for swimming — or at least it is not inviting, so I shall drive my motorcycle into town and see what the tavernas are offering by way of food and company.

With best wishes, I remain, sincerely,

Alfred de Grazia

Then years later, he lies in Stylida with a broken leg:

June 7, 1978

Foot swollen and aching this morning. Big discussion with A. M. as to cause of this "relapse." she saying my walking upon it caused it, I saying that it may be the normal effects of stressing the foot in order to get the cartilage, foot bones, muscles, tendons articulating properly. I confess, though, to a certain worry from the beginning of the case: that everything inside was thoroughly disarranged, apart from the broken bones, and may be difficult to reorder functionally. But, too, I took a long swim and that, plus walking, has markedly tightened the muscles of the calf. Wouldn't the stretch pain the tendons?

Reading in Velikovsky's *Peoples of the Sea* to recheck whether he had separated sufficiently the Egyptians' "Peoples of the Sea" from those "Peoples" alleged to be destructive elsewhere at the same time, I find that he has not and I should one day pursue the idea that "Peoples" fiction served to cover up the Martian catastrophes of the 8th and 7th century, 3-400 years before the time of which Velikovsky writes.

But the force of his arguments makes me yearn to circularize a brief questionnaire among all Egyptologists asking whether they have read the book and whether the hypothesis of Ramses III being of the 4th century is at all useful or defensible. I believe that the results would be scandalous.

Stylida evening 17 June 1978

A Swede dropped in unexpectedly. His friend is interested in buying into my land. He stayed a few minutes and left. Ami rode into town with him and brought back food and mail and news. Then we swam. I continued to hack my way with a hand ax down the bluff and back up again, as I had begun the other day. It was easier, the footholes more prominent. I slung a rope around the bush and dangled it down to steady me on the crawl up. There were 30 pieces of mail of which 2 were for Ami, one rejecting "nicely" her second novel (really the fourth she has written) and the other from a journalist who compares her in a review with Anais Nin. I received a rejection of my elaborate request for a grant from the National Endowment for the Humanities; for various reasons, I don't mind this. It's already an article or two on the "Ballroom of the Unconscious." [It is carried in *The Burning of Troy.*] I wanted the money to live on and to employ Ami who knows the literature so well, supposing that *other* means of subsistence don't come in.

Of the force that moves this varied activity through the years, there is more than a hint in a note of Deg's journal, undated but apparently of 1973, the more interesting in view of the massive narcissism that has been a-scribed to V.

Ten years ago I was induced by L. Stecchini to gaze upon the writings of I. V., catalyzed by an accidental reading of *Oedipus and Akhnaton.* This led up many different paths of philosophy and science, which I would not have had the courage or confidence to undertake, if I had not been a victim of the magnificent arrogance of R.M. Hutchins whose New Plan and own spirit of it had not pervaded the University of Chicago with an idea that man, even in this age of specialization and seemingly endless data banks, could and must master a survey of all knowledge to be educated. This happened twenty-four years beforehand.

But this would not have been enough if there had not been sixteen years before a narcissistic bending of my character in infancy and childhood, a fierce desire to keep the world in all its forms within me (to own the world) and a fierce competitiveness toward all others to enter it upon my own terms.

CHAPTER FOUR

A PROPER RESPECT FOR AUTHORITY

In the summer of 1971, Deg led a party of 300 persons, with many camp followers, up the Swiss Alps to found a college and V. came later to teach. It did not take V. long to perceive that Deg was continually in danger of falling victim to a human landslide that Deg's own explosive force had set into motion. When it came to V.'s turn to speak to the representative assembly, a beautiful contrivance of Deg which, like the French revolutionary assembly of 1789, had gone wild, V. called up Freud's *Totem and Taboo* and gravely admonished the respectful group of the danger that lay in killing their father. Deg felt embarrassed while dutifully thanking V. for his remarks, for he was a staunch republican who had always disbelieved in patriarchal leadership systems and because many of the college crowd would be all the more delighted if they could rid themselves of their father as well as a leader, killing two birds with one stone.

"I, an octogenarian, said V., stride with the young of mind. There is no cult of Velikovsky: there is only the cult of scientific and historical truth. The youths sense this, and the rebellion against the pseudoscience taught from the cathedras of the universities is not far away."

V. to Princeton Graduate Forum (Oct. 18, 1972): "Nineteen years ago I called the young ... to look for new vistas, not to be afraid of calumny and name-calling. Today I repeat my call; it's a new generation. I call you to cross the barriers between sciences ... My work is not finished ... It is in your hands. It is up to you to decide if you wish to repeat what the authorities told you or to become authorities yourselves - to grow and to be nonconformists and to take abuse and to be exonerated some day. So be courageous and don't be afraid."

If V. had been given a son, he would have wanted him to be like the astronomer, Carl Sagan, but, of course, in agreement with his ideas. Being what he was and the times being what they were, he was probably lucky to have no son. Rare these days is the child who adopts the father's views or even defends him. When V. and Sagan were appearing on the same platform at a AAAS meeting in San Francisco, V. invited Sagan to his room, and there sought, if not to persuade him of his ideas, to influence and neutralize him, perhaps in a way to hypnotize him. Sagan only redoubled his criticisms as a result; the attempt to make a son of him back-fired. Sagan regularly lectured against Velikovsky in his classes and published repeatedly his essay said to finish him off.

Still Sagan could invest himself with V. 's claims, and probably (though he would not meet with me to talk about such matters) he was convinced that the father was well dead and gone and was terrified at the feeling that V. now wished to be patriarch to him. Interviewed by Richard Baker on BBC 4 (radio) "Start the Week," 30 March 1983, he was asked, along with other guests, "the moment in your life that you've been most pleased about?" Sagan talked of the "delightful moments" when his predictions about planets were borne out by space vehicles on the spot. Pressed for a "particular discovery," he replied "Well, the discovery that the surface of Venus is extremely hot, about 380°C, [Actually it is much higher] and produced by a massive atmosphere Greenhouse Effect that keeps the heat in..." The second is a dubious theory, not at all original with him.

That he could claim the first can most charitably be regarded as a slip of the tongue, such as Sigmund Freud describes; inadvertent and often embarrassing utterances, they are usually prompted by a strong suppressed desire of the speaker to make a point otherwise prohibited by rules, morals, or truth. Sagan, one might surmise, let the claim slip out as an expression of general megalomania, but the particular claim, out of all those he might have thought of, strikes at V.'s well-established claim of predicting the high heat of Venus. There is here a hint of psychological pressure working to take for his own specifically the property of the father.

V. was fixated on authority, the higher the better; he sought out acquaintances and enemies on high levels. But he did not gather intelligent up-coming young people until late in life; he has written a book on his conversations with Einstein, yet he would never have dreamed of writing a book of his immensely richer conversations with Juergens on electricity and Stecchini on ancient languages and the history of science. Why? Because they were unknown.

His idea of arrival was naive. The great ones would recognize him on the basis of his books. The young would come along, following what their teachers say. Until late in life, he had no idea of the striking fact of intellectual history, that most geniuses and heretics start out young.

At any given moment in time, Harvard University is likely to have a couple of pets of the communists. It's a gimcrack of impeccability. Harlow Shapely was one of these — and, of course, a great deal more, too much more, member and officer of dozens of scientific associations, Director of the Lowell Observatory, and more still. In poking about, Deg discovered that he had even once invoked exoterrestrial forces to explain terrestrial phenomena.

Well, V. had thought, a man so broad in his interests and tastes would welcome a helping hand to apply legends to astronomy. V. was anti-communist and had been so since the earliest successes of the Russian Bolshevist movement had not gone so far as to efface anti-semitism in Russia. The authoritarian aspects of communism, or statism in general, did not faze him. Principles of government were foreign to him, a sharp contrast to Deg who was continuously seeking better designs for human institutions. To V., governments and men were bad or good. The Soviet leaders were bad because they acted badly. Nor should persons be forgiven evil because of the pressure of circumstances. How he would love to live quite without compromises!

The only dispute in connection with Deg's article on "The Reception System of Science" of the *ABS* issue occurred over his mentioning V.'s "respect for authority." Deg told him of the expression "the Cabots speak only to the Lodges and the Lodges speak only to God." His response was not to reform but to try more of it: he writes Deg a few months later that he knows that he is speaking like a Cabot but would Deg support him in his efforts to bring the prestigious figure of Lord Bertrand Russell over to his side?

V. was on a collision course with himself. He practiced on Aristotle, Newton and Darwin, numerous 19th century writers and then on current authorities, but impersonally and only with the slightest irony, in a situation calling for broad sarcasm.

He thought of himself as an authority but did not realize that he was undermining present authorities and that they would react as authorities invariably do, by putting him down. But, then, he was a poor sociologist. Like many a psychoanalyst (and most scientists for that matter) he barely realized that the field existed.

He was flabbergasted when his *Worlds in Collision* was attacked so vigorously and then each succeeding book was treated the same, dismissed, or ignored. It was all the more shocking because *Worlds* was a best-seller, which brought *popular* authority into play as well. Here both V. and many of his followers showed themselves unwitting victims of the market place in ideas. They did not suspect success. Deg, whose life had begun early to forge a chain of successes, had contempt for success. The concatenation of any man's successes was but a motley cluster of medals on the breast of the generalissimo of a banana republic.

V. was unhappy with the support he received. It seemed that he would get agreement and aid from exactly those sources he did not himself respect while being rebuffed by those who should flock to this banner. One had to be an anti-authoritarian to support him, but such were rarely to be found in physics, biology, astronomy and geology. Passive anti-authoritarians, yes, often erupting in personal eccentricity. Anthropology — but he knew little besides Freud's work on anthropology. Psychology — again the psychoanalytic approach, not tight empirical psychology.

So he got support from people who usually were just plain folks, intelligent (and therefore I say rare) readers, and a great many confused believers, or at least people who V. at bottom thought had no right to pass judgement on him. Like Moses, V. spent a lot of private time disliking his People. Like the barons of the Magna Carta, he wanted judgement by his peers, but in his case the peers had to be in the other sense "the peers of the realm."

* * *

Perhaps *Oedipus and Akhnaton* should have been entitled "The Oedipus Complex Unmasked," or 'The Jews were First with God." V. enjoyed thinking about titles and slogans. Deg and he would spend some off-track moments in such half-serious play. V.'s titles were exceptionally effective; *Worlds in Collision, Ages in Chaos, Earth in Upheaval,* and so were most of the titles of sections of his works: thus in *Oedipus and Akhnaton* there were "The Sphinx," "The Seven-Gated Thebes and the Hundred-Gated Thebes," "A Stranger on the Throne," "King Living in Truth," "The King's Mother and Wife," and so on.

When Deg, six years after they had met, presented him with *The Torrid Love Affair of Moon and Mars,* he had to have explained to him the Hollywood Americanism of "Torrid Love Affair" and liked the *double entendre* with the heat of a cosmic encounter, but then eventually preferred *The Disastrous Love Affair of Moon and Mars,* which denoted, if not heat, a cosmic event and catastrophe.

Later on, still, he could let himself like *Chaos and Creation,* and even *Homo Schizo,* but would not let himself contemplate *Moses and His Electric God,* but this was part of another matter, his taboo of Moses.

"You will damage me with this book," he declared solemnly to Deg. Since Deg made no reference to V.'s idea of Moses in *God's Fire,* which V. had not seen anyhow, and since V. had damaged the reputation of thousands of scholars "in the line of duty," he must have been gripped by an illusion that referred to an entirely personal problem of his own in regard to Moses. What could it have been?

Martin Sieff, a Belfast Anglo-Irish-Jewish journalist and historian — one of the cosmic heretics — spoke out in 1981 about the taboo: "The role of Moses is strangely muted in *Worlds in Collision.* Moses is mentioned only in connection with the voice of Yahweh at the flaming bush and the trumpet blasts of Sinai." Further, "in *Ages in Chaos,* one major figure who is obvious in his absence from the same historical canvas, is that same Moses."

Again, significantly, the ideas behind — not up front — in *Oedipus and Akhnaton* were instrumental in the creation of these works. V. admitted,

"This study carried me into the larger field of Egyptian history and to the concept of *Ages in Chaos,* a reconstruction of 1200 years of ancient history... More than eighteen years passed from the conception of the work and the first draft of its re-writing and preparation for the printer."

Moses was taboo to V., a subject to be turned from and skirted around, except to show that Moses came before Akhnaton and that Freud was fearful yet adulatory of Moses. Even while railing against Freud's problem with his father, V. may have seen himself as Moses and son of Moses, down to the line of succession that began with Joshua. "Velikovsky," said Livio to Deg, as they walked down the street after their first meeting with him, "will be the only man who can play Moses when they make a movie of his book." And he guffawed in his *basso profondo.*

We have, that is, two plots in *Oedipus and Akhnaton.* One is the classic scientific method and detective work. The other is the intensely private psychic world of a man whose biological father was a strong and beloved figure, Simon, and whose intellectual father, Freud, had weaknesses that must be exposed, offenses against his people for wishing to abandon them for the gentile world and for taking away and making an Egyptian of their common ancestor, Moses.

Before coming to America, V. had, in one of his few published articles, reanalyzed the dreams of Freud that were available and concluded that Freud was torn by a desire to assimilate to the gentile world. V. would have none of this. While Freud would make the Jews into gentiles, V. would make the gentiles into Jews.

Here I would quote Martin Sieff who is talking about V.'s article "The Dreams Freud Dreamed" (1941).

> Velikovsky was now using the psychoanalytic weapon his intellectual father had forged against his own creator, against Freud himself... Velikovsky went further. The initial aim of his research finally to emerge over twenty years later as *Oedipus and Akhnaton,* was to kill the Freudian father dragon in its lair. Akhnaton, the first monotheist in history, stood revealed as Oedipus. Freud's arch-saint turns out also to be his arch-sinner... Velikovsky dedicated *Ages in Chaos* to his physical father, but sought to erase the name of Freud, his intellectual father, with his *Oedipus and Akhnaton.*

At the same time, V. could not go to great lengths in redeeming Moses, the father, without incurring the danger of displaying that he himself felt the strength and mission of Moses, and that he resembled Michelangelo's "Moses" more than the other son Freud did, who went to Rome to worship the statue. Worse yet, he, too, like Freud, would have to dispossess Moses if he wrote about him, for how could a psychoanalyst have perceived Moses except as a hallucinator and manipulator of crowds? And then what of Yahweh? *Au revoir,* Adonai.

That V. was not Moses, did not pretend to be, and even denied it by refusing the question of "Who was Moses?" are not superfluous remarks. To many of his readers and followers he was a Moses of modern science and history. To himself he was one who had all that Moses possessed except the opportunity. Deg tended to agree and he had studied many men, but he was not the most devout of followers. Aside from possessing his own conceits, he did not like Moses' theocracy, nor his ambitions, nor his ruthlessness, nor his religious deception even if it was founded upon self-deception.

V. differed from his secret idol by more than he himself realized and Deg liked him better for it. If a friend, like Mel Tumin, professor of sociology at Princeton University, would say to him, as he did on the train to New York one time, I can't stand him, he's an arrogant, egomaniac bastard, *Deg* would grin tolerantly and say: "I understand what you mean, but he's not all that bad, and where do you find such minds?"

Come to think of it, this was more or less what Einstein said to an antagonist, Bernard Cohen, when asked about Velikovsky. Referring to *Worlds in Collision,* he laughed and said, "It's crazy, but it's not bad." V. could be riled up invariably by the mention of this story, and he explains carefully in *Stargazers and Gravediggers* how it was wrongly told and was used to destroy his precious relationship with Einstein, and what he conceived to be Einstein's true view and mood, and I agree with him, and so does Deg.

In this connection, a private note that Deg made in May of 1972 may be offered for what it is worth:

> I have been present on numerous occasions when V. was under pressure to be intellectually and politically dishonest. I would say he passed practically all of these tests with flying colors. The rare exceptions have practically all to do with pretending to have supporters among the authorities who did not support him so strongly. Explain. When you compare his conduct with that of scientists who had no reason to be unscrupulous, because they were already entrenched or in process of achieving established rank, he stands out like a rose from a manure pile.

Because his manner and figure were impressive and imperative, V. seems to have encouraged subconsciously the awesome stupidity of attacks upon himself. Opponents became reckless out of threat, losing their capacity to reason precisely at the moment when they were being called upon to be reasonable. This is a behavioral pattern that I take pride in having newly discovered, because Deg nor anyone else to my knowledge has ever mentioned it. Let me give an example:

In *Ages in Chaos,* V. took away five centuries that did not belong to Egyptian history, whereas in *Peoples of the Sea* V. took away three centuries that *did* belong to Egypt, at least according to Deg, who was siding with the "Glasgow Revisionists." One could not follow this important development from a reading of the great newspapers or the scholarly journals. The *New York Times* did carry a review of the latter work, antagonistic as expected, but quite irrelevant to the issue. Arthur Isenberg, an Israeli writer, addressed a reproach to the *Times* editor, containing *inter alia* a neat statistical reprimand for Thomsen's snide remark about V.'s supposed overdoing of "the first person perpendicular:"

17 July 1977

The Editor, New York Times Book Review Section
The New York Times
229 West 43rd Street
New York, N.Y. 10036 (USA)

To the Editor:
    In his reply to his critics, Dietrick Thomsen is even more unconvincing then in his (highly!) original review of Dr. Velikovsky's "People of the Sea". He begins by patronizingly awarding unsolicited certificates to some of those who take Velikovsky's book more seriously than he does: They are "fine and intelligent people, and they raise cogent points" which — alas! — "lack of space" prevents Thomsen from refuting. Next, he concedes that "in many points" Velikovsky "may be correct", an acknowledgement which he repeats (in spite of space limitations) a paragraph later. But then he dilutes the concession by means of a peculiar definition of science as a "set of mind" which, he implies, Velikovsky does not exhibit. His major objection it seems, is to the tone of Velikovsky's book — as if scientific theories should be judged by connoisseurs of tone and style to determine their adequacy.
    Tone apart, he faults Velikovsky for overdoing the use of the pronoun "I" (the "first person perpendicular" as Thomsen quaintly calls it). This prompted a little research on my own part, with the following results:
No. of Times "I" is used in 100 consecutive pages:

Author	Short Title	consecutive pages
Darwin	Origin of Species	153
Hoyle	Nature of the Universe	116
Einstein	Relativity	60
Eddington	New Pathways in Science	191
Tinbergen	Herring Gull's World	161
Von Frisch	Bees, Their Vision, etc.	132
Velikovsky	Peoples of the Sea	8

(total "I" count for the entire book, xvi-261 pages: 32)
(My counting was done hurriedly; the actual figures are likely to be

somewhat larger in all cases: Thomsen is welcome to a recount.)

A grand egotist like V. rarely lets his third person slip uncontrolled into the first person, whatever the provocation. In fact, he slips into the third person, as V. sometimes did, talking of himself as "Velikovsky."

Later on, Thomsen, the reviewer, defended himself in a letter to Clark Whelton. He was furious at the impossible task set for him by the *Times*, and for bizarre editorial cuts.

What I have tried to express here is that somehow the figure of V. made people lose their senses and self-control; rages collected and rushed about like the winds when released from the bag of Aeolus.

* * *

V. moved to Princeton from Upper Manhattan in 1952; Deg moved there from Stanford, California, in 1957. Five blocks apart, it took five years to meet, a block a year, so to speak. Deg was deeply involved in New York City and travelled to Washington. V. spent his years there in secluded study, with his wife and his daughter's family for company, his wife's musical ensemble to listen to, several meetings with Harry H. Hess, and some conversations with Albert Einstein. He did not attend conventions, or review other people's books; he did not join the network of sciences, but then how could he? There was no science of neo-catastrophism. He might have joined associations of ancient history, anthropology, philosophy and history of science, though; he did not, wisely, for he was interested in a peculiar combination, unrecognizable, except in its bits and pieces, in conventional programs of the associations. He was a special case; he would have it no other way; he wanted to sit above all of them and receive their respect.

But the ideas of an authority and heretic may be contradictory. To be a heretic is to be opposed to established authority. If V. could not be an authority, he would be a heretic. His true heroes were top authorities; his professed heroes were heretics. There were three of these, he would say to Deg.

One was Diego Pirez, also known as Schlmo Molcho. A second was Giordano Bruno. A third was Miguel Serveto (or Michael Servetus). Deg's heroes were many; he was more polytheistic, so to speak, or even antireligious. They ranged from Jesus of Nazareth to Benjamin Franklin. They would include in the Church-dominated Middle Ages William of Occam, for he was an empiricist, nominalist, anti-Aristotelian libertarian who believed that words signified only real things and events, who taught also that reason could only arrive at valid comment when talking of the real world, not the divine, which only faith could attain (thus non-religious matters were freed from church control). Occam's principle, Occam's Razor, prefers to cope with a problem using the fewest possible functions and terms, so therefore Deg would feel that his simple quantavolutionary model, Solaria Binaria to begin with, and all that spewed therefrom, was in the great tradition of the Razor.

But William was beset by the authorities, convicted of heresy, and so fled to the safety of the Emperor's jurisdiction. His influence carried down the years, and of course all who were tinged with his notions felt the hostility of authority, such as the Sorbonne Professor Jean Buridan who around 1358 was drowned (not burned) and was celebrated by the allegory of "Buridan's Ass," that starved to death because it could not decide which of two bundles of wheat to eat; the same Buridan, too, revived in the song of the student-brigand-poet Francois Villon, who in turn should have been "sanctified" as a heretical hero by the student radicals of the 1960's, but was somehow overlooked.

But Deg found heroes wherever he had gone throughout life, in India, Turkey, Italy, England, Hawaii, and so on — never mind the war heroes who were glosses on the immense rainbow of heroes — and heroines, for he found heroism came more naturally and frequently to women. Whenever one studies leadership — the movement of events, whether political or intellectual, one must first carefully dissever fame from achievement. He wrote about heroes in one of his poems, contained in *Passage of the Year,* the poetry which he published in 1967, where he said:

> ... *I shall never*
> *never understand*
> *why famous names are worshipped*
> *and writers wear their pens to nubbins on them.*
> *When they are nothing*
> *while the great ones bump*
> *our elbows and disappear in the crowd.*
> *"Wait!" "Hold on!"*
> *I call after them*
> *and they don't even turn around.*
> *They are vanished, they are dust.*
> *No cast of bronze contains them.*

One of Deg's unsung heroes would have been the man whose name I forget (naturally), the English amateur of eoliths whose protests, if harkened to rather than ridiculed, would have made the Piltdown hoax impossible. But I would not detract one whit from V.'s heroes.

Schlmo Molcho was a Kabbalist and pseudo-messiah, a Catholic convert who reverted to Judaism. Around 1529 he began to believe he was the Messiah, and Pope Clement VII granted him protection. In 1531 he was denounced, tried and condemned to burn; he was saved by the Pope and another man burned in his place. He began to counsel the Emperor Charles V but was denounced and burned at the stake in 1532 after refusing to recant and reconvert to Christianity.

Miguel Serveto (Michael Servetus) was a true Renaissance figure who discovered the pulmonary circulation system, was the originator of the science of comparative geography, and was a defender of free thought and free speech. He intimated that Christ was only human, and in his writings on Christianity preserved nothing that was merely traditional and dogmatic. Arrested in Vienne, France, and condemned for heresy, he escaped but strangely entered Geneva, heading for Italy, and was caught. All the Swiss protestant cantons were consulted and returned a recommendation that he be punished for blasphemy. Calvin, however, hated him and insisted that he be burned at the stake for heresy, for he refused to retract his dislocation of the elements of the Trinity, his argument against the validity of infant baptism, and his denial of original sin. He died on October 27, 1553.

Giordano Bruno began his career as a Dominican philosopher but was accused of heresy. He managed to teach at universities of several nations and wrote copiously in metaphysics, with excursions into satire and poetry. Finally, after fifteen years of work and wandering, he came into Venice, where he was seized, convicted of heresy, sent to Rome, and , after prolonged imprisonment, burned at the stake in 1600. Intensely anti-dogmatic, he propounded the infinity of worlds, the pantheism of matter, and the relativity of man's position in the universe.

V. seems to have put the cart before the horse: one did not need to be burned at the stake to be a heretic or a hero. And a great many heretics of history escaped the fate intended for them. Often there are ages where heretics are ignored and tolerated, as in our own time in North America and Western Europe, when practically all forms of dissent, even against the heads of state and the forms of government, except when expressed as deadly terrorism, escape severe physical sanctions. The relativity of values and practices in the "advanced" democracies of today is such that almost no definition of heresy is operative.

Notably, V.'s heretical heroes were long dead. He said once, in criticizing the magazine *Pensee* and a foundation that were working to help him, and speaking to Milton, Rose, and Wolfe, that he did not "wish, well, to carry the banners for all heretics." Waiting as he was for designation to the top rank of authorities, he meant to be wary of association with any contemporary heretic.

Deg only half listened to V.'s litany of his heroes' lives and virtues. V. would never say what really fascinated him in the human characters of these men. His was hardly the depth analysis that one might expect from a psychoanalyst. Indeed — and this must seem exceedingly strange to those who did not know him — he almost never analyzed public figures or even those who were in controversy with him. He accepted them, as if they were rational creatures and their justness or unjustness was simply a matter of fact. So it was almost always Deg who was suggesting and proposing motivations and characteristics while V. seemed to regard his opponents (and friends) as unidimensional, almost as automatons.

In this way, and others, V.'s mind and character were Mosaic and Old Testament. He did not even consider himself a member of the British Society for Interdisciplinary Studies, founded to pursue work very much along his lines. Nor did he regard his tamer organ, *Kronos* magazine, as part of himself. He consented to lecture at Deg's college in the Valaisan Alps of Switzerland one summer, but he would not go and return with the chartered aircraft carrying students and faculty, so that Deg had to authorize expensive first class tickets by way of Swissair.

He was absolutely unwilling to give anyone the slightest authority over himself. He never worked for anyone; he could barely tolerate cooperating with anyone. He had a striking inability to identify with people. He did not like to be compared with anyone alive and once exploded publicly in cutting anger when Professor Warwick, in an attempt at a supportive speech, not only seemed to make light of his claims to discovery, but dared to compare his own treatment as a doctoral student by V.'s foes of the Harvard Astronomy faculty with V.'s treatment by the same people.

This continual insistence upon treating any offensive or belittling gesture towards himself as a major event, a *casus belli*, was the facade of his immense egocentrism, perhaps of the very narcissism which, in psychoanalytic practice, he claimed, must be the first region of the unconscious to be plumbed. Again one thinks of Moses, who looked upon all opposing thoughts and practices as actions against Yahweh. But V. never called in God as lawgiver, witness, judge, or executioner. He was all of these, or all of these except the last, which he left to his supporters, and was so in the name of the rational authority of the system of science, an abstract authority, not people so much as principles, not realistic principles, but ideal principles. He expected nothing less than ideal justice.

The kind of offenses that were committed against him were commonplace in science, as in every other field of human activity. But noone dared tell him so, for if such were proclaimed, the game would be up and all the cosmic heretics of the Velikovsky camp would have to strike camp and retire. Friends left him from time to time, tiring of the game. Even if one brought up an equally nasty case, he would become suspicious that his own demand-level might be threatened. This is certainly narcissistic behavior.

Often V. would protest that he had never behaved *ad hominem* towards his critics. How could they be so personal, aggressive and vile? He said that they were incorrect, wrong, and, at worst, uniformitarian in their thinking. Hardly the invective of a mighty warrior — which he was.

But there was many another to do this job for him, and no strong or foolish critic ever escaped the lash of letters and articles from his supporters. This would be done at his urging or with his blessing. They were usually appropriate, to the point, deserved — but excessive. Noone could recall an instance when V. pulled back the reins on his steeds. He usually was playing out the reins, and slapping them; many could recall instances when V. felt that a case being made on his behalf was not forceful enough.

But why did V. maintain personally so proper a language and bearing towards scientists and publicists who were terming him a charlatan, a crackpot, a novice, and more? Partly, it was strategy: to be above the battle, to be insulted without descending to their level of retaliation. He was also restrained by his ultimate conservatism with regard to authority. Authorities might, unfairly, unjustly, without provocation, drag him through the mire, but he could not let himself do the same to them. He could unleash his minions to do so, however, and they did.

This is an achievement of a great leader — to be above the battle and yet direct it, to not lose one's dignity in a thicket of passionate verbiage, to be excommunicated and martyred without descending to the level of his opponents.

At Lethbridge University, in the prairie of the oil-rich province of Alberta, Canada, a conference on V.'s ideas was held in 1974 and Deg flew in for the event. There turned up a local professor, a German named Muller, who came down heavily upon V. in the local newspaper, and V. was outraged. He turned to his largest artillery piece to blast Muller. He would not appear at the next meeting. "You can do it," he said to Deg as he lay sulking in his tent like Achilles, "noone else is strong enough." So Deg departed from the hotel room where V. and Elisheva rested, and, when the appropriate moment came, took the floor, Muller at the rostrum, and denounced the newspaper article and impugned Muller's general competence. Deg was not especially happy at becoming a petty hero. Muller was unlikeable, true enough, and had the temerity to imply that V. was converting ethnic pride into an historical reconstruction, the type of remark that Germans had been scrupulously and correctly leaving non-Germans to make since World War II. Yet, when it appeared that Muller was excessively disliked, and on his way to becoming a whipping-boy, Deg felt sorry for the person, a feeling that returned a couple of years later when the same Muller was murdered by a jealous colleague on a matter of adultery.

I doubt that Deg bothered to tell V. half the horror-stories he knew of academic and publishing crimes, let alone the sixteenth century heretics. In one case — it happened to be his own — Deg went off to World War II as a co-author and came back to find the book, half of it his composition, published under a single name, this not his own. "Well, I'll be damned!" he said, when sent a copy of the book, and was soon busy with other matters, nor was his friendship with his co-author more than temporarily bruised.

More annoying, Deg believed, was a case when his *Politics for Better or Worse* was published in 1973. Three young women instructors from different universities did a study of textbooks on American politics to prove how demeaning were their authors toward women, how indifferent, how ignorant. Then, at the last minute, Deg's book appeared on the market, was snatched up and thrown into the bonfire in an appendix to the report that they caused to be distributed widely at the national convention of the American Political Science Association. That is, they flagrantly lied about, distorted, ignored or did not read the book which, had they known, he had deliberately planned and executed as a radical exposure of the situation of women and of the need for reforms leading to sexual equality. When he composed an indignant letter to the culprits, weeks after the damage was done, he showed it to his learned daughters, Victoria and Jessica. Their advice: don't get so excited, Daddy! (How willing are children to sacrifice their parents!) He wrote a note of gentle chiding and that was the last heard of the matter. I wonder whether he should have introduced a thunderous denunciatory resolution on the floor of the Convention. After all, his book might have sold tens of thousands more copies if it had been properly contrasted with other textbooks.

V. could never understand that the crime against him was not horrendous nor uncommon. It was remarkable in the evidence being so clear and the subject being in principle so important. It was especially remarkable because he was his own biographer. Every slip of paper — every insult and complaint — was treasured. Since he succeeded in finding a great audience, in publishing his other works without difficulty, and in attracting to his areas of interest several dozen excellent scholars (a most rare achievement for even the most famous and successful scientists) he might just as well have been amused, scornful, and satisfied. Albert Einstein actually wrote him just this, after reading an account of the insulting opposition to his work: "I would be happy if you, too, could enjoy the whole episode from its humorous side."

That was asking too much, especially from V. For V. only the respectful conversion of heads of science would suffice. He respired authority and power; therefore only authority could legitimately crown him. Crowds were fine, because they were pleasing in themselves but always, too, they were used by him as a measure, such as of the pressure that his views must be exerting on the experts and unbelievers. Crowds were not authoritative in themselves.

Deg often hinted, remonstrated, and harangued: "You must not pin your hopes on conversion of the leaders," and would list the reasons why the leaders would not budge, the "sunk costs" of their lives, the unavailability of heavy sanctions against their retaining conventional views, etc. and sometimes Deg would say: "Tell me if there is a single reason why an establishment leader *should* side with you on any controversial point of yours. What's in it for him?" V. would rather not answer. He realized that he could not say "Because I am right," although that is what he would have liked to say. This would betray narcissism.

For over thirty years, V. suffered this situation, in which he was inextricably trapped. Not in full awareness, not as a strategy — because they could not be fully acknowledged as such — he adapted in several ways to the implacability of the scholars.

He claimed the understanding and sympathy of the young; uncorrupted by old ideas, they would see his ideas without prejudice or jealousy.

Becoming a champion of youth did not come easily to him, but it was an acceptable line of public argument, a stereotype of the culture. He was never an active advocate of the young, certainly not during the critical years of student rebellions.

He diagnosed the problem of the established authorities as "collective amnesia." Again, this argument came later. Deg does not recall V. having advanced it when in 1963 they had long conversations on the motivations of his opponents, but the argument is prominent in *Mankind in Amnesia,* posthumously published. As we shall see, the concept itself falls into doubt when it is used without specific valid tests to label or unlabel the behavior of persons or groups.

He watched for, sought to encounter, and carefully tended any maverick from the respectable herd of scientists. When he heard that an Australian astrophysicist, Bailey, had announced calculations showing the sun to carry an immense electrical charge, V. corresponded with him, and hosted him on a visit to Princeton; Bailey received acclaim from the heretic circle that he could not receive from the scientific world. V. corresponded with and visited Claude Schaeffer in Europe when he came to read Schaeffer's *Stratigraphie Comparée,* but, as in the case of Bailey, there was a warmth of shared sentiments without noticeable movements of these men to the Velikovsky camp. Trainor, Michelson, Santillana, Hadas, Kallen, M. Cook, Sagan, Einstein, Dyson, Bigelow, Hess, Kaufman, and others were approached, responded in greater or lesser degree and withdrew to their proper spheres.

Robert H. Pfeiffer, Harvard Semitic Scholar, appears to have accepted V.'s *Ages in Chaos,* without carrying out substantial work that his approval might logically have entailed. There was also in the seventies the category of scholars who were outside of academia, or young, or still unfulfilled who had, like Deg, entered the full stream of V.'s work, men like Ransom, Milton, Juergens, Cardona, Sieff, Greenberg, Dave Talbott, Reade, Crew, Rose, James, Lowery, and Gammon. C.J. Ransom was, V. confided to several supporters, "for a while the only physicist who saw something in my work and followed it."

The ideal supporter, to V.'s mind, would have been a fully accepting astronomer of renown, who could announce the success of an indisputable test of a near-encounter of Venus and Earth 3500 years ago. Astrophysicist Robert Bass made an effective sally in the seventies. When two British astronomers, Clube and Napier, entered wholesale upon V.'s terrain with a model of recent cometary encounters, they hardly mentioned him. Yet they possessed foreknowledge of his work and they could have used it legitimately as a foil, contrasting his planetary theory with their own cometary theory, and accepting openly much of his historical and legendary reconstruction in place of their own, which was weak. Once more we have an authority problem: though expecting a spanking, they hoped to avoid a trouncing. They received two spankings, one conventional, the other heretical; are two spanks less than one trounce?

Actually, when one goes to the heart of the matter, Deg was the only scholar of considerable previous reputation who accepted most of Velikovsky's work in the natural and historical sciences, absorbed it, and carried on with it. Most friendly or tolerant scholars of established reputation acted like a trapeze artist who pauses for a moment on his swing to watch an especially neat trick being executed by a tightrope walker in the next ring of the circus.

CHAPTER FIVE

THE BRITISH CONNECTION

For many years Velikovsky's books had been popular in Britain but his supporters were out of touch. Recalling the early days, Brian Moore wrote:

> The popular science writers occupy an important place in the communications system which links the scientist and the public, and they have played a major role in propagating the unfavourable image of Velikovsky. Having been officially declared a heretic by the scientific Inquisition, Velikovsky has been handed over to the secular arm of the scientific popularisers for public torment. Some readers may think this an extravagant metaphor, but any objective examination of the available evidence on the "Affair" will lead to this conclusion. My own interest in Velikovsky stemmed in part from the hysterical scientific reaction to his ideas — a reaction unique in this century when books proposing unorthodox ideas swarm, are ignored and sink without a trace.

I am led once more to remark upon how vulnerable the public opponents of quantavolution, particularly of Velikovsky, are made by their arrogant certainty. A full generation of repetitive experiences has hardly affected their effrontery nor hence mitigated their discomfiture.

I would point out a feature of the ridicule not elsewhere commented upon. The scientific community will have its jokes: enough to say "Velikovsky" in a group of scientists and there would arise that ineffable combination of good humor, snarls, titters, knowing glances, intellectual nudging that tie people together, like mention of a joke would other groups: "Remember the story of Pat and Mike at the wake?" (laughter in the tavern) or "'They're reprinting the Bible in a plain wrapper for the Alabama schools," or "Did you see where Ronald Reagan has gotten the Nobel Peace Prize?" (laughter and snarls). There is comfort, mutual solace, malice, subconscious fear, a bonding of spirits in possessing a few names to which phrases and epithets can acceptably be applied.

In these times Deg visited England without knowing Brian Moore or the many others who came together ultimately and with whom he later associated happily. He would visit old friends from the Eighth Army of World War II like Rayburn Heycock of the BBC or of politics, like Michael Fraser, and go about his business. In London on June 16, 1968, he is writing in his journal:

> Russell Square is green in the cool of morning and the fountain may be heard to play now that Sunday has stopped the motors. Four small boys have come out early to play a frightening game with the taxicabs. They run out in front of them just as the signal light is about to turn green. They put their faith in accurate timing of machines, just as their elders.
>
> Last night I dreamed that Velikovsky died, and was much disturbed. I wept. I felt there was a terrible loss. He died suddenly, as an old man will. I confessed that I knew nothing, that I could reconstruct nothing of his work. Just bits and pieces that meant nothing.
>
> It must have come from my walk through the British Museum yesterday afternoon. I read so many inscriptions, all flatly against his ideas of dates. One bore the suspicious rendering that I have remarked before — "Pharaoh 'A': name borne both by 'Q' in the 12th century and 'R' of the sixth century." The same man with the centuries so wrong?
>
> I searched for Greeks and Assyrians with horned helmets to correspond with those of the 'People of the Sea' whom Velikovsky places with the fourth century Greeks and noticed several features on statues and vases. Braids that look like horns, short plumes (?); Athena of Pergamon with two horned projections towards the front of her helmet (baby wings out of a crown?)
>
> The airplane ride from N.Y. had seemed short to me. Nothing had been fully solved by departure time — I left several highly important matters in the hands of others — collecting my debt from Simulmatics, the merger of our company PIT with "3is", the contract for my American government textbooks, the fate of the expedition to El Arish (permission for which has been denied by Israel), John's case at court conveniently and perhaps forever postponed and summer itinerary awry, my contract with Simon and Schuster for both "Republic in Crisis" and "Velikovsky and his Critics" pending — but in all cases the formula of the execution is assigned to someone. (Little did he know, alas, that all would proceed according to Murphy's Law: "If anything can go wrong, it will."]

The early 1970's witnessed the founding in England of the Society for Interdisciplinary Studies (SIS), conceived by a gang of four, and on a Halloween night. The first issue of their *Review,* later to be attractively printed, was in mimeography and, at that, barely readable, but its contents were of excellent quality. The founders, and those who signed up, many of them American, settled into a flexible oligarchy. The dominant members have been, on the whole, Brian Moore, Malcolm Lowery, Peter James, Harold Tresman, Martin Sieff, Euan McKie, Ralph Amelan, Geoffrey Gammon, John J. Bimson, Eric Crew, Hyam Maccoby, Michael Reade, Bernard Newgrosh, and Bernard Prescott, with possibly others, but obviously enough in number to forbid an easy sociometric diagram of the networks of cross-influencing, not to mention the differentiation between those who were primarily organizers and those who were intellectual contributors. With two exceptions, they never met or heard Velikovsky in person, although his work inspired their organization; by contrast, all of the involved Americans knew him personally.

The Constitution of the Society adopted in 1978 declared as its principal objectives:

(a) to promote a multi-disciplinary approach to scientific and scholarly problems and in particular to promote the active consideration by scientists, scholars, and students of alternatives to the theory of uniformity in astronomy and earth history;

(b) to promote a better understanding of the nature of the earth, the solar system and human history, through the combined use of historical and contemporary evidence of all kinds, and to encourage a continuous reassessment of the validity of the basic assumptions of the discipline concerned by testing these against evidence;

(c) to promote better co-operation between workers in specialized fields of learning in the belief that isolated study is sterile;

(d) to foster research among scientists and scholars towards achieving these aims.

It was not at all the American condition, where years before, following only upon occasional bulletins that supporters of V. issued in the 1960's, there came *Pensee,* a production of the young Talbott brothers, Stephen and David, whose enthusiasm for his work crystallized into a conversion of their small magazine on human rights into a forum on the Velikovsky Affair, at least for ten issues. Stephen Talbott was a brilliant editor and organizer, bent upon opening the world to quantavolutionary ideas, but also to criticism of them. After spectacular successes, *Pensee* collapsed under a load of debt and overwork. As it was ending, it promised to broaden its interests beyond Velikovsky and to discuss ideas irreconcilable with his.

V. would have no part of this, and several of his Eastern supporters — with Lewis Greenberg and Warner Sizemore leading — issued the first number of *Kronos. Kronos* became editorially the child of Lewis Greenberg, a young art historian of the faculty of Moore College of Art in Philadelphia. He recruited a group of convinced supporters of V. who contributed articles and evaluations, and who, being the closest to a prestigious academic group he could put together, he should have called "Board of Advisors," but whom he called "Staff," and he set up grades of Senior Editors, Associate Editors, Contributing Editors, and Staff, hoping to build a respectable latticework of authority such as is conventional among scientific journals.

Financing, production, and management fell to Warner Sizemore, who by virtue of his faculty status at Glassboro State College, was enabled to establish an academic connection for the journal, a public relations device of no small value for a new review with a disreputable and controversial perspective in science. *Kronos* remained essentially and in many details under V.'s thumb until his death, performing very much the function of *Imago* for Freud.

This is not to say that the directors of *Kronos* were uncritical; in the very first issue, Zvi Rix ventured ominously upon weak points in an article upon the origins of anti-semitism and the Ankh. They simply had to acknowledge V.'s power, his help, his thesaurus of notes and materials, even on occasion his financial aid, and above all — what men such as Stecchini, Motz, Jastrow, Sagan, Hadas, Gordon, and Deg, especially, had in their own way to bow to — his well-nigh complete erudition and orderly mental inventory on the matters at issue.

Early in 1976, Deg appeared at the British Library Association in London to speak to the Society; first contact between the Americans and British was made. About a hundred persons were present and Deg talked informally but to good effect on subjects both sociological and quantavolutionary. Questions from the floor were numerous and only a sense of decorum brought the meeting to a close. Afterwards the ringleaders adjourned to an English approximation of a cafe and carried on a conversation for hours.

The high competence of the British group was manifest; if they were strongest and at "state of the art" level in history, they evidenced also in abundance the imprecisely defined general background in the sciences and humanities which is so necessary in facing up to questions excited from all quarters of knowledge when exoterrestrial encounters are at issue.

I wish that I might now introduce some of the many letters that the heretics exchanged over the years; they would display the interweaving of ideas, the reportage, the delicate personal relations and the ramified research and life activities that inevitably and essentially occur in an intellectual movement. Even a single instance — a letter from Deg to Malcolm Lowery — may lend the flavor of it all.

Naxos, July 16, 1976

Dear Malcolm;
    Thank you so much for you letter and the transcript. It was excellent work and my best compliment is to edit it immediately and return it to you.

So here it is. I probably have been imprudent in letting everything stand, as you hoped I might. But it is fair, I think, and fairness is one up on prudence. I have made a number of technical corrections, clarified words, and introduced a euphemism or two. I understand that you intend to split the presentation and leave the operation to your discretion... Your article on Kugler was most intriguing. Have you sent Stecchini a copy? (...) The material is rich and your commentaries and presentations of the source matter referred to by Kugler valuable. I would expect the whole, amplified even to the extent of a complete translation, would constitute a welcome book. Perhaps one for *Kronos* Press... Was the *Atlantis* item really August '61, as you write? I'd like to see it; perhaps you can confirm the citation next time around. The Tuareg are a mysterious people, you know, of undefined race and origins. The Fabrizio Mori reports, if locatable, would be more valuable... You do bring up surprises *re* Velikovsky. No, I've only heard of original work he's done in electroencephalography, that he may have been the first to propound it. What you quote is fascinating. It does relate to the suppression of instincts, of which I make much in the transition from hominid to man... it gives us time to think, but heightens general anxiety at not being able to respond. My general theory of the subject is being prepared for limited distribution prior to the long haul on publishing the book, so I shall hope to send you a copy. Meanwhile, I would suppose you could readily do the translation yourself. Rix has a lot of trouble with English. (I try not to distinguish 'lower' from 'higher' species. In my present lonely spot, I am compelled to admit the many superiorities of the ants)... I haven't received the T.L.S. review of *Velikovsky Reconsidered.* I've gone through Temple's work on *Sirius* hurriedly. He moves into his theme backwards — first the Africans, then the Egyptians, then spacemen. Dr. V. in his "Chronology and Astronomy" found Sirius (Sothis) a yardstick for measuring the Venus-cycle. The one item (well-known) of the tribal recognition of the invisible star goes along with other ancient knowledge of the skies that was lost and recently recaptured by telescope *(cf.* my brief article — Did I leave a copy with you? — on the rings of Saturn and bonds of Jupiter). Better eyes, magnifying atmosphere, closer proximity, ancient telescopes?? — we'll have to make up our minds in the light of a total well-developed theory of Revolutionary Primevalogy... I wish that we had transcripts of the many additional hours that we spent in discussion. Which leads me to say how much I enjoyed the whole of my visit with you all. I'm due to fly back in haste...

So went the messages, back and forth and around. In the States, Deg worked closely now with Earl Milton of Lethbridge, Canada on *Solaria Binaria.* He saw Sizemore regularly in Princeton. He visited with Velikovsky. Most of the American network communications in these days funnelled into Greenberg, with whom Deg had only an annual telephone conversation but about whom he received information from Sizemore.

Kronos magazine sponsored two meetings at a Motel in the Princeton area; Sizemore exhausted himself to pull them off successfully. One was before V. died in November, 1979, the second later on, and Elisheva dropped in upon it.

Deg missed both meetings for being abroad. The second was unexciting, save for wrangling between Greenberg and Whelton. So far as I can understand the causes, there were none of substance. Clark Whelton spoke up in general criticism of the proceedings as lacklustre and Lewis Greenberg tore into him from the Chair with *ad personam* indignation which was incomprehensible unless, as I was told, "You know Lew..." Few friendly heretics — never mind the unfriendly larger participation — had no occasion over the years to receive his uncomplimentary remarks and the consoling words from others, "You know Lew..."

Greenberg's correspondance with the British was equally a mixture of rationality, abuse, and threats, and since he never would fly, he did not appear in England and only Peter James had a pleasant encounter with him. But that was once. When Greenberg invited James to become of the "Staff" of *Kronos,* Peter accepted. He was almost bumped from it when he wrote an early piece of criticism of V. and V., in a fit of anger, told Sizemore and Greenberg that they had to get rid of him or else he would withdraw his support from *Kronos.* Then according to Sizemore, V. reconsidered, recalling no doubt his own reputation as a champion of freedom of speech and press, and called up to withdraw his demand. Nevertheless, not too long afterwards, what V. had wished came about, when Greenberg and James quarreled and James resigned, as will be explained later.

In the Spring of 1980 Deg reappeared in London to address the Society. By this time his agenda was full of friends of catastrophist persuasion. *The Velikovsky Affair* had appeared in a British edition in paperback with a new preface. Earl Milton was coming in from Alberta, Canada, to speak, after which, with his wife Joan and his little son Davin, he was to join up with Deg for a heavy workout on *Solaria Binaria* at Naxos on the Aegean Sea.

On Deg's list of telephone numbers in London for the occasion we find Peter James, his primary host, informant, and contact man, a slender scintillating young and blonde man who seemed to be everywhere and into everything in London, who lived on vegetables and beer in a collectivity, and who had surpassed intellectually the university degree he was arranging to pick up. He supplied Deg and Ami with an apartment, perfect in every regard save its price and lack of telephone, of which the latter was the more serious. Hotel prices were prohibitive. Food was expensive and as always bad, except in the oriental and European restaurants.

Luckily down the street was the Baeck Hebrew center, school and library, tended over by Hyam Maccoby who took to reading Deg's *Moses* manuscript while Deg stuck heavy coins in unending numbers into the hallway telephone. For, on the aforesaid phone list were all those he wished he might see: Geoffrey Gammon, Malcolm Lowery, Brian Moore, Peter Warlow, Harold Tresman, John Bimson, Martin Sieff, Eric Crew, Robert Temple, Fred Freeman, Redmond Mullin, Rayburn Heycock, Margaret Willes, Nick Austin, and Chloe and Mike Fraser. There were thereupon added in a confused network the names and numbers of all the people who were contacted in order to contact others and the temporary, supplementary, changed, disconnected and "try-him-at" numbers.

And on his "to-do" list for the two weeks were to write his paper for delivery to the Society, to have his novel *Ronalds Norm* typed up and copied, to read the latest exchanges on *Solaria Binaria* and discuss them with Milton, to discuss with Sphere Books the *Velikovsky Affair* and his manuscripts (the same with Margaret Willes of Sidgwick and Jackson), to discuss "Aphrodite's true identity" with James and explain the ideas of an Encyclopedia and the possibility of a Quantavolution Institute, to open a bank account at Barclay's, to edit finally and send *Chaos and Creation* to the Indian printers, to visit the headquarters of Amnesty International, to visit the Temples in the countryside to see how their garden was growing and where Robert's mind was in the aftermath of his book on the *Sirius Mystery,* to write his son Chris in Rotterdam and send him some money, to meet Fred Freeman of Liverpool whose ideas on independent welfare action and tax reforms were *simpatico.* And much more, but of course, much was not done, bogged down in conflicts of time and logistical difficulties like the telephone and vainly-searched-for typist.

When his plane took off from London, he entered some lines in his journal, captioned

Failures of a trip to England — England in the Spring — Oh to be in England when...A book yet to be published jests at my ability to concoct surprising numbers. Here are more (on time expenditures]:

Trying to find a good place to eat	12.5%
Discussing the food and service	12.0%
Writing the talk that should have been written beforehand	23.9%
Futile Communications with Publishers	4.0%
Walks and visits: external sociability	29.0%
Management and commuting	10.5%
Eyeball-to-eyeball discussion about quantavolution	5.6%
Listen to others perform and performing	8.0%
All others	9.4%
	114.9%

Adds to over 100% because of doing more than one thing at one time, e.g. "No, I don't think we passed the restaurant; that was a good piece you did with O'Geoghan," or "Carter's foray into Iran was foredoomed; why did Dayton (author of a magnificent book on ancient ceramics and minerals) waste so much time decrying the mentality of archaeologists?"

Now what more would I have wanted to do? Talk to Bimson *re* opinion of natural disasters at Megiddo
Dolby *re* ice ages
Moore *re* poetry
Lowery *re* linguistics
Sieff *re*... James *re*.... etc. etc.

I am diverging and must return and repeat: the British and their magazine were more of a free association and farther removed from V.'s hulking figure. Hence it would be more likely that opposition should arise successfully there. First it happened when Euan Mackie, a proverbial tall dour Scot, a Glasgow Museum curator and co-founder of SIS, began to place monuments that were seemingly oriented to the present directions of the compass, such as Stonehenge, in the period before the Venusian catastrophe of around -1450 BC when the Earth was said by the V. scenario to have changed its axis of rotation and orbit, hence its orientations and its calendar. Further, when Deg appeared in England in 1976 and presented his thesis of "the Disastrous Love Affair of Moon and Mars," he found that the English view, led by Peter James, rejected his, and V.'s, and Robert Graves' identification of Homer's Aphrodite with the Moon, insisting that the goddess stood for the planet Venus, not Moon. James published more criticism, and Deg was given to understand that he had been worsted — Rix, Cardona, Gordon and others espoused the James thesis and Deg was driven back to the stack shelves. V. said to Deg that he had more material for the defense somewhere in his files, but he never produced it.

But then the heavy onslaught came with the long-awaited publication of *Peoples of the Sea* and *Ramses II and His Times*. After intimating dissent for some time, the British now mobilized at a conference in Glasglow in April, 1978, and delivered a set of papers that confirmed V.'s worst fears. The British — or let me say, the historical fraction of the SIS elite — while affirming their support of V.'s reconstruction of Egyptian (and hence total Mediterranean and Near East) chronology until the end of the 18th Dynasty said in effect "Stop. Disposing of 500 years is enough." The rest of the Egyptian historical sequence is in respectable order: Ramses III was not 4th century, he was also moved back to the 8th century. The Hittites did have their Empire before the Chaldeans and were not a side-show or a double for them. The end result was to cut V.'s immense loaf in half and to reassure him that "Half a loaf is better than none at all."

One might see the pattern emerging. By 1983, when Brian Moore had been elected President and Peter James Editor, much more emphatically than in 1978 might it be said that the "essential purpose" of the Society was "to promote active consideration by scientists, scholars and students, of alternatives to the theory of uniformity in astronomy and Earth history." This could only mean the general approach of revolutionary primevalogy and quantavolution. The lines of advance would move outward from Velikovsky but SIS would deny that it "is committed to any specific catastrophic theory." The *Review* would not become involved *ad hominem* and in emotionally charged wrangling but "will concentrate on the *real* issues at stake, as for example the occurrence of exoterrestrial catastrophes and the reconstruction of ancient chronology." The *"SIS Review* offers the broadest spectrum of opinion and the most objective approach..."

By this time, however, signs of a broader movement were also emanating from its elder, *Kronos,* triennially printed in America, and the younger *Catastrophism and Ancient History,* a biennial magazine founded and published by Marvin Luckerman at Los Angeles, California.

There was still no broad monthly of the type of *Science 83* (an AAAS publication) which Deg had been advocating on both sides of the ocean. He would have liked to see a published magazine *"Quanta"* and an *Encyclopedia of Quantavolution and Catastrophe,* so he caused to be sent around to hundreds of persons interested in the field a circular describing the projects as follows:

PLEASE GIVE US YOUR VALUED OPINIONS ON TWO QUESTIONS.

Project I. Quanta.

A monthly magazine, large format, dedicated to presenting to a wider public all current news and developments in the sciences and the humanities related to the theory of quantavolution: the theory that the major sources of change in the history of the world, both in the natural sciences (all fields) and in the humanities (all fields) and including human nature and behavior, have come from sudden, high-powered, and large-scale events.

It is an idea with a rich past, of famous writers, but, too, of writers whose works have long been submerged beneath the conventional tides of uniformitarian, evolutionary, and gradualist thought. We must pull out and bring forward into contemporary review the greatest of these ancient, medieval and early modern writings from all over the world, ranging through legend, through religion, through literature, through science, in all their diversity and format, so that once again they become part of our civilized heritage. Simultaneously, we must select, from the enormous volume of indifferent but carefully prepared scientific and humanistic work that is oblivious to the quantavolutionary idea, the remarkable findings, the nuggets, the truths and reality that are buried there.

Finally, *Quanta* should publish the best of the new generation of writers who are ready to tackle and overthrow old images of science and philosophy, the old idols of thought, and to discover in the world of nature and life, including human conduct and behavior, the validity of the quantavolutionary vision of the world. *Quanta* will preach and practice objectivity.

We are presently in a most disorderly state of publishing, whether of books or magazines. In this confusion of the age, there must be a place for a modest but forthright publication, and that is what *Quanta* seeks to be, that publishes for a certain critical mass of readers the facts, theories and news about a general and liberal approach to the phenomena of geology, psychology, astronomy, biology, and other science.

Project 2, The Encyclopaedia of Quantavolution.

A person who is interested in the quantavolutionary modes of change in natural and life history is often frustrated when he searches for information about a writer, a river, an animal, a myth, a phenomenon, a period of time, a place, an excavation, a planet, a concept, or a philosophy; indeed, just about anything that one looks up becomes a source of frustration. Why? Because practically every subject treated in conventional reference books has been passed through two centuries of suppression of the quantavolutionary, of the sudden, intense jumps that have been responsible for the largest proportion of change in the universe.

What has been written has not been referred to and has been actively lost. Begin with the letter "alpha", go to "Aaron", and proceed; every article has a missing slant, a missing theory, absent evidence. But so much is left out, and so many useless things are included for the quantavolutionary scholar, student, active reader, whatever the realm of inquiry, that there is a pressing need for a new encyclopaedia, so new indeed that one has to go back to the Encyclopaedia of Diderot in the Eighteenth Century to conceive of such an innovation and advance in the history of science and the humanities.

The present tight capital situation is not favorable to investments in publishing projects. Orthodox foundation channels are clearly closed. Nevertheless, given that the shortage of financial aid has not impeded thought and progress in quantavolution, the initiative and participation of scores of competent scholars in all fields of learning can be counted on to carry the project along. A cooperative organization, headed by an international editorial committee, can produce alphabetically a series of fascicules that would in three years range from A to Z. Then the total product would be bound in cloth and paper for public sale. During the interim, individuals, libraries and institutions would subscribe to the fascicules to provide operating capital, receiving in the end a sizeable discount on the final Encyclopaedia, which would cost at present prices about $90.00.

The returns were not encouraging. It appeared that the costs of finding a sufficient market for the magazine and encyclopedia would exceed the costs of production. That is, if a quarter of a million dollars were to be spent in development and first publication, not counting contributed and compensated time, at least that much money would be required to carry the message through the dense thicket of mass book and magazine advertising. The competition among the *National Geographic* magazine, *Science 83, Discovery, Museum, Geo, Science Digest,* the *Smithsonian Magazine,* and other journals was so severe, their struggle for survival and expansion so costly, that a small voice, no matter how sharply contrasting, would be overwhelmed.

The situation of an encyclopedia could be different. Here Deg discussed with Jeremiah Kaplan, an acquaintance of some 35 years and Chairman of the Board of the Macmillan Company, a possible participation of Macmillan. Kaplan had put through the great *International Encyclopedia of the Social Sciences* and was now directing the preparation of an *Encyclopedia of Religion*. The question of the controversial nature of the Encyclopedia arose not directly but indirectly. With Charley Smith, the appropriate Macmillan editor, they put together a scenario, a typical setting for the use of the Encyclopedia.

> A high school girl walks into her school library and asks the Librarian where she can find material for a short theme on evolution. The Librarian advises her to consult the *Britannica* and the *Encyclopedia of Quantavolution and Catastrophe*. The "Ev" volume of the first is being used by another student, so the girl studies the article on "Evolution" in the new Encyclopedia, writes her paper, gets a failing grade from her teacher, complains, embroils the librarian, and the librarian is told by the science teacher never to refer anyone to *that* book again.

The librarians, it is concluded, want or must buy encyclopedias that provide "unbiased" conventional articles in the name of prominent authorities; there is only one truth in science. Deg thanks his host for the fine lunch and walks out whistling upon windy Third Avenue thinking "Macmillan *has* changed since 1950. The customers now exercise pre-censorship." He did not, of course, agree, and could offer other scenarios — but what was the use?

The great one-world society was a large handicap for the movement. Creative workers were spread around the world. Far from each other, their communications were poor, and relatively expensive, given that at least half of them had disposable incomes at the official USA poverty boundary; few were well-to-do. Deg made Peter James an offer of a subsistence and "pie in the sky" if he would collaborate, but James was working and studying in a combination of a job and studies designed to extract a higher degree from the University of London. Deg talked also to Martin Sieff, who from time to time, like most Northern Irish, wondered whether he should move out before he was blown out by a bomb. On May 18, 1981, he was writing to Sieff at the "Belfast Telegraph":

Dear Martin,

I do regret that I cannot plot some position for you that would enable you to carry on your valuable work in quantavolution and history, both social and natural. We have, I believe, the phenomenon of an emergent new general paradigm for science and philosophy, and you should be on hand as parent and midwife (the parthenogenetic simile is not amiss in ancient age-breaking and age-making, as you know).

We need to publish many books. We need a magazine building upon the extant ones — *Quanta, I* call it. We need an Encyclopaedia of Quantavolution. We need an information storage and retrieval system that is set for quick production and dissemination of old and new materials. When done, our progress will be rapid, and we will generate a much larger supporting group from scientists, public, and science reporters. I cannot be blamed if I see you highly productive and influential in this state of affairs. Your journalistic experience adds to your potential.

Besides yourself are the others and I feel strongly sympathetic, too, towards James, Lowery, and a dozen more.

But visions without resources may be blameworthy. The great research centers are situated where costs of living are high and life complicated — New York, Princeton, Washington, London, Paris, Israel, Amsterdam, Basel, etc. Only a commune would work in these places. I cannot accredit the hope for large donors or, these times, a university that would accept a new institute in its budget, much less one such as ours in spirit. I tried indeed with the University of Maryland, New York University, and elsewhere; the answer, even when friendly, is "Bring in your own funds." Velikovsky's resources went into a family shop, supporting additionally Jan [Sammer] and Richard [Heinberg] for the time being, whence all products carry the brand name "made by Velikovsky." What Elisheva is doing is wonderful. Greenberg is hopelessly guarded in his *Kronos* den. Noone, however, can say it is the beginning and end of quantavolution in science, history and philosophy.

So what can be done? We are frustrated. My own income is cut deliberately to the subsistence level in order to pursue my studies, precisely at the time in life when I could be enjoying the highest earnings. But if not Quantavolution, then Kalos, the World Order movement, would occupy me ungainfully. Only a bonanza of some type, whose chance is perhaps one in ten, would let us set up some type of communal operation or institute on Quantavolution. A five year lease on an appropriate property near a good library; subsistence for perhaps eight persons, about $20,000 for materials, expenses, and initial publications: we are approaching $100,000 a year of minimal costs. Sources of funds: grants, donations, side earnings, correspondance courses, conferences, publications. Should you have any ideas

I would be eager to receive them. Meanwhile I shall brood and watch, like a demiurge, grasp at whatever creativity I can, and pounce upon any larger opportunity....

On Dec. 21, 1981, as it seems that Sieff may be enticed onto Yankee territory, Deg writes again:

Dear Martin:

There is small occasion for cheering you on to these shores, except for my wish that you might come and succeed and be nearby. Several major dailies have folded up recently. The *New York Daily News* is on the block. There is a new market for papers and talents in suburbia around the land, catering to shopping centers and a semi-literate public. Magazines are plentiful, unprofitable and short-lived. The economy is in a recession, whose end I do not see because it is shrouded in an apparently bottomless pit of world and domestic problems into which politics refuses even to peer much less descend. Book publishing, too, is floundering in the muck. Great talents, such as your own, are of little advantage; mediocrity, with unflagging snuffling in all corners, would stand you better. I don't doubt that you'll get along; that you'll be at home with your dreams, I doubt.

With all this, ought I to say, also, that the teaching field is in poor shape? The lower schools are emptying and entering into their biggest crisis since the dawn of free schooling. College and university budgets are all in poor shape. There are scores of applicants for every small opening. That still does not mean that very fine candidates are being hired for the few jobs available. Back to coda: you may find something, but you won't like it very much.

May I suggest this: If you come, come to stay; choose the spot where you want to live beyond all other; once there take on any kind of work to make ends meet and begin the aforesaid snuffling around; sooner or later, you'll find something better than most, which will give you a little freedom and cash. If you don't have friends to begin with, you'll find them everywhere at about the same level of intercourse. No matter whether Tampa or San Francisco, not any more. If we had the kind of society we wished for, I wouldn't need to write this letter because there would be a community of persons digging our sort of interest and you would make your way here naturally, and there would be a place for you without saying. The University of Chicago was that sort of area in the 1930's; almost everyone was a genius or considered himself such, and most were broke, and most were into what they thought might be the new world.

Here in Trenton, I'm isolated in a way. I have to go long distances to see people and they to see me. My little old house bears no resemblance to the fine and spacious house I once had in Princeton. The Princeton libraries are only twenty-minutes drive from here, but you cannot afford the car and gasoline, were you to crowd in with us. We'll probably be leaving for Greece in March for several months, so there is a possibility of arranging for you to stay here while we're gone. But I can see no advantage to this, since you'll be having to travel by train or by car to wherever you might be needing to go to seek a position, or to get together with people. No, it would make no sense to stay here unless I were here and then only for so long as a couple of days for an exchange of views. Even for this, I'd try to find some friend around here who could accommodate you comfortably while we visit together. I'll give you all the names I can think of, with all the compliments to accompany them, anywhere in the country you may wish to go. I'm not optimistic about this procedure, but I'll be glad to oblige. Do you remember how costly it is to travel? And wherever you go, the way Americans live in their far-flung warrens, you'll not be where you want to be even for the moment. The distances are an enemy, especially for the poor. How, by the way, do you expect to get a job without a work visa? I think you have to find an employer who will make a special request before coming. Or else, come, find a job, return and be called back. Isn't that the way it works, unless you come as an independent writer without a wage or salary paid you here.

If I had even a little money to pay expenses, I would invite you here to join in preparing the Encyclopedia of Quantavolution, a project that I think would move our cause forward greatly and sooner or later pay off financially. My idea would be to provide alphabetic fascicules every month or two until the job would be complete, financing the venture largely from subscriptions to these (with a large discount on the ultimate bound volumes), do it all in 2000 pages, all fields, half written by five editors (e.g. besides myself and you, say Brian, Bimson, Milton, Lowery and other good colleagues who might want to come aboard) and half by about 100 other contributors, taking three years in all, appearing in three volumes in 2,000,000 words and selling at a low $89. I think Princeton would be a good place to center it but I wonder about Cambridge, Eng. (with occasional editorial conferences in Naxos). I would readily contemplate a move to Cambridge if there were a few enthusiastic souls about and a minimal cooperation by the Cambridge Library authorities. Couldn't we lease an old house big enough to barrack visitors for a reasonably small sum for three years and have a go at it? The production should be done in-house on a word-processing system that would provide print-out for the fascicules during the whole creative period and then feed floppy discs to the automatic typesetter for the final production of the bound volumes. We would attach a newsletter, perhaps the Newsletter of "Workshop," to the fascicules and when the Encyclopedia comes out continue the publication of a wide-public magazine *Quanta*.

I was going into Manhattan today but am glad that I changed my mind and could therefore get this letter off to you, among other things. Holidays don't turn me on; I make my own, as often as possible. Concluding, let me not give the impression that I have ceased to think about what you might do and where, but give me feedback and encouragement and I'll do better next time.

>Cordially yours,
>*Alfred*

Martin Sieff came like a whirlwind, and came again not much later, a short, dark counterpart of Peter James, a comic book buff, friendly and grateful, darting brown eyes through heavy glasses, missing nothing, spewing out accounts of college days at Oxford, the dire internal politics of Israel, the latest bombing of his Belfast newspaper, the psychology of Velikovsky, the girls of Long Island-Belfast-Jerusalem, the personalities of the cosmic heretics of Britain, the confusion of the British Society for Interdisciplinary Studies ("Nothing at all like the big way you do things here, no support..." "What do you mean? We are disaster-stricken. Out of touch, nasty little arguments and all of that..." "Not really, I thought that was us!" "Not so, I thought that was us!")

Martin wants to see Clark Whelton and he and Deg hear of Clark's longing for an Association where we can all get together on a regular basis. Alas, Clark is assistant to Mayor Koch, on 24-hour alert; he is writing a novel; he is going through the trauma of kids readying for college. How, when, with what means and who? Everyone looks blank and slightly pained. But the outer world must have something in mind when they speak of the "underground" the "well-organized tactics" of the catastrophists, the invariable sharp attacks greeting an offensive remark about Velikovsky or against short chronology or for exoterrestrial eternal peace, as, for instance the London *Times Literary Supplement* of 26 June 1976 murmuring about "a powerful force in the underground of academe."

Not long afterwards, dodging about the streets of Belfast (he has spent most of his thirtyish years in two civil emergencies, of Belfast and of Israel), Martin rifles a letter to Clark Whelton at the Mayor's Office in New York, expressing fear of the collapse of the Society for Interdisciplinary Studies journal.

Belfast, 9 August 1983

(...)
"There is only one solution that I can see — the appointment of an Editor-in-Chief with full authority over production, and over all SIS copy — both Workshop and Review, able to appoint and fire editorial staff at his discretion, responsible for deadlines, and responsible himself directly to the SIS Chairman, creating a workable Publisher-Editor relationship. Should you succeed in launching a U.S. version of the Society, this is the only way

to get the thing done. Government by committee is a wash-out. As long as Lowery was on form it served as a useful camouflage for him to operate under, while he actually put out a high quality product. But once he pulled out, the whole cumbersome system of referees and editorial committee responsible in its turn to Council, another committee under a mini-Lowery in its turn, just fell apart. Peter James is an outstanding scholar. But he doesn't know the meaning of the word "deadline". Brian Moore put an immense amount of effort into the Review's production — and had nothing to show for it at the end of the day...

There was of course no money to pay an Editor. Sieff feared a collapse of the Society, and could only pray that its membership would be patient with the leadership a little longer. [In a letter to Deg later on he expresses surprise that the phoenix is arising from its ashes.]

And then, horror of horrors, Martin announces re-re-revisionism of ancient Egyptian chronology: I am becoming convinced that everything that happened in the Exodus and in the crisis of the Ipuwer Papyrus may well have been at the end of the Old Kingdom. At this point Deg's mental vision shutters down like a toad's eyelids. When the revolution comes, nothing is spared, and then it feeds upon itself. No, you don't, Martin! That's too much!

Here is how Sieff declared the consensus again to Whelton: *"Ages in Chaos,* Vol. I still stands. Minor corrections and improvements, yes" — but the Hyksos are the Amalekites; El Amarna tablets fall in the time of the prophet Elisha; Queen Hatshepsut of Egypt is the Queen of Sheba; Thutmose III is biblical Shishak. "To which I will add the correlation — Ramses III in Jeroboam II's time; Merneptah kicked out by Azru = Uzziah/Azariah; Ramses II = Late Bronze-Iron interchange." In these words, 30 years after *Ages in Chaos* first appeared, Sieff is pronouncing the validating results of thirty years' work, practically none of which was done by anti-heretics, and which, whatever else happens, in cosmology and chronology, are sufficient to bring the rewriting of much of ancient Egyptian, Hebrew, Syrian, Anatolian, Greek, and Roman history. But Martin is part of "whatever else happens" and so are Peter James, David Rohl, John Bimson, and Jim Clarke who are energetically taking V. apart and putting him together again. The old chronology is gone but there is yet no tongue-in-groove replacement.

* * *

In April 1983, Deg and Ami, after two months in France to promote her just published novel, *Le Pigeon d'Argile,* go to London from Paris and he speaks on Homo Schizo, on the gestalt of creation that in short order makes a cultured person out of hominid. This time they have the apartment (and telephone) of Stimson, Peter James' friend, with a monster bed embracing its room, from which everything is reachable with levers and buttons and on which all is do-able, apparently including dining, for there is no dining space.

There is a fine celebration after the meeting, proverbial homemade English pastry playing a nostalgic part; drink flows freely and the survivors end up at the pub nearby. Deg meets Jill Abery so can tell her that he admires her snippets on fossil assemblages and many other mini-reviews of the quantavolutionary literature. Again he misses John Bimson and, too, Bernard Newgrosh, the medical doctor who edits *Workshop* for the SIS.

He does a fast trip to Brian Moore's Cleveland haunts and the two of them ascend the Observatory hill in Edinburgh to spend hours with Victor Clube and William Napier who have published their *Cosmic Serpent,* which Deg had read, but they have not read *Chaos and Creation* so he gives them that and they give him a reprint and all are full of talk and trying for a common ground while sniffling about a bit doggishly.

Clube and Napier call their quantavolutionary scenario "the disintegrating comet theory." They set themselves to showing that at great intervals of time the Solar System encounters galactic clouds of cometary material and suffers heavy destruction from collisions. Residual comets accompany the Solar System, and their periodic visitations, on rare occasion, end in disaster. Like many others working on catastrophism, the two Edinburgh astronomers find themselves isolated, both because of the extremity of their ideas and because they need much material from fields like mythology and linguistics that they cannot grasp themselves nor command expert consultants to provide for them.

The crux of the matter is that, while both groups grant catastrophes in human times, the Scottish astronomers want to read "comets" where the Deg-V. contingent read "planets" and they bring out reams of calculations on Encke's, Halley's, and more to come, while Deg is confident by now of Solaria Binaria and cannot wait for the book, which, if not calculation-full, is calculation-proofed, and he feels good about some tag-wrestling matches to come, where with much better historical reconstruction and with Milton at his side, well, we shall see, he thought happily, as they stepped out upon the Observatory site overlooking beautifully the fine sombre city with the sea beyond, and they took their jovial leave.

Deg was pondering, wasn't this setting where Comyns Beaumont placed the world of the Bible and was Edinburgh Jerusalem and it was all transferred to the New Palestine after the comet struck — nonsense, of course — to what lengths will not subconscious ethnocentricity lead one, but how far and how near was Beaumont to William Blake the mystic poet and painter who envisioned Jerusalem as England, pathetic genius, lost soul amidst the steam and soot of his century.

* * *

Time had come to leave England for New York, but two matters had to be settled. After much thinking and talking, Deg decided he could entrust the manuscript of *Solaria Binaria,* which he had been hoarding all the while, to Rosemary Burnard of the Society for composition on the IBM type-setting machine that the Society had scraped up the funds to buy and use for its publications. A type-font was chosen, the format designed. Within three months all would be done and the pasted-up camera-ready copy would be sent to Milton and Deg for final correction and printing. Not so: July stretched to January before the job was done. Shall I stop to explain the six months delay, Deg's fortnightly fury, the sweet, bold abstracted character of Rosemary, the trials of the intellectual underground in Britain, speaking of how things don't get done and finally maybe do get done in the perennial bohemia of generation after generation of the Western World intelligentsia? Of course not. I cannot allow myself a Proustian self-indulgence in prose. If there is a page to spare, it must go to the heroic efforts of it seemed everybody to penetrate the U.S. Immigration Service just enough to get Ami aboard a plane to New York.

Excepting the several millions of Indians who already were on hand, the vast majority of individuals (and I use this term significantly) who came to the shores of the New World were driven away from their old haunts — by the Old World authorities, by famine, by failure of one kind or another — and half of them came within the past century. And they are coming now, in vast numbers, such that the system of restraints has broken down, and the question now is how to legitimize millions of persons as Americans without setting in motion a similar advent of millions more. At work, of course, is the U.S. Immigration and Naturalization Service which, you must understand, is separate and distinct from the Department of State, but shares this with the Department of State: that they live a life out of Kafka's *Castle,* full of resounding laws, rules and regulations, and of textbook principles of administration.

Now, as in Kafka's books, the people most removed from the intent of the laws are bedevilled by them. So it is that an apolitical, well-behaved French writer, who is married to an American, unrecognized for the troublemaker he is, can have more difficulty getting in and out of the country than anyone of the mob of persons whom the agencies are instructed and exhorted to screen, examine, and order into various categories. So it happened, that the aforesaid French novelist, female, law-abiding, with a stamp on her passport letting her in but stuck with a paper not letting her out beyond a certain time, can be prevented from coming in and must begin at the beginning — lines, forms, physical examinations, faceless officials, and time without apparent end.

Here then enters Professor de Grazia, professionally, fully, sceptically, ironically, indignantly aware of what imbecility *ad infinitum* bureaucracies historically display, whether in science or in travel, yet who still imagines that a minor delay in the return of his wife, for good reason (for the good of the USA, too) will not cause much of a problem, if he addresses the Immigration Service in London properly and in good time. One week of good time goes by, and a second week. Ordinary communications, cables, phone calls are not enough. Interchangeable faceless beings turn on and off. The system cannot cope with the request to reenter; a ping-pong game is set up, with the US offices on the one side and on the other side of the Big Pond reluctantly striking the ball, after resting in-between shots.

I cannot be sure of what finally happened, except that at a certain point Deg stopped acting like a proper ordinary citizen trying to get his wife back home and began acting like a politician and a border-runner. Ultimately are mobilized the good offices of a U.S. Minister, a Consul, a U.S. Senator, several U.S. lawyers, and a politically prominent British Lord, coupled with a partially blocked presumptuous entry upon a British Airways plane with the baggage flying solo, until somehow something cracks in the system at the New York Airport, and the message gets through to the airline that if Anne-Marie de Grazia were to be aboard a certain plane no objection to her coming home to America would be raised by the Inspector at the immigration counter. Nor was there.

PART TWO

CHAPTER SIX

HOLOCAUST AND AMNESIA

As his last year begins, Dr Zvi Rix is writing to Deg from Rechovot, Israel. It is January 9, 1980 and he sends New Year's greetings, and hopes that they might meet before long. "I am very cut off at the place where I am living now. This does not only concern libraries, but other matters too..." for the mails are slow and books arrive late in the shops. He is in touch with Christoph Marx. They travelled together to Glasgow... "He was quite obliging... So far I have not formed a final opinion of him."

I would nominate Zvi Rix to be the hero of this chapter, but it is up to the reader to find his own heroes in this book. Rix was a man who Velikovsky would have liked to write *Mankind in Amnesia* in his place. He was a medical man, deep into psychiatry, and a refugee from Nazi Germany. Deg knew him only through their correspondance. Deg was glad to get a description of him from his widow, whom he met shortly afterwards at the home of Christoph Marx near Basle. She wrote to Deg on January 23, 1981:

Dear Prof. de Grazia,

My husband died very recently; as is customary for Jews, even not practising religious commandments, we stay at home at least a week. In this time I went through his many letters and found also yours.

I have the impression that you were very friendly and very much appreciating his work. Therefore I write to you that I am very thankful to you. He was a very lonely man and every encouragement was a help to him. Here he had nobody to talk to, I myself am much too obtuse to understand half of what he was talking about and as he was also very shy he had no contacts; besides that, his ideas were not exactly what people here would like to hear. It is a semi-theocratic world. Ruled by a conglomeration of

Zealots (...) they call themselves socialists or rightwingers, its all the same. Our dreams went awry.

> Yours very respectfully,
> *Melitta Rix*

Rix, whose scrambled writings are being kept by Christoph Marx, was hard in pursuit of evidence that the cometary destruction of civilizations around 3500 years ago had warped the human mind in the Near East, inciting human destructiveness, religious excesses, and sexual deviations.

* * *

Christoph Marx was a computer expert from Basle, and an amateur of Velikovsky's work and all that it connected with. He circulated an invitation to whomever he knew to meet in Iceland, a typical groping, logical yet mad, of cosmic heretics for a way of expressing themselves and their message. Logical: let us assemble in Iceland between America and Europe, a catastrophically threatened land even now, set athwart the great catastrophic Atlantic Ridge; mad: Marx was teetering on the edge of interdiction by everyone, the British, the Americans, the Europeans, Deg included, a heretic practically excommunicated from the heretics. The conference did not materialize. Marx tried again in 1980, this time in his home city, and found a few communicants.

> The *minimum* consensus of all people positively involved with the work of Immanuel Velikovsky may well be characterized as *an interest in the true reconstruction of mankind's genetic history,* and thus also of geologic and, in part, cosmic history.... Developing Velikovsky's psychological inceptions, the goal — of bringing home to collective consciousness the realistic conception of the world, as opposed by the present mania holding sway over cultural evolution — would include nothing less than safeguarding mankind's life on earth, imperiled by (1) by the acute danger of self-destruction, and (2) by not attempting to prepare against some future chaos in the solar system. However, whether some of us are attributing such healing powers to the recognition of true history, or whether others would simply consider it as a value in itself, does not seem all-important: both parties will equally perform a supporting function in repelling collective irrationality and fanaticism, the worst effects of which are mass killings through war and murder. We know that Velikovsky comprehended his own striving for the true picture of history in this perspective....

The consensus among cosmic heretics of which Marx spoke in his announcement did not really exist; however, it is certain that V.'s unique and original way of searching for the roots of anti-semitism was a revelation to many thousands of people who would otherwise have not even considered the problem or would have lived with a few, often anti-semitic, stereotypes. Measuring such influences is impossible, but, by any standard, V. was a great Jew who disabused the minds of many incipient anti-semites.

Deg's journal Paris, August 19, 1968

    V. keeps two secrets, or doctrines half-hidden. He has expressed himself to me so often that the "secrets" are apparent. He would perhaps deny them. I am sure of them. He does not believe in God. He is a Hebrew, therefore Israeli, imperialist. Both doctrines, if publicized or known, would involve him in a whole new line of controversies, would make new enemies and unwanted new friends.

    Evidence, examples:

    Of 1: direct statements; writings; philosophy of psychoanalysis; his theory of "great fear" as bringing religion; belief that Jews were even in Biblical times polytheistic.

    Of 2: works of his life — Zionism; gift of income from his property to Israel in June 67; written works analysis; conversations; hatred of anti-zionism even at cost of other values (e.g. El-Arish incident and Brandeis professor).

After a long trip following V.'s death, Deg returned to 78 Hartley Avenue (he could never remember the house number, but would send his letters to 34 or 85 or another number, any number, and V. was puzzled — What significance could forgetting it have for Deg? "You can address me just at Naxos, Greece and I get you alright at Hartley Avenue, Princeton!" "I have gotten letters just to 'Princeton, NJ' " — "So there you are!") to see Elisheva. The parlor was little changed. V.'s unimpressive chair stood facing the two stiff couches and the coffee table between. Deg thought, "Should the chair be sat in, moved, replaced, bound across with a museum belt, what?" It struck one with incompleteness, an uncertain quaver. He would slip some books and papers upon it. Elisheva and her assistants Jan and Richard lined up with Deg on the couches. Like a cordial committee they sat, drank tea, and reported to each other: health, manuscripts in progress, people seen; and they passed papers and books around.

    Thus went the meetings in the years thereafter. Sheva would at some point ask: "Did you see Marx?" and Deg would say no or yes, and she would say "How can you see him when you know how bad I feel about him," but she was curious nevertheless, while Deg tried to evade the subject and one time she said "I will not speak to you again if you see Marx" and Deg threw his arms around her jovially and said, I tell you what, if you don't see Greenberg, I won't see Marx, and she was taken aback and all laughed because she had mixed feelings on that subject too and knew that

Greenberg was not his favorite among the cosmic heretics, but setting up proscription lists in the Roman style was pointless.

It was on one of his earlier returns from abroad, in 1977, that Deg heard about Christoph Marx. V. spoke of a visitor, almost in religious tones, who had lifted weighty burdens from his shoulders, and would establish his rightful fame in Central Europe. He gave Deg a copy of a well-executed chart of his reconstructed chronology of Egypt, in color, which Marx had drawn. "Good, good," commented Deg, who was surprised, bemused, and sceptical at the same time. "What's happened?" he asked Sizemore and others when he met them aside. They seemed confused and uneasy.

What happened is this. A Christoph Marx had telephoned Velikovsky to pledge his allegiance to his ideas and to offer support. There was much he could do: he could help with the translation of V.'s books into German, working out of his more respectable (in V.'s eyes) Switzerland; he could launch a campaign to bring the Germans to their senses, so that they would remember the horrible Nazi past and thus cleanse themselves of the pest of comfortable oblivion, with its eventual compulsion to repeat the past again; he could organize study circles to confront the establishment with Velikovksy's ideas.

On April 14, 1977, V. wrote Marx, confirming in most cordial terms an invitation to visit. For ten days, Marx settled into Princeton. Professor Lynn Rose, who V. said at various times would be his literary executor, came down from Buffalo for some of the discussions. Marx departed on Mayday. V. writes him: "Dear Marx: You left on Sunday, you called from home on Monday, and today is Friday — and very many things did happen in those few days... Earl Milton from Lethbridge, Canada, is with us since yesterday and leaves tomorrow morning together with Alfred de Grazia — who just now spent with us some time — and left copies of letters he wrote to Enc[cyclopedia] Br[itannica] and to NY Times. Sagan sent me a new book of his inscribed with all good wishes and a day apart arrived the tape of this year's lecture on the yearly theme — Venus and V. — in which he indoctrinates future astronomers in their first year with derision toward me and my work..."

Three days later V. is writing about turning over rights to the royalties from various foreign translations to members of his family. He says he is turning over the management of worldwide Spanish language rights to his recently acquired agents, Scott Meredith. He says "I reconsidered and wish to suggest the following plan: your share is one eighth (12 1/2%); but you retain countries not 'gifted' an additional 7 1/2% for work that furthers our goals — at our common discretion (such will be the case with Germany),..."

V. writes also to Lynn Rose on May 11 that "I let him (Marx) have broad powers to act, and have already the first report from him. He will take over most of the European Continent for contracting my books with publishers, and be a rather central figure in organizing groups of interdisciplinary synthesis, and in opposition to the Establishment." He mentions other rights to be bestowed upon individuals and adds "Christoph Marx will be in charge of these and many other activities."

On May 16, Marx replies that he will proceed as desired. He wonders whether the gifting of "income" rather than "rights" is not the better procedure, and suggests that the literary estate should be kept centralized and managed efficiently. His idea is of a Velikovsky Institute, a foundation not-for-profit, with an office in Switzerland and another in America.

V. seems to be in a manic phase. He sends off sundry "Notes to my Collaborators," a newsletter in fact. *Inter alia* he mentions lending Marx his unpublished manuscripts and writes that "I gave him wide powers to represent me in academic contacts and arrange for the publication of translations of my books."

In August, V. visited the office of Scott-Meredith Literary Agency in New York and met the head of their foreign rights department, Mr. Vicinanza, who "showed great eagerness to represent me on a broader basis." An offer was made to enter the greater European market. Vicinanza estimated that $750.000,00 could be obtained in advances worldwide for *Worlds in Collision* in 18 months: so V. reported to Marx, adding, "Against such figures the offers made to you appear minuscule,..."

A month later Marx reports to V. with several offers and expresses doubts (as did V.) about the high figures. Marx would like to sign in the name of the "Velikovsky Institute." In any event, he would like to draw upon the expected advances to begin microfilming and indexing V.'s archives.

Then suddenly, V. telegraphs "Please don't sign agreement with Umschau. Wait my explanatory letter. Greetings." Something has happened. There is a flurry of letters and telegrams. In a telegram, V. says that his books are being returned by the thousands due to the book *Scientists Confront Velikovsky* (by Asimov, Sagan and others) and "other adverse publicity." Marx appeals by telegram for confidence and trust, to no avail. They also talk on the telephone. Marx is seeking to give "rational" answers to all objections, but says "I have legally signed the agreement as your proxy within the frame of German and Swiss law. At this point I again wish to thank you for the powers you have entrusted to me, which I consider as a wide obligation toward you and your family."

I suspect that around this moment, Marx had been hit by the inevitable reaction to the Grand Vision. V., always a procrastinator in decision-making, facing opposition from his family and the lack of enthusiasm of friends such as Rose and Sizemore, could not overcome his profound aversion to things German, including now spending resources "to help reeducate them." Marx might as well proceed; V. would never have returned to the Great Vision; his idea of therapy would have to be applied by others, if at all.

Marx has signed the contract on November 22; the Umschau Verlag signs on November 29. He reports that he is putting the money in a special account in German Marks, which are moving upwards against the dollar. He continues to report editorial activities.

Now young Jan Sammer, who has come from Canada to live and work with the Velikovsky's, writes to Marx. Without expressing his authorization, he relates that V. is upset with the disapproved signing, that Doubleday Company will probably insist upon 25% of the proceeds, that V. does not favor the Velikovsky Institute idea, that Marx has "overstepped the powers that V. granted" him, and that he could negotiate but not sign an agreement without the author's approval. Marx is told to stay out of affairs in Holland. Marx replies both to Jan and to V., avoiding a confrontation.

Jan writes again repeating himself more forcibly, adding a warning to Marx not to pretend to represent V. in speaking to any scholars. He repeats words written earlier by Marx: "Umschau in due course will wish to have proper signatures to the contract. You would have to empower me accordingly." How, asks V., through Jan, can you now say you had power to sign.

Marx argues at length to this point: V. had orally and even in writing granted the power to sign. Marx speaks of a further consideration being "my understanding of how distasteful Dr. Velikovsky would regard a duty to sign a German contract personally." (Deg remembered that V. had considered even not permitting his books to appear in German.) Marx states that V. had told him not to worry about any claim of Doubleday to the subsidiary rights.

Finally on March 1, 1978, Mrs. Elisheva Velikovsky writes to Marx, repeating that Marx had himself said that further empowering authority was needed, insisting that he not present himself anymore as V.'s agent, and condemning the idea of an Institute. Marx rebuts this, and indicates a desire to visit Princeton to settle matters.

The visit is declined by Mrs. V. Marx inquires about V.'s health. His letters continue to carry news of books and meetings. Jan says in the middle of a letter of May 17, regarding Marx's expenses of purchasing books, that "in any case, they would have to be paid by you from the 7 1/2% designated for expenses connected with your efforts to arrange for translations." More reports. V. telegraphs for an accounting twice in the same month, the second message being misaddressed to "Immanuel Marx." And a third cable demands the transfer of funds to America. Marx sidesteps these and writes of his work on the Dutch contract, which he had been called away from, and of his dislike of entitling the German translation of *The Velikovsky Affair* (Deg's Book) *Immanuel Velikovsky, Die Theorie der Kosmischen Katastrophen,* a publisher's presumptuousness that one might find annoying.

On August 15 goes to Marx the first letter by V. in two years. It asks the transfer of money, and that V. be informed of all negotiations from the beginning and that no contract be signed without written approval; if not, any authority will be revoked. Marx on August 24 refuses the "fundamental change," acknowledges the end of the agreement is inevitable therefore, and suggests he be allowed his 20% of receipts from books signed up and be given all German language rights. '...Such German monies are not going toward an enrichment of myself... no other people in the world need your works as urgently than the German speaking peoples.' On September 5, V. signs a handwritten message, witnessed by his lawyer; it "terminates our business relationship." Further, Marx is accused of having been in California and Washington, D.C., "but did not give a ring to Princeton."

Marx retorted that he had too many rebuffs to continue telephoning. He protests that, in V.'s name, the *Kronos* magazine group was denying him permission to publish in German various of its articles. He also received in due course damning letters from Lynn Rose and Warner Sizemore. Rose adds a postscript calling "a deliberate misrepresentation" a letter from Marx to the *Times* which asserted that "Velikovsky saw the Holocaust in terms of collective amnesia."

Matters had been sliding into the hands of Robert Pinto, Velikovsky's attorney and, with V.'s death, attorney for his Executor, Elisheva Velikovsky. The ensuing fol-de-rol among Estate, Publishers and Marx went on and on and is of little interest here.

So a kind of love affair ended, brutally, with injury to all concerned. Sizemore wrote to Marx April 3, 1980 that "the last year of Dr. Velikovsky's life was almost totally taken up with the question of how to put a stop to your activities. He rued the day he ever met you." This may be so, but is it rightfully so, and is it all? Velikovsky was not working well for years. Further, in the last week of his life, Deg had him smartly discussing substantive topics of quantavolution. (Marx went unmentioned.) Yes, in a way, Marx was V.'s Waterloo, his last grandiose effort to launch himself against an opposing world. He loved Marx for the vision, even if Sheva and Warner and Rose and Deg and all the others could not share the vision nor needed it. Deg had not yet met Marx.

On May 9, 1980 Deg is writing to Mrs. Velikovsky:

Naxos, Kyklades, Greece, 9 May 1980

Dear Sheva:

When I called to say 'good-bye' before going to Greece, you had already gone to Israel. I hope that you enjoyed your visit and are well at home now. Ami and I spent a month here and then three weeks in Western Europe, two in London. The Society held a day of meetings on April 26. Talks were given by Dayton, Warlow, Milton, and myself— I spoke on "Ten Propositions concerning the Quantavolution of around 1450 BC," or something like that. About 150 persons were present. There seems to be a continuing high interest in Immanuel's work.

C. Marx came from Switzerland for the occasion. Somehow he had learned of my coming and had written Sizemore to pass along any messages *via* myself. Isn't that interesting — implying that I was in contact with him. Furthermore, he had been sending to the British group letters presenting his case to represent Velikovsky, including even Immanuel's will, which I therefore had occasion to read, and which fortunately is simple and clear and free of any embarrassing detail.

After my talk, which was the last, Marx introduced himself. I exchanged a few words with him. As you say, he is disarmingly mild and inspires immediate sympathy, to the point of affection. I advised him first (after commenting that he should not have tried to give an essay by himself a ride on my book of the *Velikovsky Affair* without consulting me, by trying to put it in through the publisher) that he was all wrong about you and that you had been kindly disposed towards him in the beginning and that he should write you a letter of apology. Second, I advised him not to perpetuate a controversy that would only damage him and cause everyone great costs, and rather to put his case up for arbitration by three persons, not including myself, to determine what, if anything, was and is due to him for his work and achievements. He didn't seem to care for the advice, but my last words to him were to think it all over. Probably you have heard that he is hoping to gather a conference in Rekjavik, Iceland, soon. I have no idea who will come.

While in London, I stayed at an apartment only a few meters away from the Jewish Synagogue and college where Hyam Maccoby works, and we had several meetings and a lunch at the best Jewish restaurant in London, Reuben's. He read most of my book on *Moses and His Electric God* and found it plausible and interesting. He knows the sources very well. I have heard nothing from Charles Lieber in New York, who is supposed to be finding a publisher for the book.

We shall probably be leaving Naxos for Athens and New York at the end of June and thus be mainly in Princeton during the summer. Is Richard still with you? — I suppose so. Please give him our regards — also Ruth, and Warner when you see him. I look forward, then, to seeing you again before too long. Best wishes meanwhile.

> Affectionately,
> *Alfred*

On May 11, Marx addresses Deg, expressing pleasure at their brief meeting:

14 Years ago you pointed to the Velikovsky affair and its implications, and still good scientific form seems to require that even Velikovsky's main theses together with the principal view whether the reconstruction gives a

true picture of mankind's past cannot be considered as fact, from which to proceed to new work. In spite of all the experiences of these 14 years a rather naive opinion also seems to persevere, that if only one persistently kept to so-called scientific method, in the final analysis everything will turn out just fine. For the disastrous non-success of Velikovsky's ideas in science a Scientific Mafia is found responsible, but science itself, the field that many Velikovskians are employed in or would like to be part of (if just for status only), and which from its beginning has allowed the most irrational large-scale delusions to grow *(Grosswahnbildungen* I call them in German), is glorified by naming our hero one of her greatest representatives. After I've seen science destroy the more important of these delusions, such as ancient history or some myths of physics, by its own methods, perhaps I'll be ready to call Velikovsky a scientist: until that time, which I don't really expect to really come true, I prefer to know Velikovsky, along with Freud, as the brilliant analyst he was; to withdraw him and his work from the clutch of science; and thus remain free to expose science wherever necessary or as a whole as one of the great systems of thought (after classical philosophy and religion) shielding the collective from its memories.

He complains of "the most unfortunate job Mrs. Velikovsky is doing in ordering an about-face of her husband's approach to the Nazi Holocaust." He thanks Deg for suggesting arbitration and will, he says, essay a move in that direction.

On June 4, Deg replies:

Dear Mr. Marx:

Thank you for your letter. The Breasted citation and pages are welcome. I will seek the hieroglyphics, now. Concerning your last paragraph on the 'arbitration,' I have already written to Mrs. V. of my suggestions to you, so certainly you may refer to them if you wish. I am glad that I was never part of your complicated and difficult relationship with the Velikovsky's, else I would feel responsible at least in part and therefore more sad than I am.

Any impression that the whole story has been told would be incorrect. The major issue is hardly reflected in it. The more one considers the affair, the more one senses an underlying tension. Would it be the pronounced incapacity of either V. or Marx to work with others? Certainly Deg's original scepticism of the relationship was based upon his acute awareness of V.'s tendencies to call his troops forward, only to have them halt before commitment and forever be frozen there. V. called himself a procrastinator.

But Marx was a patient and loyal and demonstrative person. He could have gone along indefinitely and, given the neat bind trapping both parties, the relationship, hot or frozen, would have persisted.

The crux was the Holocaust. It was deeply disturbing. The matter could be put syllogistically: historic catastrophes resulted in severe collective amnesia; the world's peoples, having suppressed their memories of catastrophe, are compelled psychologically to recreate the conditions for reliving them; thus emerge warfare, massacre, self-destruction and the destruction of others, man-made holocausts. Whereupon one reasons: the Germans, like all peoples, have undergone the ancient catastrophes; like all peoples, they have suppressed the memories of them; like all other peoples, they are prone to recapitulate them, and do so on occasion, as during the Nazi period.

Now the process implies a therapy. To cure the penchant for human destruction, the victims of collective amnesia (practically everyone) must be led to confront and appreciate the extent to which their minds contain the experience of past catastrophe and hence the seeds of future ones; once this is done, the human will realize the meaning of his conduct and control it so as to break the endless chain of disaster. What is good for all peoples must therefore be good for the Germans. Hence any effort to cure the Germans of their collective amnesia is to be commended and supported.

This, in brief and with such defects as I shall point out, was Velikovsky's social philosophy, and this everyone who paid any attention to V. knew to be his philosophy, and Marx clearly saw this, too, and was fully persuaded of it from his reading and from his early communications with V. He was deadly serious about it.

Long before all of this, on December 18, 1963, we find V. writing to Dr. Zvi Rix in Jerusalem: "I found two of your ideas magnificent, the hatred of the Jews because they claim of having the upheaval made for their benefit (the Hyksos actually profited); and the words of the Gospels about the fiery furnaces and Hitler's accomplishing such vision and doom (by extrapolarizing his own hateful traits)." Again in a letter of January 7, 1964, he calls the idea "stupendous." He "wished that somebody else should write 'The Great Fear'," because he is so busy, but suggests a cooperative book to which he might also contribute. Nothing came of this highly unusual disposition to engage in collaborative work.

In 1974, V. journeyed to the University of Lethbridge in Alberta, Canada, to receive an honorary doctorate. The Conference in which he starred was devoted to the topic of collective amnesia. His own address was subtitled "The Submergence of Terrifying Events in the Racial Memory and Their Later Emergence." There he commented that "the inability to accept the catastrophic past is the source of man's aggression... Warfare has its origin in the same terror." Leaders imitate what they perceive to be the gods in action. Nobel Peace Prizes have been futile. Freud, V.'s predecessor, first developed the theory that each individual desires subconsciously to repeat the catastrophe or trauma, which he believed to be the murder of the father, the Oedipus Complex.

In place of collective amnesia from the murder of the father, V.

substituted collective amnesia from the trauma of natural disaster. His therapy, like Freud's, was to get the patient to realize the origin of his trauma. With Freud, the aim was not to realize the primordial murder, but to realize the oedipal complex operative in infancy. With V. it could not be this easy; catastrophes do not occur with every generation; therefore natural and human history required exposition in the light of catastrophism.

Velikovsky accused many scientists of functional blindness, psychic scatoma, which he would probably assign in large part to collective functional amnesia of the anciently experienced disorders of the solar system.

Thus, on November 2, 1974, he was saying to a Philosophy of Science Conference at Notre-Dame:

> Astronomers do not like interference from other sciences, and certainly not from what could be called 'legends and old wives tales...' The ancients tried desperately to tell us what was going on... We wish not to know anything of this. We wish to believe we are living in a peaceful world.

As a psychoanalyst, he was professionally unable then to accuse them of sin. They could not help themselves. He could not denounce them even if they refused to see when the truth was explained to them. He had simply to grant that their therapy was incomplete. The excesses of their attacks upon the analyst were to be expected and treated by inducing self-understanding.

But he was personally involved, which is an impropriety. He became a kind of Catholic psychiatrist, who has to tell his patients that they are sinners. Worse, since he is sinned against, he became inevitably angry with the sinners. There was no "Forgive them, Father, for they know not what they do." The German national case of psychic scatoma was, of course, much more deadly than the case of the scientists.

V. writes, "You cannot put the human race on the couch." And then he looks at his own fate. "Without preparation, without giving the patient a chance to prepare himself you cannot slowly release from his subconscious mind the necessary recognition of the traumatic past, and so, the patient has experienced great paroxysms and has rebelled against my revelations." But now, by patients, V. means specifically the scientific community that opposed his ideas, which like humanity as a whole, rejects bringing to the surface memories of natural catastrophe.

Many of V.'s supporters agreed with these propositions, Christoph Marx certainly did, and some, like Marx, wanted to devote themselves to its application. Not so Deg, who found both the theory and the therapy grossly simplistic. Having spent most of his life in examining human ideologies and devising techniques of changing, controlling, and accommodating them, Deg had long since abandoned hope of finding a quick fix for human destructiveness.

V. hardly recognized in his psychological theory what was so obvious in his history and in the reception of his book, that over all of history and today, the vast majority of humans and their religions actually demand that we recognize, denominate, and respond in every sphere of life to the occurrence of ancient catastrophes of fire, flood, wind and earthquake.

Destructiveness seemed to Deg "normal," "intrinsically human," ineradicable without genetic engineering and breeding. It could only, by known political means, be diverted, shaped, made to play games with itself, rendered innocuous, and displaced in a hundred ways. Destructiveness was neither more nor less created by natural catastrophe than human nature in its other behaviors, including an abstract active concern for the human race as a whole. Further there was probably a genetic switch, prompted by catastrophe as were most mutations and primary behaviors, that had changed a primate quickly into a human. These ideas were developing in his mind throughout the seventies as the theory of Homo Schizo.

When, after V.'s death, I passed along to Deg a copy of the posthumously edited work, *Mankind in Amnesia,* that Jan had given to me, widely advertised as V.'s great testament, called by himself his most important work, Deg was prepared to be disappointed. When I said "How did you like it?" he said "Even more disappointing than I had expected it to be. Simplism is still the hallmark of the theory. Systematic development is entirely absent. The evidence is second-hand and commonsensical for the greater part. The recommendation for social therapy is nil."

Deg felt a deep chagrin. "The work is true only on the most general level, and therefore unoperational and inoperative. It contains jottings and exclamations. It reads like a string of notes. Its publication could only have been justified as 'notes and stories,' or 'Velikovsky's Lament.' Dr. V.'s claim to be a 'citizen of the world' is inacceptable, unless any person's declared wish that the world not be blown up by nuclear bombs makes the person a 'citizen of the world'." Nor was V., in fact, for all his high qualities, ever such.

The work is too brief for its purported task. Still it wanders; it contains extraneous matter. Too, the work had been long in the making; on July 2, 1967, V. had written Deg that he had "decided to concentrate upon it," at the urging of his publisher. He concluded the same letter: "Keep well, write again, and infuse yourself with impressions that will make out of you a ringing advocate of a need to understand the racial hidden springs of hatred." No need for exhortation: Deg had been such a resounding advocate since childhood.

In reading the new book, Deg had to reflect upon the fact that V. and he had never discussed the work, whether because there was nothing to discuss or because V. wanted to talk of less important matters or because Deg was uninterested in the theory beyond the basic fact, with which he accredited V., the fact that ancient natural catastrophes have played a large role in human and natural history. As much as he believed in the high value of introspection and of the deep interplay of honest minds, Deg had long before meeting V. assigned only a limited potential for good to a knowledge of true history.

"Psychological revelation" would help the world, commented Deg. "Philosophy and anthropology well insist upon this point, but the means for such are not given by V. (see p.207 of *Mankind and Amnesia*) and therefore the statement will hardly perform the miracle. I can hardly believe that he says psychology and sociology had nothing to say about the Jonestown (Guyana) massacre and mass-suicide, yet he does say so, whereas the dynamics of this event were crystal clear to the ordinary social psychologist."

Where is his evidence of a 'racial inheritance' of an experienced fear, an attitude, no less. This is a Lamarckian genetics that I cannot accept. I asked V. once, in the 1960's, for his idea of what physiological process memories could use to ensconce themselves in the racial soma, to which he gave no response. He didn't show me what, if anything, he was writing. I would have been most critical. He read my Lethbridge lecture on fear and memory. I gave him my first sketch of Homo Schizo theory, but I doubt he paid any attention to it, although there I made explicit the only dynamic by which Freud and Lamarck might be married, through psychosomatism. Yet V., who was repelled by Jung's complaisance with the Nazis, would not admit to being a Jungian. Moreover, his ethnocentrism is again apparent. He attributes significance to the presence of the five-pointed star of Venus on the helmets of American, Soviet and Chinese soldiers (only an American general officer is in fact authorized to wear the emblem), but he does not mention the ubiquity of the Star of David in the Israeli army (p.201); did V. or his editors delete the "Mogen David of ancient Israel or even of Israel of today" that he had joined with the others in his Lethbridge lecture (p.27 of *Recollections of a Fallen Sky)?* He indulges freely in anti-Arab statements (p. 150 *et passim).*

In his vagaries, he does not however mention any of his close associates; Stecchini is found in a footnote (p.67), also A.M. Paterson (p.66), and the mention of Rose was a post-mortem insertion. He mentions several correspondents; a temporary assistant, Cathy Guido; a New York City teacher; a jail inmate; a man from Topeka, Kansas, writing on tornadoes, and a conversation with St. Clair Drake, which meeting he places in the Swiss Alps without acknowledging that the two were there at Deg's invitation as part of a revolutionary experiment in higher education aimed at diminishing destructiveness and creating a beneficent and benevolent world order (p. 111). But the most striking omission in the rambling work is that it sidles past the Nazi Holocaust. Of the purest, and best-documented case in history of the working of his theory of aggression and amnesia, not a word is said! [Actually there were a very few words alluding to the German case, and these were excised by Mrs. V. before publication.]

And Deg wanted to go on, but I stopped him. The question of

anti-semitism interested me more, so I got him onto this track. In Deg's opinion anti-Semites define Jews and Jews define anti-semitism, both in their many forms.

As to how many types of Jews there are, I know of no classification. First you have to grade Jewishness as a subjective feeling, an intensity, say, of five grades. Then these are role-operative, transactional, that is. If I feel somewhat Jewish, this is fully or moderately or little sensed, depending upon whether I am transacting socially and psychologically in a setting dominated by the perspective: much, some, or little Jewishness - in a Jewish family, friends or club, or a totally gentile setting, and in the gentile setting, the reverse sets in and I can be made to feel much, some or little of my ordinary moderate Jewish sentiment by the objectification of Jews that the gentile setting exudes. So at any point in time or space, I am liable to be in any one of hundreds of states of Jewishness. Moreover, my character possesses 'X' degrees of stability, but is never so stable that my sense of Jewishness cannot be stepped up or stepped down by my hormonal balance, or some other physiological or sensory balance, as, for instance, when depressed, I may feel more Jewish, and so, too, when manic, but less so in-between. And, of course, all that I say about my type and other types of Jews are averages of quantities.

But now you must go farther. The historical knowledge and life experiences of Jews differ greatly, hence the symbols and references to which we respond, which are so varied. The physical signals of Jewishness are of course symbols, too. To some Jews I "look Jewish," to others rather so, to some not at all, and so to gentiles. There is a Jewish look, which is a combination of a culture-look and a genetic-look. It has a set of grades of attractiveness and repulsion, one set among Jews, another among gentiles, depending of course upon which Jewish or gentile culture and sub-culture you are using as the standard. And with all of these possibilities the area of Jewishness and gentileness and their interrelations is most complex and varied.

This very state of complexity, in which no Jewish race, or culture, or religion, or nationality, or historicity, can be said to aggregate more than a small fraction of those who think themselves some kind of Jew or are regarded as a Jew, fosters anti-semitism, because among strongly authoritarian and dogmatic characters, perhaps 10% of any population, the tolerance of ambiguity and variation is low. Objects and people must be pigeon-holed; they cannot help themselves; that's the way they are and they are eager for any distinction that will discriminate, any line that can be drawn, "a drop of Jewish blood" or "a Jewish grandparent," or, on the other hand (and this is often forgotten), sometimes a thoroughly rigid character will accept as such any person who says "I am a Jew" and then also any person who says "I am not a Jew," like not questioning a person who says "I am a Chicago Cubs fan," or "I am a Dallas Cowboys fan." Since the same authoritarian or discriminating character is also inclined to penalize ambiguities, he is at one and the same time eager to define a Jew and to penalize the Jews for being so difficult to define.

Velikovsky, I should say, and even more so Mrs. Velikovsky, perceived the world strongly as Jew and gentile. Mrs. V. was a fine artist, a fully acculturated Judeo-Christian as a musician and a sculptor, but voted the straight party line, so to speak, when it came to Jewishness on most other matters, including holidays, diet, and intimacy. The big chasm in V's tradition of Jewishness was opened up by modern western science; he lacked belief in the substance of Judaism, whatever his participation in its rites and routines and despite his refusal to discuss religious preferences with anyone.

The Velikovskys were among the "most Jewish Jews" whom Deg had known, even though he had from childhood held Jews among his closest friends and, while he had something of the heart of a Catholic and the culture of a Protestant, he had the mind of a Jew, a twentieth century "assimilated" midwestern American Jew, that is. That was what his wife of thirty years was too, except that she originated in New York. He was more a Jew than an Italian, although his descent was purely Italian, even of certain Sicilians who had been the most nationalist of Italians, but this line had practically stopped at birth with a father who was chauvinistically determined upon the Americanization of everyone (except musicians, it sometimes seemed).

V. couldn't comprehend this very well. He tended to stereotypes and would conspire up an ethnic image of everyone. When once he wrote to Matthew Harris of Doubleday Publishers, upon his own insistence, a letter advancing a book scheme of Deg, he said, "You know, of course, who Professor Alfred de Grazia is. He is a fierce fighter for causes he thinks just; thus he fought for my cause but occasionally we disagree. I would think that born in a different place and time he would have become a Sicilian captain roaming the seas; then Medicean Florence put an aura around him even before he first visited the country of his ancestors...' (Dec. 28, 1968). Perhaps so, but Deg's great dream as a boy of the prairies was "riding off into the Golden West."

Stecchini was Italian by birth and upbringing, but that was not all of it. He had studied in Germany for one of his several degrees and picked up another at Harvard. "Did you know that Stecchini was of a Jewish father?" Deg asked V. one time, to observe his reaction. "No." "His father was a prominent Italian anti-Fascist named Levi who had finally to flee the country. And his mother was a countess." V. was surprised, and Deg was surprised at his surprise, for V. had now known Stecchini for some years, and they had been together scores of hours.

V. was certainly able to work well with gentiles. With Freud, who was an assimilationist, there had been concerns and crises over the role of gentiles in psychoanalytic circles; nothing could be observed of a tension of conflict along such lines in V.'s circle, no more than there had ever been in Deg's circles. Time after time, Deg was asked about V.'s religious beliefs by members of an audience, but remarkably, there was no hint of anti-semitism in the question, nor did he ever perceive any among V.'s many acquaintances.

Deg surmised that Christoph Marx was a Jew for various reasons (despite his Christian name, which was not heard in the Velikovsky household or correspondance) for V. had a tendency, in matters familial and financial, to draw into Jewishness. Deliberately one day, when Elisheva was remonstrating against Marx, Deg said he supposed that Immanuel thought he might have confidence in a Jewish representative when dealing with Germans. She was astonished — Marx Jewish? — not at all. Nor did Immanuel ever think so. Deg convinced her he was so, or perhaps of Jewish and Christian parentage, and she said, "That must be it. They are the worst." And then she telephoned Deg who had been laughing at her to say of course she didn't mean that, meaning of course that she recalled that Deg's children were all of mixed Jewish-Christian parentage. As it turned out, when Deg told him the story, Marx confirmed that he was not Jewish.

When, after V.'s death, Warner Sizemore ("to get money for the cause") ventured into Amway consumer-business circles and into the formation of a "far-out" protestant church, he told Deg how surprised he was at the manifestations of anti-semitism among folk in such circles. That's to be expected, Deg advised, for the world of the aspiring small businessmen and millennialists, with its rural, radical protestant, and poorer base, held large contingents of anti-semites in America and Europe. Yet, also, this same base provided, at least among its more educated elements, many enthusiastic readers of *Worlds in Collision* and *Ages in Chaos*. Since the first Puritans, America has attracted the "true Israelites," the "Christians who had been persecuted by the Jews and Romans."

CHAPTER SEVEN

FROM VENUS WITH LOVE

When Deg was proofreading *Chaos and Creation* in 1981, he recalled a half-century earlier overhearing Bob, his Scoutmaster, confide to a deacon of St. Chrysostom's Episcopal Church in Chicago, "Sex rears its ugly head everywhere." The recollection was triggered because among innumerable problems foreseen and unforeseen there occurred in remote India the castration of Geb. As illustrated in the book (p. 125) Nut the Egyptian Sky Goddess reaches down to embrace pronouncedly ithyphallic Geb the Earth God. But the printer's proof of the illustration that was sent back by Popular Prakashan Pvt. Ltd. reached Deg *sans phallus*. I quote now Deg's admirably restrained letter of January 29, 1981, p. 2, point 3:

> I note that the phallus of the god of earth on figure 15a has been removed. This drawing is a famous archaeological figure and should not be tampered with. Was the excision made for fear of censorship or customs and prolonged controversy? I had no idea that there would be a problem. I don't want to delay the books by even a day. But it takes two sexes to mate, even Sky and Earth in mythology, so a semblance of masculinity has to be restored. I will be criticized as an unreliable author by many people as matters stand (unless directly beneath the caption 15a on page 125 there is printed in parentheses — "Earth's exaggerated phallus has been removed — reduced? — by the printer to conform to Indian government censorship regulations").

Back comes the reply of Mr. M.G. Shirali, Production Manager, dated February 2, p. 1:

Re: 'the mystery of the missing phallus' — figure 15a, page 125 — let me explain. You will recall this drawing was traced out by our artist from the original xeroxed sheet you had sent, which you will remember, contained a lot of other things such as minute specks. This being possibly photographed from a stone mural or some such thing. So while tracing out just bare outlines as you desired, this somehow just got lost in the maze of specks. Believe me, never for a moment did we think of tampering with, nor was the excision made in deference to the customs, nor for fear of censorship. Pure and simple it was an unintentional slip. Please accept my sincere apology for the lapse on our side and also my thanks to you for pointing it out. And now it has been 'arranged to be restored to the rightful place'!!!, as you will see when the final proofs come to you.

The new proof returns. The phallus was restored — by half. Persisting, and because he fears that his original has been mutilated beyond use, Deg writes on March 28, 1981:

Enclosed is a copy of the famous Nut and Geb picture. It occurs to me that, without any redrawing, a cut should be made of this as it is, leaving the shading, *which is from the original papyrus,* and thus the picture will not appear so prominent. I think this would indeed be an improvement, it is, after all, only a detail in an immense work. *To repeat, photograph the new drawing exactly as it is here, and thus keep the shading in the final cut.*

Indeed sex does pop out of all corners in the material of human history and is especially illuminating in regard to catastrophic events. It is remarkable how V. managed to suppress sexuality from becoming a major theme of his circles. It would have been easy to follow a path similar to the one of Wilhelm Reich who found in a kind of electromagnetic life force, expressible in sexuality, the beginnings of an answer to all things, including a kind of communism for which he was evicted from the communist party in Germany.

Elsewhere, in *The Burning of Troy* and in related pages of the *SISR,* a story is told of how V., following Cicero, claimed the root of Venus to be the word *venire,* meaning 'to come', and therefore the planet must be newly arrived, but Lowery, analyzing the words, finds them unrelated, nor is this the first time Lowery and the tribe of linguists dashed cold water against the heated claims of catastrophists.

Christoph Marx and Deg independently found a subtle connection that Lowery missed and I take leave to quote from a paper circulated by Marx, dated May 8, 1982:

Easy to see now how Venus from 'venire' is quite equal to Venus standing for 'love' because to love — if successful — is the same as to come (as any body past adolescence may experience). The dream-like efficiency of the term 'ven' may easily be judged by those with the faculty of imagination and an analytical turn of mind. To make visible the tradition of violence embedded in the term I would only add the example of a French porno movie, in which 'to come' produces "The End of the World" (the film title). It shows, of course, the love-making while the atomic rockets are on their way, but only in the end we see how they were released in the first place: Merrily, the president of the United States and the General Secretary in the Kremlin over the Hot Line are exchanging their experiences while being serviced by their beautiful private secretaries; the President of God's Own Country comes, and in his ecstasy hits the red button, leaving mankind with a movie's length of final lovemaking=coming.

Etymology must begin with the study of Arno Schmidt and James Joyce who purposefully used and analyzed etym addressing. Etymology is not at all the successful tool Lowery makes it out to be when, e.g., he points to the reconstruction of the ancient Egyptian language: the decipherment of the hieroglyphs was not an achievement of etymology, and whoever has read a translation, say, of a literary text such as the Book of the Dead cannot but agree that there is hardly anything more senseless in the way of expensive books — understandable perhaps to the translator's analyst, but certainly not to the ancient author. Etymology for the present is not more than a systematized part of established science, the mechanism for the continued repression of the past.

Electricity has in folklore been connected with sexuality, just as has the coinage and usage of words. Jerry Ziegler, a physicist, in the 1970's circulated his work on ancient knowledge of electrostatics and a copy came to Deg who got in touch with Ziegler and recommended his study to V. who ignored it, but Deg began to develop it in a number of ways. This was not uncommon; V.'s closest associates moved in their own way; Sizemore was aware of a world of marginal sciences that he would not discuss with V.; so Stephanos, as will be seen; so Juergens who moved from the Princeton area as he had moved toward it, because of V., first to be near him, then to be away from him; so Stecchini; so Bill Mullen; and the British heretics, so devoted yet so independent of thought.

Ziegler found many associations of ancient religion with electrical practices, and persuasively in his *YHWH* informs us of what interested so persistently and for so long the ancient sects in their mountaintop ceremonies. To be near to the gods, yes, but to be near the sources of enhanced electrical stimulation, too. The people, led by priests, went up the mountains for ecstatic purposes where religious rites and sexual experience were joined. Electrical discharge was supposed to enhance the sexual libido.

Significantly, when in modern times there began many experiments with electricity, following the invention of the Leyden Jar, the scientist Sigaud tried to pass an electric shock through a company of grounded men, a trick that others had achieved, and when the attempt failed, he suspected that one of the company was "less than a man," a eunuch or *castrato,* that is; but then, as Heilbron's history tells the story, it developed these, too, jumped where discharge was passed, and were electrically conductive.

But Zvi Rix, of all the cosmic heretics, went farthest into the exploration of correlations among ancient religious practices, sexuality, and cometary disasters. Marx took over his manuscripts from his widow, but the task of disentangling them and reformulating them into fairly conventional prose proved to be arduous.

When he was a boy, Deg believed that sex was a simple function: a male found a female, like an arrow shot from a bow pierces the bulls-eye of a target. For the several years that he was confined to autoeroticism, his fantasies and exercises, occurring privately, aimed at real female acquaintances and attractive female images in equal proportions. By increments of experience and learning, before he was forty, he could publish the article of a friend in Psychology at the University of Minnesota, arguing that sample surveys might be improved if they solicited information that would place the respondent on scales of masculinity-femininity, allowing sex to be a finer variable, capable of more meaningful correlations with other behavioral variables like "candidate preferences."

By the time he was sixty, though still an active heterosexual, the image of the arrow and the bulls-eye had resolved into the image of a fragmentation bomb, striking promiscuously and erratically in all directions. Homo Schizo, it seemed, from his beginnings and forever after, had lost, sexually as with all drives, close instinctual guidance and gained an uncontrollable but vast world. The modern theory is that if you don't find indications of homosexuality in a man and lesbianism in a woman, you have an unusual person who is rigid and lacking in affect.

Roger Peyrefitte, a French writer, ex-diplomat and professed homosexual, discussed and wrote about what he regarded as the homosexuality of Jesus and his apostles. He was challenged to a duel by a fiery Spanish psychiatrist, but refused the test. The same understandably underground theory was shared by V. but Deg was unimpressed, not needing V.'s innuendos, meaningful glance and obvious reluctance to say so, but still V. had to let the cat out of the bag, like "you know, there is much to be said in this regard about Jesus." But Deg had no doubt that the tradition went back to the nasty circumstances surrounding the trial of Jesus. I'm sure they called him everything, he said, not disagreeing but not caring at the time to plumb V.'s data base on the question. There was little Deg could not find a place for in his mind, ranging from Jean Genet to Don Juan, and all the ambiguous feelings, attitudes and practices in between.

The closest V. comes to offering a theory of sexuality occurs in *Mankind in Amnesia*. There he asserts that neurosis is based upon narcissism, ultimately, the autistic libido that has to be located and treated first of all (p. 162). This done, the therapist must move to the treatment of homosexual problems and then into alleviation of the Oedipus complex. The theory is rather directly one of Freud's many, and V. generally arrived at these several stages quickly with his psychiatric patients. Fifteen minutes is often enough, he said to Deg, to understand what is going on with a patient. Repeated visits and phonecalls were to be expected, of course.

V. was remarkably prudish. Over the years, he gave Deg the impression which actually was obvious at first but scarcely believable in a psychiatrist, that he operated on the idea that "men are men" and "women are women," a simplistic notion. He seemed not to notice that several of his most brilliant and active supporters might have been homosexuals of one kind or another. Fight off the homosexual urge, he seemed to be saying, and stamp out the narcissism that stands beneath it. Laius, father of Oedipus, had introduced, according to legend, the practice of "unnatural love" (V.'s term) in Ancient Greece (which, insists V., is at the origin of the terrible curse upon his house).

Onetime in America and once in England, Deg was asked with a certain wonder about homosexuals in the movement. Their participation was not surprising, he answered; no movement is a rational and random selection from the population, no more than the establishment it stems from; homosexuals are more active in innovative and intellectual movements; all that we know of the sources of creativity and cultural change would be contradicted if they were not. New movements, whether scientific, cultural, political, religious, or social do not come from the average norms and normals of a culture.

Deg ought to have explained fully, right out of his reading of *Oedipus and Akhnaton,* which so impressed him. There, on pages 48 to 50, is told the story of Amenhotep III, father of Akhnaton, and of the Roman Emperor Hadrian, and of the Greek's and Oriental's indulgence of homosexuality and the Hebrews' condemning of it. In a delicate lacework of wide-sweeping history V. manages the following pejoratives regarding homosexuality: "Greek love," "invert," "iniquity," "spoiled by," "contemptible," "work their will (on Lot's guests)," "horrible punishment," "aberration," "unnatural urge," "sinful," "horrible retribution" (Laius' descendant at Thebes); throughout the passage, luxury, splendor, power, idleness, extravagances, high culture and civic freedom are dwelt upon as the ambiance of homosexual inversion. No wonder, thinks the innocent reader, that Akhnaton was so queer. But Akhnaton is not the issue here. Three features emerge from the passage: V. absolutely rejects homosexuality; homosexuality is portrayed as an exotic and attractive luxury of high cultures; V. does not, here or elsewhere, appear outwardly punitive to homosexuality.

Deg could name a half-dozen of his acquaintances, all of V.'s circle and on at least three sides of any argument that came up — not a clique, that is — who were homosexuals, but he never thought of what might be the seductiveness of V. both at close hand and at a distance. For my part, being more distant from the scene, I would guess that V. subtly presented the image which homosexuals in those years (not the present liberationist gays) could best accommodate to: a stern attitude exuding a luxuriant bath of guilt and a seeming tolerance, delicacy and understanding precluding any but the most "delicious" punition, which was necessary for the enjoyment of their homosexual feelings. (Nor, to be fully aware, have we of Western culture quite learned to enjoy heterosexuality without guilt and fear of punition.)

V. liked Nina, Deg's second wife, who was at the Swiss college on and off. Deg recalls an especially vivid image of the two of them silhouetted in the sunshine and snow against the Alps on the road to Haute-Nendez, talking volubly in Russian. Long after, Deg was reporting to him that Nina had gone to Berlin to marry Peter Bockelmann — a fine musicologist said Deg, and a fine man. Whereupon V. began to speak of Tolstoi's "Kreutzer Sonata," a story in which a husband, according to V., enjoys sexuality homosexually by turning his wife over to another man. Deg was amused at this. He had been happy that she had found so good a friend after their separation. What were V.'s motives for the story — his liking for Nina, his dislike of Germans, his need to carry a dubious theory into every human relation, a jealousy of Deg's philandering, a homosexual impulse of his own? That is to say, when it came to conjecturing and examining motives, Deg was unwilling to let others escape. Or perhaps V. just had not gotten the story straight; they were separated and still friends: it was a plot not to be found in V.'s manual.

One of the sillier passages in V.'s *Mankind in Amnesia* propounds the idea that nations have a masculine or feminine character, Germany and France being among his examples (pp140-2). This kind of social psychology is not only unproductive but also false (like Mussolini once in anger calling the Germans a "nation of barbarians and pederasts") and only made Deg more irritated at V.'s pretentiously published book.

For the infant college in the Alps, Deg had invented a concept which he called "rapport psychology" that was intended to be a form of group encounter usable for his "kalotic" world order, he wrote in the *Bulletin* of the School:

> The basic rapport group usually consists of eight to fourteen members and the leader or facilitator. The group uses verbal and non-verbal exercises and encounters, and typically has no set agenda. It uses the feelings and interactions of group members as the focus of attention. This allows for maximum of freedom for personal expression, the getting in touch with feelings, and interpersonal communication. Emphasis is on open, honest and direct interactions among members in an atmosphere that supports the dropping of defenses and social masks characteristic of normal academic relationships. Rapport group members come to know themselves and each other more quickly, deeply, and fully than is possible in the usual academic situations; ordinarily, a strong feeling of group solidarity develops. The resulting climate of openness, risk-taking, honesty, and trust displaces feelings of

defensiveness, rigidity, and mistrust. Members can identify and alter self-defeating attitudes and behavior patterns, and explore and adopt more innovative and constructive ones. In the end, most members can experience daily life and work more pleasurably than before, on campus and off.

Deg was trying to connect the personal to the universal without the usual intervening madness. Amidst the continual hubbub of hand-to-hand struggle at the new school, he could not operationalize the theory of the Rapport Center. He left it to the attention of his brother Edward and B.J., a group leader whom Ed had recruited from his experience at the famed center for group therapy at Esalen, and to the students, aged 18 to 28. At one moment in a group session, on the way to the brave new world, two men decided that they would make love to each other and went off, after which one, a virgin in such matters, "tossed his cookies" in a rush of shame and disgust.

The word got to Deg and to V. as well, who accosted Deg on an alpine pathway and denounced such conduct nor, said he, will I stay on these mountains with this going on. Deg solemnly and reassuringly listened, and told Ed "What the hell happened there anyhow?" He didn't expect much of an answer, nor got one. The Rapport Center remained popular and undirected to the new world order, whence I remind my readers of two axioms: few truly wish and are psychically prepared to address themselves to the necessary new world, and "bringing life into the classroom" is a beloved pedagogical expression with absurd possibilities.

V. stuck it out on the mountains — actually he enjoyed his stay — but he could not help but slip a reminder of the incident, camouflaged, into his notes and ultimately into *Mankind in Amnesia,* where, in a diatribe against both the old and the new, he says (p. 185):

> The rebellion of the young was full of hope — the millennium was about to begin. The hair was grown long, John the Baptist was imitated in appearance, but the rebellion was against asceticism as well as against materialism; regulations were to be violated, young and not-so-young flocked to 'rapport-psychology' which struck out Freud and the rest of the 'schools'; orgies were practiced as curriculum in some campus classrooms as the call came for tearing down all inhibitions.

But V. did not pursue sexual investigations of Jung or Marx, contenting himself with stressing the obvious resentment of Jung at being regarded as a son. Bronson Feldman, a Velikovsky acquaintance and supporter, introduced sexual analysis to back up V.'s claims, but we must remember how chary was V. to let anyone claim to represent his several views, with every excellent reason. Feldman, who became understandably mad and confused when dealing with Central European anti-semitism, added little to historical reconstruction.

He did point out, for instance, that V. had misstated a famous report of Freud's swooning in the presence of Jung and others. V. forgot to mention that not only had Jung been defending the efforts of Akhnaton to erase his father's memory but had just been hotly accused by Freud of the great academic crime of non-citation of authority — namely himself, Freud — in his writings. Thus Freud had taken two blows from his disciple and son, Jung, and probably a third unmentioned blow, a Christian effort (at least a suspicion thereof) to bury a Jew's contribution to knowledge; of this suspicion we have ample evidence, and of the fact, too, whether in Jung or in Nazism, that the contributions of Heine, Mendelssohn, Einstein and many another Jew to German high culture were buried. And, incidentally, Deg spoke in *Politics for Better of Worse* of the recent era in America, "of those highly skilled and creative people who had built the arts and sciences, half of them Jews;" for he was irritated that in whatsoever history book or sociological work on America no such statement, even the approximation of such a statement, is to be found. But Jews are divided in their minds and amongst themselves whether to lay claim to their achievements or to play them down to avoid envy and resentment.

The sexual verges upon the political and the political, I must now make the point, verges upon the sexual. I mentioned that V. was a prude — or was he canny, realizing that scientists and scholars are sexually repressed and in our civilization will not respect an authority who ties in the sexual link too closely with the processes of the intellect? I would say V. was publicly rather priggish, and privately more so. He did not like at all Stecchini's introducing Peter Tompkins to his circle, nor did Peter visit more than once, although a war hero, a man of some fame then (and more to come), of great personal attractiveness, and a potentially influential supporter: why? Because Tompkins had written on cults and practices of eunuchs and virgins and saw in the history of the planet Venus, which he credited to V., the mad unfolding of the human mind into sexualized institutions.

With perhaps more reason, V. was exceedingly wary of a "hippy bookman" in Manhattan, Theodore Lazar, adorative of V.'s books, who wrote a pamphlet about Venusian-derived phallicism, the cometary image as it entered so many ways into the brain and behavior of mankind. V. was wrought up at Robert Stephanos, a Philadelphia school system psychologist, the most faithful, pleasant and helpful of disciples, for pushing favorably the work of the New Yorker. And, later on, he was angry to hear that Stephanos had been flirtatiously corresponding with a Southern devotee and, not long afterwards, in a paranoiac mood, came to suspect that Stephanos might even be purloining papers of his. You must remove him from the Board of Trustees of the Foundation for Studies of Modern Science, he told Deg, the President, and others.

"Politics makes strange bedfellows," but so does science when it strikes out in new directions. Whoever wants to sleep with the partner of his choice or to sleep alone must give up creative dreams. V. sought hard to deny his bedfellows, but they were with him from the moment his book struck a popular chord, attracting many who were looking for bedfellows. Not so strange, he or his fellows, I hasten to stress. Just variegated.

CHAPTER EIGHT

HOMO SCHIZO MEETS GOD

Great mysteries of existence such as human nature, divinity, time and governance are intimidating. The ordinary person is content with a few slogans about them, a kind of catechism, and to be allowed to make off with a piece of one of them — so small as to be indistinguishable, therefore safe — to play with for life. There are also those few persons who, emboldened by a successful encounter with a great mystery, become drunk with the genre and go on a rampage, knocking over distinctions and laying claim to new territory extravagantly. You can tell the type, if by no other sign, then by the way they have of looking upon the universe as a cabbage patch and treating great historical figures as their neighbors.

One could see it long ago in Deg, who after taking the worst and the best of the army for four years, came back finally and managed a Chicago election where, introducing his distinguished professor Charles E. Merriam to a mass meeting (luckily the Fifth Ward had the greatest concentration of intellectuals in the world) he said enthusiastically that he had studied with Merriam 'like Aristotle at the feet of Plato' and then was ribbed by friends and poignantly embarrassed, so that, as you see, even now he can remember to tell me about it.

Therefore it is no surprise that thirty five years later he can be treating Charles Darwin and everyone else familiarly, even arrogantly. "What is your opinion of Darwin?" was, of course, the question. The tape spun; Deg picked up his notes and spoke at the machine:

> Charles Darwin was an apt hero for nineteenth century biology and the public and scientific mentalities of the nineteenth century. He came from an expanding empire, did his "field work" young; he lived for many years quietly, gestating his ideas; he published at the right moment for coalescing the views of the scientific and cultural world; his theory of natural selection

was simple, vague, and in line with what the secular person thought was his own idea.

Now that his ideas are wearing out, the psychiatrists, methodologists, and philosophers have picked him to pieces. He was an uncertain person, never a fully convinced Darwinist. In the contemporary vein, R.C. Lewontin writes that "Darwin's work is filled with ambiguities, contradictions, and theoretical revisions." Velikovsky once pointed out that if Darwin had followed some of his own observations while on the voyage of the *Beagle* he would have become a catastrophist. He almost became a Lamarckian at one point, so fetching is it when one's own theory is indefinite, to imagine that the soma can be changed permanently by a forceful environment.

"Darwin was ambitious, courted success and successful men, and cared for their approval:" again these are Lewontin's words. So too was Velikovsky. In 1858 just before Darwin published the *Origin of Species by Natural Selection,* he wrote that he did not yet feel set on the truth of any point of his theory, and was in this state of mind when Alfred Wallace wrote from far away to tell him about his own theory of natural selection.

When he consulted his friends, their solution was to hustle him into publishing his manuscripts along with the essay of Wallace. What else could they do? Otherwise, Wallace would have priority. As Darwin said, "All my originality, whatever it may amount to, will be smashed ... It seems hard on me that I should be thus compelled to lose my priority of many years standing."

But let us be clear...

Ignoring the machine, Deg produced a statement out of his drawer of epigrams; "I used to hate epigrams," he said, but now I collect a few, "especially my own." He read: "Priority in science is a political claim. It is of no interest to scientific advancement that A or B captured a strongpoint first, so long as it was taken. A proposition is denuded of its generator. It ends life as it began, in anonymity."

He spoke feelingly because a continual annoyance of a generation of the Velikovsky affair was the bickering about claims and predictions.

> The lead was unfortunately provided by Princeton physicist Valentin Bargmann and Columbia astronomer Lloyd Motz when they assigned V. a priority on the heat of Venus and the radio noises of Jupiter (upon his instigation) and recommended reading his work for further clues as to what to expect. Such words from an astronomer and a physicist were naughty; they excited V. and his followers and angered other scientists, all the more because they were involved themselves in this racket.
>
> The ideas of 'priority', 'prediction,' and 'claim' are more political than scientific. The word 'claim' connotes possessiveness — not a happy human quality. V. liked the term; the press liked it; ambitious scientists like it, and long years of struggle have gone on in such fields as physics and

psychology to try to assure people's claims to discovery, as if all of knowledge is of little bits, ever-diminishing bits as well, that are owned by an individual forever.

Darwin need not have worried; his location, his friends, and the ample, ambiguous, diffident qualities of his writing, pitched at the consensus of all-who-mattered, the 'happy few' of the day, would assure his work 'priority.'

Velikovsky's work found no such consensus. Perhaps it deserved no such consensus. Perhaps it earned at that point precisely what it deserved, and what Darwin's work deserved — an audience, a hearing, a turning of minds, a refurbishing of hypotheses, some of the patient, indulgent, reflective, detailed processing that is supposed to characterize science but does not markedly do so.

Deg's un-darwinian *Homo Schizo* was present for many years and began with the conviction that man was essentially non-rational. When Deg first joined the faculty of Stanford University in 1952, he was working on the phrasing of Lasswell's law: political man displaces private motives onto public objects and rationalizes them in terms of the public advantage. This conception had burst upon political science in the 1930's, joined with pragmatism and neo-machiavellism, and overran the 2300-year-old positions of rational-legal-institutional political science.

Deg radicalized the concept. He could not see anything extraordinary about Lasswell's political man except in the intensity of his involvement with power. Too, he was critical of the notion of rationalization, for since boyhood he had found everybody doing nothing but rationalization. So he suspected that reason and rationalism and rationality were real processes of rationalization. When he came in the seventies to ponder the nature of man, he could now perceive a brain structure and personality altogether of the schizoid type. His newer concept was of instinct-delay, blocking, and displacement of response to a stimulus, forcing terrible self-reflection, and in the control of these reflections — the polyego — there occurred the human character. The essential polyego assured an eternal existential fear, whose high level, being constant, goes generally unnoticed.

Homo sapiens, whom he finally termed *homo sapiens schizotypus,* is most rational when he is acting (thinking being a form of acting) pragmatically, that is, calculating and adjusting to the consequences of his behavior while transacting with an environment, both human and natural. Logic, and hence science, and hence most of what is ordinarily called reason, develops as a means of most efficiently connecting an entering stimulus with an effective response. In this sense, man, seemingly farthest removed from the animal kingdom, finds his triumph in emulating instinctive response. He aims at reducing his high level of existential fear by logical, "rational," and scientific conduct.

But as the underlaid instinctual apparatus of the animal does not guarantee it against the multiform assaults of nature, whether represented intraspecies or in the transactions with other species and inorganic nature and whether uniformitarian or disastrous, so too man's efforts at reconstructing and reinforcing his less genetic, delayed instinctual apparatus, are continuously ineffective. All the achievements of the calculating and even scientific homo schizo cannot win control over the self, others, and the natural world.

As in the beginning and even in the most rationalistic technical ages, homo schizo continues to rely upon the organization of his far-flung displacements for adjustment and control of himself and the world, so that religion, culture, and the arts are, if not preponderantly his road to "happiness," most useful and welcome companions of pragmatic scientific conduct. Alone or together, the sciences and the arts, cannot create a creature other than homo schizo. Even if they could, the monster would be limited to some portion of their own envisioned ideal that they could agree upon, and they would promptly regret having made such a substitute for the unrealized larger portion of their ideal.

I should not try to explain the full theory here, not when two volumes about it are available elsewhere. However, it is appropriate to comment that Deg began his development of the model of homo schizo to test the Freud-V. theory that historical traumas produced a character who simply had memory problems but was otherwise "rational" by nature. As I said, Deg was already prejudiced against this idea, and it was no accident that he almost immediately placed the idea of the intelligent evolving savage into a restricted enclosure. He searched instead for the larger meaning of catastrophe, now quantavolution, that formed a different creature to begin with. Primordial man was now catastrophized in two senses, first genetically and second in the sense of reinforcement through repeated catastrophic experiences.

The latter, the reinforcement process, gave Deg no trouble; there was ample evidence of a "law" operating whereby the intensity and duration of an experience (read "catastrophe") determined and varied directly with the amnesia and compulsive sublimated recapitulations of the experience. Further, therapy of such a condition (control over it, that is) was exceedingly difficult, whether of the individual or of the collectivity.

More difficult was the establishment of the genetic basis of human nature. Here Deg found his way, first by undermining the case for gradual darwinian and anthropological evolution, and second by discovering uniquely human variances in current research on the structure and operation of the central nervous system. He came to attribute humanness to a brief glitch in the stimulus-response system, which I mentioned above. How he visualized it becomes crudely clear in a note from his files, entitled "Making a Chimp Talk: a Suggested Research Project on a key element of *Homo Schizo.*"

MAKING A CHIMP TALK

Premises
1. Homo Schizo theory says that mankind became human and is human today in connection with a millisecond delay interfering with instinctive response.
2. The delay a) diffuses (displaces) percepts, concepts, and memories widely because of lack of immediate response, b) forces the being to sense itself, that is, at least two selves, c) activates existential fear mechanisms because of lack of control of a) and terror from lack of control of b).
3. To tie itself (itselves) together, the being communicates with itself and the result of this communication is inner language, the basis for external language.
4. External or social language occurs as the being continues its inner operations by external means, employing whatever it can, such as gestures, utterances, and other signs and signals.
5. All of 1. to 4. occurs with little relation to the size of the brain, with some relation to hemispheric symmetry, and with relation to other possible delaying mechanisms. A person can be raised to behave normally in speech and behavior with 1/10 of the brain matter normally encased in the cranium provided that all elements of the brain are represented by proportional fractions.
6. A chimpanzee brain is within the human functional limits so far as size is concerned. Its vocal apparatus and other symbolizing mechanisms are adequate. It is a highly sociable animal, so 'presumably would like to communicate. Chimpanzees and other non-humans can learn many isolated symbols..., but they show no unequivocal evidence of mastering the conversational, semantic, or syntactic organization of language.' (H.S. Terrace *et al.*, 206 *Science* 23 Nov. 1979, 891).

Thesis: Chimpanzees do not speak because they do not undergo an internal electro-mechanical compulsion to speak.

Corollary: Chimpanzees would speak if their instinctive brain operations were continuously and unconsciously blocked for milliseconds, [thus supplying the compulsion].

Experiment
Baby chimpanzee Abel is subjected to partial commissurectomy; insulin injections to arrive at constant 10% higher blood level; and background human videotape television plus human handling as of normal human babies of up to 26 months of age.
Hypothesis: Abel will at the age of 26 months emit 50% (rather than 20%) of the expansive adjacent utterances of human infants of the same age (and proportionately more than chimpanzee 'Nein' of that age — in the

Terrace *et al* experiment).
Corollary hypothesis: Availability of the conditioned animal will permit application of a full range of tests of humanism, including intelligence, self-awareness, self-images, aggressiveness, persistency (obsession) in task performances, memory and recall, with special attention to the generation of the several components of schizotypicality, including various tests of insanity.

Here I think that Deg is downright ignorant regarding the possibilities of Dr. Frankenstein experimentation with apes. The ape is a massive system of unique organic connections and resultant behaviors: unless you get into the gene system and perform a systemic mutation there, you will get nowhere by monkeying (excuse the expression) with the post-natal resultant. He proposes to cause artificially a totally ramifying system of displacements, fear, and ego split when all the settings of the ape's organism are deadset against alteration. The animal will simply die. That is a much more logical and simple response than to undertake the enormous burden of behaving like a human.

Deg's archive carries many another note of different kinds — sketches, designs, critiques. They begin as a broadly spread-out and miscellaneous aggregate, and then come together as the book is written, but many of them are locked out in the end. Here are three of the excluded ones, let to view:

Deg's Journal, December 20, 1968

In pregnancy, especially during the last three months, when the placenta is largest, the placenta manufactures a large amount of blood ceruloplasmin.
1. Ceruloplasmin alleviates many cases of schizophrenia
2. Women with schizophrenia are alleviated towards end of pregnancy
3. Relapses and initiation into schizophrenia may occur following pregnancy, i.e. post-natal schizophrenia is common
4. Schizophrenia is 'split personality' disease traditionally, although Hoffer and Osmond deny this definition, saying there are not two persons, despite hallucinations and feelings of persecution. They are in a major sense right.
5. The correspondence of high C production with the period at which a woman faces the traumatic need to split her baby from herself makes me think that the body protects itself (or the 'mind') from the effects of this traumatic experience by exuding into the blood a specific defense against schizophrenia.

About this time there occur also various petulant scribbles on his readings, *viz.:*

Glancing through *The Scientific American's* handsome volume on

Human Variations and Origins, I see many errors behind the skillful graphics. There is Eiseley's idiotic article on Lyell, for example. The 'distinguished' academician knows much about his man's surface and nothing about his dynamics, nor does he understand the real conflict between uniformitarian and catastrophic evolution. Eiseley's reputation comes from a deadhead riding the commonplace, uttering mystic words.

Later in the book I see all manner of speculations treated as facts, simply because they come from scientists. Man's spotty history is given a coherence by rhetoric, not data or even good theory.

I see a picture. I read a caption. It shows an extremely tall negro and a short, chunky Eskimo. The first's height is supposed to be an adaptation to heat, more surface per pound; the latter's chunkiness is supposed to conserve heat. But whence the Swede? Whence the many fleshy African negroes? The Ibos, Pygmies, etc. Doesn't moisture and dryness of the air matter, etc.? I have seen pictures of chunky short Indians of the Amazon and Orinoco tropical jungles.

The theory of evolution is full of hopeful guesses. I am working with a sample survey of attitudes and experiences of the U.S. population right now. I am, as always, acutely impressed by how the first relating of variables can mean nothing and *always* means nothing unless one is satisfied that all the other factors are interpreted and counted. Women have the same accident incidence as men: fine, but that's the end; afterwards all manner of crosscutting forces change the character of their accidents and incidence when compared in sub-groups.

The defensive scientist retorts irritably: 'But this is only *popular* science! We don't make such errors in our *real* inside work.' Nonsense. Every specialist is carried along on these so-called popular currents, not to mention that he likes to call 'popular' anything that he doesn't find agreeable or true. There is the beautiful image Merton and other students of science, who are admirers of the image, employ: 'We are but pygmies, standing on the shoulders of giants.' We should also say, 'We are giants standing on the shoulders of pygmies.' Or better, 'We are monkeys, swinging carelessly along a dizzying network of vines mysteriously placed and oriented.'

Sometime in 1970, Deg met biologist Dr. Karl Schildkraut of the Albert Einstein Medical School through Dr. Annette Tobia. He was interested in Deg's University scheme and Deg talked with them a couple of times about heredity. Perhaps these contacts brought about a note foreshadowing some of his passages on evolution:

...Unless one resorts to an immense number of mutations (practically begging the question whether uniformitarian or catastrophic), it is impossible to conceive of the complex intra-organism adjustments (changes) that must accompany an organic innovation, that is, 2^n where n = affected parts; if brain convolutes by mutation, then how many elements of the body must adapt immediately?

If all chromosomes and genes are linked, then there must be a chemical 'universal element,' bringing about a total viable system change.

Note, too, the received evolutionary doctrine offers in evidence the numerous similarities of all living cells. The same fact of universal similarity is applicable to the doctrine of simultaneous systemic mutation, both regressive and progressive.

Deg sent an early version of the theory of homo schizo to Lawrence Zelic Freedman of the Institute of Social and Behavioral Pathology at the University of Chicago at the suggestion of Harold Lasswell. Freedman raised two issues with the theory, issues that Deg addressed in the final work: Could man have been catastrophized other than by natural disaster and could a catastrophe strike into the hominids *en masse*. Freedman wrote:

> ...The notion of contemporary man as a schizotypicalis is one which I find easy to accept, and your adumbration of the contemporary social and psychological dilemmas of knowing — if not understanding — man, magnificently expressed... the elemental catastrophe of separation and confrontation with hostile elements during the process of birth might be the individual equivalent of the massive confrontation with overwhelming stress which the model catastrophe hypothesis demands.

Deg considered that human birth is not much more traumatic that anthropoid birth, hence, if it has a greater psychic effect, that is because of a prior genetic constitution which has to be explained. Freedman raised a second major issue: "the high probability that significant elements in the general population would escape the pathogenic influences of the hypothesized catastrophe."

Deg worked out of this dilemma by devising a primordial scenario in which a radiation turbulence, causing millions of mutations, altered the physiology of a given hominid such that full schizophrenic behavior was promptly induced in its descendent and, by virtue of the powerful capabilities of the individual, within a thousand years produced a multitude of operative humans spread over a large territory. Alternatively, owing to a catastrophic turbulence, a changed atmospheric constant might have constituted in effect a genetic change by continuously, "ever after," conditioning a new hormonal state in a pre-potentiated hominid species, in which event, the humanization process would have been speedier. That both processes, genetic mutation and a changed critical gaseous constant, could operate sympathetically was also foreseen.

Deg sent the same early booklet to his friend at the University of Haifa in Israel, Professor Ernst Wreschner, who found the homo schizo theory especially vulnerable in regards to its catastrophic scenario and the short time allowed for humanization:

> I accept that Pleistocene upheavals, cosmic tectonic — a combination of fire and water — must have been for generations of homo erectus, Acheulean man, Anteneanderthals, Neanderthals as well as for some Cromagnon, and whatever names archaeologists give to them, an experience of realities that were outside their powers of coping with mentally. It is feasible that by these very experiences mechanisms could have been developed which enabled men to survive more or less sane during times of the twilight of the gods. But I also believe that the very principle of natural selection could and did cope with the possible influences of catastrophes or cosmic radiation escalations. Either in the mutational sense or in the mentally adaptive or both. Which would mean in the biological and cultural fields. (...)
> The postulation that catastrophes were always global and had overall consequences is untenable, as is the date expounded for a decisive point in human history such as 13,000 B.C. (...) The deep dualism in the human make up developed and existed in their "animal context" becoming mentally or psychologically pronounced when self-awareness could fathom them. But this happened in a process of culturisation and this forced men to deal with them, even without catastrophic catalysts. (...) And language is also not a sudden creation. Many factors worked towards it, biological (anatomical and cultural ones). Man is by nature an experimenter, based on mammalian trait of curiosity. It was 400,000 years ago that he experimented with fire and limonite to get a result which was the red color mineral hematite. Many others after him, either independently or by diffusion hit on the same. Many thousands of years passed between these experiments. And those with the developed brains put the red color to symbolic use, when other beliefs needed a carrier for associations connected with life and death. Thus with the first burials the red color in the form of ochre appears and afterwards red color symbolism in many forms spread and you find it ever since in variegated ideational meanings, in burial practices, myths, rituals, legends and ceremonials.

In reply, Deg seeks to explain their basically different ways of looking at human evolution:

25 December 1977

Dear Ernst:
   Don't look now, but it's Christmas Day. It's cold and rainy. Saturn has come down with his disastrous reindeer from the North Pole. I am hiding out, for a couple of hours, nursing my cold, which is true, but also releasing my soul from the desperate festivities, which I shall rejoin soon enough,

and my appetite for turkey will be sated. I shall try to behave with the appropriate jollity. I shall try not to be ironic, and not to make too many anti-materialistic or even learned remarks. I have become incapable of joy "on order" though I am quite eager for joy when I am in the mood. The holidays in our current world have become twistings and turnings of human relations in an attempt to find some traditional form that is quite alien to the form that they assume during the rest of the year. Ah, well, for the moment it looks as if we might have peace in the Near East this next year, owing to that remarkable Sadat who is neither Jew nor Christian, and probably not even a member of the CIA.

Both *Kronos* and the review of the Society for Interdisciplinary Studies (England) have asked me to publish my *Homo Sapiens Schizotyicalis* and I think it will be done. I am suggesting to them that they ask you to prepare a commentary from your letters and other thoughts, if your time permits, thinking that you will have half done the job already. Strangely, I think you have understood my theory very well but you have not understood the weaknesses of your own conventional flooring quite as well. If you will permit me to say so, I would assert that time after time you (and that means a flock of learned gentry of evolutionary persuasion) will employ sloganized concepts and terms to bridge whatever has to be crossed. Like the word "cope" as "the principle of natural selection could and did cope with the possible influences of catastrophes and cosmic radiation escalations." or employ the phrase "decisively influence" in place of "created" to deal with the change in mind. That is, you have no mechanism for the changes that occurred, but rather words that are accepted and unquestioned. And you say that symbolism is created by the adequate faculties of man — then and now — to explaining things rationally. But why does he have to explain? Why doesn't he just let the matter go by? Noone demands that he explain, except himself, and this he does because he must control himself, and thence the gods and others. That is, the reason for human reason is not reasonable, that is, functional in the sense you put it, but he is compelled to a certain kind of reason by his very being that has been changed, and the change is not reasonable but is simply the kind of change that produced the new kind of being.

I have been reading the book by Walter Fairservis, called *The Threshold of Civilization,* as I have thought about your letters, and I can see him to be unconsciously evading all of my major points. He systemically lays out the division of societies into hunting-gathering, agricultural, and civilized (using useless terms), prettying up the old evolutionary sequence. But how much hard evidence exists that hunting came before agriculture? I think that they came together and that later on perhaps when a society became strikingly one or the other, secondary differences occurred. To me, it seems logical that the earliest homo schizo went on for a moment of time grabbing at all the bugs, carrion, and plants he would find, but discovering right away that by escalated sign behavior and organization he could do immeasureably better than before. That is, the gestalt of the creation permitted breakthroughs culturally along the whole front of life. Think of what the Renaissance in Tuscany did with a few ideas; it penetrated every shore of culture and did it within a few years. This was the Renaissance Gestalt. From time to time, too, you mention long temporal periods as elapsing between events and I can see that unless one frees himself mentally from the long-term evolutionary frame of mind, the aggregate of events that I say happened almost simultaneously cannot by definition have happened. So one must hypothesize the collapse of time, understand the dynamic that would then be possible, and thereafter go back and look at time to see whether it is conceivable that we are wrong in believing it to have been so stretched out. I realize that the odds seem impossibly great against a short-time measuring rod. All I can say at this stage is that I have spent some time with every method of measuring time that exists and in every case maybe found some Achilles Heel. To give one instance, it is possible to make a case for Olduvai events to have been contemporaneous with the destruction of the Cities of the Plain — geophysically, anthropologically and in legend. Not a good case, to be sure, but there has never been a study with this hypothesis in mind. And what I have discovered is that the whole world of rocks, skies, nature, and culture can be twisted into a short-term frame, hypothetically, scientifically, to where a whole series of studies could without fantastic efforts give the "yea" or "nay" to the general theories at stake....

* * *

Given so heretical an idea of man's origins and nature, we cannot expect less heresy in Deg's religious views.

I think that Deg's troubles with religion and his carping at gods was because God is a Hero. Deg did not like heroes, saying "Heroes are foreseeable accidents that befall a following." Let us say that at the least he wanted a hero he should control, which is at least an ambivalence if not a contradiction. This in turn had something to do with his early childhood, when there was a benevolent, authoritative father and a brother older by a couple of years who was always excelling, frustrating, lending help diffidently. Harold Lasswell in an impromtu speech at a banquet one time, when both brothers were present, referred to 'Al' as generated out of 'sibling rivalry.' I suppose that Deg had tried to manage Lasswell, that great god of many social scientists, over the years and did the same with Velikovsky. There were other gods as well, and probably he escaped being some great man's Boswell or Harry Hopkins because of his persisting ambivalence; it is not an uncommon trait, especially among women, with whom Deg always felt at ease and in touch.

At one time he made the following note:

It should be an offense for anyone to *speak in the name of gods,* or *to say that gods speak to him,* or *to call upon gods to intervene in the world,* or *to treat anyone in the name of gods,* or *to assign to gods human traits.*

V. and Deg talked little about, and hardly searched for, religion and god. V. had no religion and had never intended to possess one. Deg had no religion, always intended to discover one, but seemed never fully to get down to the search; meanwhile he was forever peering into the crevices where people kept their sacred idols and their firm or faltering notions, and he acknowledged the value of religious discussions. V.'s indifference to religion annoyed him. "God is an open question" was Deg's saying, and he stuck it into lectures and books and conversation, meaning not only that God is in doubt but that God *was* in essence an Open Question.

In November 1972, he makes a note to himself: "Reconcile V.'s intense jealousy of God as a Jewish invention and V.'s expression to me of his belief in plural gods, and Yahweh as Saturn." [Actually V. thought Yahweh was Zeus, and Elohim was Saturn.] "I do reconcile them by saying that V. changed too. His original belief changed even though the momentum of his original routine drove him on. Compare him with the creationists, for example, Bass, Ransom, and others not known except through writings (e.g. Donald Patten) who became quite good and imaginative in scientific and humanistic work on a new secular plane." Here Deg is saying in effect that he was sympathetic to and enjoyed the creationists, whereas V. thought that they were wasting their time. Judaism was the tool of Zionism, so far as V. was concerned. It had little other value but to claim additional authority for Israel skywards as well as landwards. Martin Sieff, studying V. from a distance, came to the same conclusions, which he expounded at an SIS meeting:

> Velikovsky's life's record clearly identifies him as a Jewish cultural nationalist, his youthful experience in the Moscow Free University, his great work in producing the *Scriptae Universitatis* in Jerusalem and in Berlin, his pioneering in the settlement of Palestine in the 1920's, all fit firmly into this pattern. It is likely that he was early influenced by the Russian Jewish Zionist writer Ahad Ha'am, who died in Tel Aviv in 1925, shortly after Velikovsky himself had moved to Palestine. It is important to note here that such a cultural nationalist identity stood very well clear of any religious commitment. Believers may search Velikovsky's published works in vain for any mention or acknowledgement of God. The most they will come up with is in the *Theophany* section of *Worlds in Collision,* a carefully oblique reference which may be taken different ways, to "the great architect of the universe". This is what makes the pseudo-scientific attacks on Velikovsky, by people who have not troubled to read his books, so ironic. Velikovsky himself is in no sense a fundamentalist. His tampering with the biblical texts as they stand and his antipathy to several of the major biblical heroes,

as well as major strands of the Hebraic religion, testify to that. Did Velikovsky believe in God? In his very revealing 1967 interview with the *Yale Scientific Magazine,* one of the few occasions when Velikovsky really lets his hair down, he stayed very well clear of this issue, stating: "people are looking for something in my works, and they cannot find it." It is doubtful, I would speculate, that Velikovsky was an agnostic, and I very much doubt that he was an atheist. The sense of moral destiny, of right and wrong is too strong in his books for that. At the same time, however, just as Freud quailed before Moses, Velikovsky gives us the imagery of Ahab and Saul quaking before the prophets of God, and his sympathies are clearly with the sinner kings.... Velikovsky kept some orthodox Jewish practices rigorously, but insisted that he only did so for the sake of his wife. As they enjoyed 57 years of sympathetic accord in their marriage, this may seem somewhat spurious rationalization... as George Orwell wrote of Tolstoy, for both men, Freud and the latter figure who was so influenced by him, their attitude towards God was rather that of two birds in a cage, suspicious of God as posing a rivalry to their own dominance. Psychoanalysis was God, cast for Freud in the image of Oedipus, and the devil-reflection of his own repressed frustrations. For Velikovsky, God was in the image of the planet god that brought purpose and terror, judgment and fire, to the peoples of the earth.

Deg recollected, when he read a copy of Sieff's speech, a remark that V. had made at Lethbridge. He found that it had been kept through several revisions that delayed its publication for several years. "The noises caused by the folding and twisting of strata, noises of the screeching Earth described also by Hesiod — the Israelites heard in them a voice giving ethical commands." There can be little doubt on the matter. In this work, which Milton happily entitled "Recollections of a Fallen Sky" (V. did not like the title, but Deg ran interference for Milton on its behalf), V. speaks from his view of all manifestations of divinity, that they are natural, material, and that they promote delusions.

His few passages on religion in the posthumously published *Stargazers and Gravediggers* are scarcely revealing. He lumps together religious and scientific dogmatists; melo-dramatically, he writes "were it possible to burn my books and their author publicly, then most probably the councils of the church and of the scientific collegium would have fought for the privilege of taking hold of me and would have dragged me, each out of the grasp of the other, to its own stake."

In the same work, he declares that "to my way of thinking, these books of the Old Testament are of human origin; though inspired, they are not infallible and must be handled in a scientific manner as other literary documents of great antiquity." Well, one man's "inspiration" is another man's "delusion."

His public stance on religion is disclosed in an interview for *Science and Mechanics* magazine (July 1968):

> ...I answered only once when a group from prison in Illinois wrote to me that this occupies their minds very much and they debated and would like to know how I stand. To men in such a distressful situation, I felt that I owed an answer and I wrote to them. But generally, I keep such things to myself because it's just the same as asking whether William Conrad Roentgen, who discovered X-rays, believed that X-rays were created by God or not. The problem is not whether he was a churchgoer or an atheist; this is not the question at all. The fact is that he discovered X-rays. Now you can approach it from the philosophical viewpoint and say "this is the creation of the Lord," and you would be perfectly right. If you are a disbeliever and claim that X-rays are the result of a soulless Nature, you are consequently correct. But you should not confuse historical and scientific questions with theological considerations.

There was, incidentally, little of moment in the letter to the prisoners. Try as he would, Deg could not remember anything in it. When I checked with the Velikovsky Estate to verify the letter, Sammer and Heinberg denied its existence. They agreed that it was written in longhand and no copy was preserved. Possibly Deg remembered V. telling him what was in it, and there being nothing tangible, forgot what it was. We can be sure that V. did not send the prisoners to the Bible, and one of the most persistent and risible of canards raised against V., especially by the humanist movement, was that he was an anti-scientific Biblical revivalist. Many scientists picked up this idea, too. That he was often used by evangelists cannot be disputed, but in such cases Velikovsky was not a Velikovskian.

V. could not be pinned down on God (Deg noted in 1972: "I am certain that he does not believe in God.") but he would use the Hebrew Lord to belay others. The most revealing passages of V.'s view came at the end of *Oedipus and Akhnaton* at the expense of Freud, whose book on *Moses and Monotheism* he denounced; Freud, he declared, had done his people a great disservice by taking monotheism from them as an original invention (again the idea of a "claim"), making of Moses an Egyptian, and of Yahwism a primitive cult; Freud, he actually wrote, was neurotic. His anger at Freud overflowed onto Akhnaton so that this magnificent free-thinking Pharaoh, who tried to liberate a great culture from priestly and traditional thralldom, became now psychotic, deformed, a nudist, monolatrous (not monotheistic), incestuous, homosexual (bisexual), a pacifist bungler of his country's affairs, and, if not a wife-beater, a wife-banisher.

V. harbored the thought that Moses was not a monotheist, that true monotheism did not come to the Jews until the time of Jeremiah, whom he regarded as the first to formulate the idea. He never expressed publicly his view, for the same reason that he had criticized Freud for publishing *Moses and Monotheism*. Too many Jews would be upset, he said onetime privately to Wolfe, Milton, and Rose. He believed that late editors of the Bible and Jewish rulers had refashioned Moses into a monotheist, and that not until a few years before the Babylonian Captivity did the Jews become officially and fully committed as a group to monotheism.

V.'s secret can be deciphered in *Worlds in Collision,* however, where, although he mentions the facts behind his theory, he gilds them by speaking of a striving to attain monotheism from the time of Moses onwards. Like other honest scholars, and ordinary people too, V. could not conceal his discoveries of "truth" even though he felt morally justified in doing so, and actually believed, with some guilt feelings, that he had succeeded. Still, his attempts at concealment had also a political angle, for he was enabled to deny that Akhnaton was a monotheist, and to call him an idolator of the sun, while letting stand the convenient notion that Moses, who came before Akhnaton in his reconstructed chronology, was a monotheist.

The reader will readily recognize in the Illinois prisoners incident that V. had picked up the typical American pose to avoid trouble: keep religion out of discussion — separation of church and state carried to ridiculous lengths. Elisheva was telling Deg proudly of V.'s position; evidently she, too, not only used the excuse, but was self-congratulatory about it. She was taken aback when Deg said that it was irresponsible: how can a person write so much about religion, realizing full well that defenseless people are being affected by what he is saying, and then shut up like a clam when the consequences of his statements are under inquiry? This is especially the case in a free country, where, unlike in police states, one loses little by honesty.

I agree, and it is proper to say that V. lacked original ideas about contemporary religion. He was a materialist, a proto-marxist (rebuffed by persistent anti-semitism), a Jewish nationalist who had to reconcile himself to the powerful Judaic orthodoxy within the state of Israel and within his family, an orthodox freudian believer that psychoanalysis can free the mind, a believer in science as a realistic and rational ordering of the universe, and a shrewd evader of religious controversy, which, if he had entered upon it, would have alienated half of his public support.

* * *

Deg's position was quite different. He was a pro-Jewish anti-mosaist, even though a profound sympathy for Moses is apparent in his book on *God's Fire,* and, I might add, he felt, too, profound sympathy for Karl Marx as a mind bursting with social reality and grim wild hopes, even while being a life-long antimarxist. He felt dreadfully sorry (remember what I said earlier about his empathy with historical figures) for those Jews, often in the majority, who tried to wrest human and civil rights from Moses — Aaron, Miriam, the Golden Calf rebels, the wanderers who heard "the call of Egypt," the Korah rebels, the Scouts, and the intercultural revellers of Beth Peor.

Deg's idea of religion could not develop fully until he had successfully framed the problem of historical religions and satisfied himself of the essence of human nature. You have to find these two keys to the history of religion and man. The first key he discovered by pursuing man's interest in things sacred back as far as possible, back to humanization or creation it seemed. It appeared that all gods were alike, that all men were religious even when atheist, that all religions were psychologically at least polytheistic, and that a succession of changing gods was a reflection of catastrophic cycles of nature and culture. All religions were basically similar: they ritualized celestial and natural phenomena in human terms; they sacrificed, they slaughtered people; and they secured and protected them. Their historical behavior was basically schizoid.

There were two ways of finding the divine, both almost inaccessible to homo schizo; one was to open up oneself to one's innermost depths in order to know whether some part of oneself is divine. The other was to examine the universe outside to see whether the divine must exist there and whether it is manifesting itself. This was a futuristic theology, to be sure. It was anti-rationalistic, that is, anti-Aristotelian. If more words need be applied, it was a phenomenological, pragmatic, existential approach.

In 1965, there occurs a mention of the idea of entropy, and Deg's view of religion may be said to have emerged from his reaction to this "law of nature."

> The world of the second law of thermodynamics — the dying world — is the product of a dying mind. When the mind ceases to die and begins to live, the second law of thermodynamics will be replaced by an equally valid and scientifically acceptable law of creative evolution or creative condensation or creative intensification of specialized activity. [This ultimately ended in the theory of theotropy, thirteen years later.]

He remembers, of course, the aura of publicity that had attended the work of Norbert Wiener and cybernetics, and a kind of gloominess associated with the notion of entropy, merged with the character of Wiener who, he thought, might have committed suicide in Stockholm. Not long afterwards he came upon a book of Melvin Cook in the New York University library stacks; published in 1966, this difficult technical work on geophysics was by all odds the most competent and confident assault upon the premises of long-time geochronometry to be found. Cook's model of crashing ice caps and splitting continents set up the basis for Deg's geology. The main problem was to reconcile his own exoterrestrial first causes with Cook's Earth-based scenario. Besides this, Cook, in a few paragraphs on negative entropy, rendered Deg sensitive to a possible place in theology for a new process. As the time approached to write *The Divine Succession*, the negativism inherent in his destruction of history was unexpectedly counteracted by a positivism from this source.

Deg's Journal, July 10, 1979

End of my generation begins. [I cannot deduce what he means by this.]

NEW PROOF OF THE EXISTENCE OF GOD
If our model of the solar system is correct, with therefore a time of 1 to 15 million years and if the universe is large and populated as it presently seems to be, the manufacture of negative entropic features of short duration should be occurring with much greater frequency than now conceived (although if time is infinitely regressive then the speed of their creation is inconsequential). However, in either case, the probability of say 10^{20} 'intelligent' (negatively entropic) worlds is very high. Now, there is no reason to use mankind as the measure of the 10^{20} intelligent worlds. Whereupon I postulate an X number of worlds where the creative dynamics of negative entropy produce beings of such intelligence and power that they may be called 'gods.' If these are defined as 'beings with n times the intelligence and power of mankind (and they may be aggregates as well as individuals), *one* of them may be considered to be of such Intelligence and Power that it may establish control over the universal process. In that case, we have the traditional concept of god exercised in a new form of proof of omniscience and omnipotence — that is, one who is created by the universe working towards that goal (by its essence) and who ultimately turns around and controls the Universe. If the chances of such a One having appeared are low, and of such a One surviving temporally in addition to all his other powers (i.e. 'God is eternal') sets up a chance that One existed but no longer does, then the Universe may still go on and on in the expectation that sooner or later it will create its eternal, omniscient and omnipotent master, whereupon truly the universe will be intelligently (as *vs.* the present chaos) ordered and in which the far-flung parts will be compelled to cooperate.

However, ideas were converging from all quarters. The theories of *Homo Schizo* and *Divine Succession* went along together and interlocked without difficulty or even awareness.

September 9, 1972

I am going to Princeton today, expect to see Velikovsky. Have continued to probe his work though I have a mountain of tasks before me for the Fall. Am continuously tempted to rewrite his theories in my own language, to test them, to add to them if they test out, *to explain their importance,* and to put them into a logical psychological historical framework that cannot be ignored. I am scarcely prepared for the task, in time, resources, information, so keep nibbling at the edges (one would hope like the Martian rats that destroyed the army of Sennacherib, according to the Egyptians).

At this moment, am reading the scarifying Babylonian poem to Ishtar (*W. in C.* p. 200). I note the line 'O furious Ishtar, summoner of armies,' that concludes the poem. Again, this works two ways: Ishtar causes the people to wander and fight; V. says catastrophes engender migrations, flight, armies clashing in the dark. Agreed. Many corresponding events in Greece, Near East, etc. ca. 1500 and 8th-7th century.

But comes another reason for the armies and the clashes. When people are fearful, they assemble. In numbers there is strength and comfort. They do not disperse as 'logic' would tell them to. Any combat officer will tell you how difficult it is to get men to scatter for cover when under attack; they want to huddle together, even though the collective 'good' lies in spreading out.

The rationalization of 'huddling', the assembly of armies, the summoning, is that the enemy is One, its intentions are unknown, the collective judgement of the tribe or people is needed (the greater the roll-call the better, the more secure the judgement) and the enemy may be the friend, who, it is desperately hoped, will be impressed by one's forces or *lead* one's forces against our enemies, indeed, demand to lead them. "I am your god, your leader. Why are you not gathered to greet me. Why do you run away; your running is suspicious. I demand that you assemble for My Coming!" All of this is notwithstanding that in *some* places and areas people would in fact scatter to the caves and clefts, as the premonition of disaster came to them. *(cf. W. in C.* 212-3).

Deg's journal, Oct. 10, 1972

I showed Sebastian several pages of V. dealing with ancient China. He was moderately impressed. I asked about Tao. Sebastian holds the unconventional belief that the Chinese notion of 'heaven' is animated. It is a Being. I have that hook to hold on to. What set me to thinking was this: Tao seems like a refutation of catastrophism; no bloody gods. But in the beginning it relates the stories of heavenly conflicts. I was baffled. Tao seems so benign, calm, apathetic. Then the thought came: but perhaps Tao became Chinese uniformitarianism! Centuries ahead of the West. Perhaps Tao came to soothe minds and restore calm to the heavens. Really it wasn't long after Mars-Ares-Huizilopochtli-Nergal that Plato clamored for laws vs. disbelievers in celestial harmony. But now see: the West remained unsettled of mind. The gods did not go away carrying catastrophic theory with them. Humanists, historians and scientists interrupted the movement towards uniformity and celestial serenity until the 19th century and then the latter triumphed for only a century. Is it that Judaic Christianity carried the Bible, whose catastrophism would not be denied or effaced, right down through the centuries in the face of all amnesiac needs in religion, society, and science? Is this why the Western world (including the Muslim) has been so turbulent and aggressive? What is behind Tao? Do we now have a third amnesiac development out of catastrophe: Greek pantheons, Judaic chosen tribe and monotheism, and Tao calm reflectiveness?

Deg's Journal, New York City, 1 A.M., May 24, 1973

Just awakened by a call from Jack Martin, Baptist Missionary in Bangkok, regarding Paul. You cannot give up hope for man or woman, knowing that, if you do, the next moment will bring you a person who will reveal that you are wrong.

EPILOGUE TO THE SETTLING OF HEAVEN

If one has stood amidst a burning city, been shaken in an earthquake, or watched the throes of death, or looked down yawning chasms or into the ocean depths, or heard artillery shells scream and strike, each 'with my name written on it,' — then one can better ponder the awful predicament of our ancestors who over thousands of years suffered disaster manyfold and many times over. They cannot be gainsaid their fears and plaints, and the qualities of their gods, those deeply involved companions of humans who became ever more human as they took the gods into themselves and ever more diabolic as they sought to master the games of the gods.

The gods have retired into new forms. But they still operate through the busy humans whom the poet Rilke called 'the bees of the invisible.' They are everywhere and scarcely as remote as our scientific texts would have us believe. They are in astrology, in fortune-telling, in magic. They fly to the scenes of disaster. They augment the forces of authority. They heal and console. They scare. They make anxious. They set the rituals for many as they have done since the age of Ouranos.

They assume their own negations: for they argue with themselves in Natural Law, in Bureaucracy, in Dogmatic Materialism, in Reified Words, in Mummified Heroes, in Time and Worlds without end. They let themselves be molded into One, and the One obliges his necessities by becoming Many. Beyond all, they stand at ease waiting for Armageddon and the Day of Judgement. Then they will don their armor and rally their hosts.

The gods have retired, yes, but it still takes rare courage to contemplate all of their continuing manifestations and to resist the invention of their negations. There is yet nowhere else to go. And few who would follow.

By skating along on the thin ice of the cerebral cortex, mathematical astrophysics or another such exercise may sublimate the gods. Dumb bestiality may be equally functional in sublimating them. We think that of all ways of facing them, the best is to look at them everywhere, contemplate their every manifestation, anticipate their reappearance, but do no more. If there is any question of human madness, it is erased when one pretends to be divine. Our human destiny is an open question. We deny our humanity if we try to close it. We belittle ourselves if we plead with the gods to answer it at any cost. Here we shall have to leave the matter rest.

Deg's Journal, Stylida, Naxos, July 3, 1978

The Old Testament of the Bible has been much on my mind this summer, because of my study of Moses and the Exodus, because of several interesting articles dealing with it by Sizemore, Greenberg, *et al.* that have come to hand and because Ami reveals herself in a new light as once a child who has remembered prodigious amounts of the Bible from the nuns' school in Mulhouse that she attended.

I have come to look upon the Old Testament as a great mountain range that has yet to be explored in regards to its effects upon the human mind, history, education, anti-semitism, politics and society in general. Just as there is no good book on the Jews — sociological, psychological, and behavioral — so there is none on the Bible.

The early scientific rationalists of the Enlightenment (and their socialist successors) thought that merely to expose the Bible as a typical unscientific and superstitious document would be enough to put it onto the shelves of dead religions, anthropology, myth. They treated it as a discrete entity that could be taken off like a suit of clothes.

* * *

What did our homo schizo Deg do socially with his polyego while inventing it? Personal affairs were not easy with him over much of the seventies. The daughters peeled off the family stalk into Bryn Mawr, Smith, and the University of Chicago. The four boys broke off prematurely. They split in every direction. Only Carl went through a university, held at the Peabody School of Johns Hopkins by a devotion to music and a character too irritable to knock about abroad. The others went here and there in the world; wherever the newspapers were speaking of "endless Summers," of places where the action was, of Denver, Bangkok, Florence, Amsterdam, Australia, Cuba, Morocco, Istanbul, and San Francisco, word would also come from them.

Jill decided upon a separation or, perhaps more accurately, redefined her relationship with Deg around 1970 and Deg came thereafter as a visitor to Linden Lane in Princeton and then to his mother, on which occasions he would also see Velikovsky and Sebastian and maybe Tom and Rosalyn Frelinghuysen. The split was not abrupt or devastating; it was a drifting away that he felt less distressing because he was immersed in tides of preoccupation. It was like a pattern that stretched until unrecognizable, and then tore, or like the string tricks people do with their fingers, when with a single movement of the fingers the strings slip into a new form.

Following upon his relatively flushed income of the sixties, when what he wanted to do coincided with what agencies with money wanted him to do — investment brokers, publishers, Bill Baroody's American Enterprise Institute, the war establishment — his finances fell into poor shape during the seventies. Despite ordinary and extraordinary family expenses, and his contributions to his mother's welfare, he took leave from his University and spent all of his savings and gave his library to the Alpine college. He gave up trying to publish his works on world government in America and published them in Bombay, where his friend, Dr. Rashmi Mayur, was building an Institute. Deg was insisting that a Kalotic World Order movement should come out of Bombay or Istanbul, not the United States.

He stayed at Washington Square when in New York, became intimate friends with Nina Mavridis who lived in his building, he taught his courses, wrote steadily, and put together the college in Switzerland with the help of several students. Nina was generous, but could hold her professorship at La Guardia College for only a year. They married, but separated after several years of their being together and she moved to Berlin. He moved from Washington Square Village to 110 Bleecker Street, where he spent little time. He stayed with Dick Cornuelle, he moved into Ken Olson's loft in Little Italy, and he visited happily with Donna Welensky for a while.

In Europe he lived in Switzerland and in Naxos. He was close to many people during the seventies. Although a gypsy he gave the impression of being fixed somewhere and of soberly pursuing a reasonable plan — people knew not exactly where — except that the where was not where they were. One month he would be in Viet Nam, then he would be staying for a week at a little hotel in Sion where the barmaid and he became fast friends and at odd hours he would tell her of many things and she would tell him of her Algerian mother and what the people of Valais were like and how they regarded her. Then he would be in Naxos, building without the means to build, fixing with crude tools, and writing. Friendship would be struck up with those who came by his isolated place and people would come from town and he would go to town. Sandy came from Australia and might even have swum from there, and he heard of the culture and society "Down Under," and they travelled together to America, and he laughed to watch her tap-dance. Sigrid Schwartz came from the Black Forest with her little boy who carved the surface of his marble table with a neolithic flint while Sigrid told of her mother who asked to be carried to the grave with a jazz band playing "The Saints Come Marching Home," and so it was done. He spent a good deal of time underwater in a diving mask and knew the bottom like his own land, and could pluck a bit of pottery out of its rock fastenings any time and give it to a pleased Hamburgian, Londoner, or Trondheimer.

Wherever he went in the world, he never truly wandered, but was always bent upon something to do with study, business, politics, education, and everything else seemed to be related. He was sometimes impatient, pressed by perceived obligations, but never at odds with himself. And wherever he went, half of his baggage consisted of folders, full of reprints, chapters in progress, manuscripts, proofs, correspondence and notes, never less than thirty pounds of these, including the folders that dealt with the job he was on. Hence he was never bored, nor even idle when he wanted to be idle, for he could hardly wait for the day to dawn in New York, London, Tokyo, Saigon, Bangkok, Bombay, Cochin, or Paris so that he could write and read in order to write.

Many were the occasions, though, when the needed piece of paper had been left behind or a needed book was on a faraway shelf. Nor could he half control the crazy-quilt appearance of his work in progress, paper of different sizes and quality made in different countries; handwriting altered by different writing surfaces, some of vehicles in motion; writing in pencils and pens of blue, black, red, and green.

His psychological counterpart, Jean-Yves Beigbeder, would turn up or he would find Jean in Paris or at Nevis in the West Indies, and they would celebrate life and make great plans, until one day Jean slipped into the sea from a stalled motorboat off of St. Kitts to swim ashore for help and was lost into the night and forever. So he had many friends, good friends, he thought, most of them going unnamed, like Carl Stover, Rashmi Mayur, Kevin Cleary and his gang who hated their enemies more than they loved him and wounded the college, Jay Hall, Barbara Schmidt, Christine Ressa, Peter and Annette Tobia, Charles Billings, Carl Martinson, Phil Jacob, Ken Olson, Levi Fournier, Dick Cornuelle, Jay Hall, Savvas Camvissis, Ilse Lackenbauer, Rosalyn Frelinghuysen, Susan Weyerhauser, and always Stephanie Neuman. Even to mention them is not fair to his wishes, for he will complain bitterly that each person means everything to him when they are together so that he cannot stand seeing them on a list, where they may seem like numbers of the days on the calendar of a long-gone year, deprived of all the riches that they presented to each day.

Life carved its channel more narrowly after Anne-Marie Hueber came upon the Naxos scene. They lived in comfortable poverty, travelling irregularly and eccentrically, along the path of Washington, New York, London, Paris, Alsace, Florence, Athens, and Naxos. Great energy now went into the Quantavolution Series, while she wrote her novels and lent him a hand.

All this I wanted to say, though briefly; creativity is always in context — whether Marco Polo in his vast Asia or Immanuel Kant in his little garden — and I fear not so much being irrelevant as that I will convey neither the context nor the created substance, whether in themselves or as they meshed together. Whatever he was up to and wherever he was, by the late sixties, Deg, like many another but in his personal style, was radicalized. He no longer believed in small solutions — whether laissez-faire in economics, gradualism in politics, or incrementalism in biological and cultural development. Pursuant to many early signs, holospheric quantavolution took possession of him.

PART THREE

CHAPTER NINE

NEW FASHIONS IN CATASTROPHISM

Deg's Journal, November 24, 1967

Rereading carefully V.'s *Earth in Upheaval*, I read the sections on the age of waterfalls this morning and, as I poured coffee beans into the coffee grinder just now I wondered at the marvellous parallelisms or analogies of force — an old observation of course — cascades great and small, all the same — what makes them "different"? Man's size? Which separates everything in the world into big and small? Time is such too. Easy to see and believe the existence of gods who pour Victoria Falls as I pour coffee beans.

Think if all the world would be reduced to the same proportion, would we then get a marvellous set of insights into hitherto baffling problems? Would suddenly the rich world become dross and dull?

Another entry, several days later:

Velikovsky came by for a few minutes, left a couple of items, and loped off saying: "I have left too much for the last mile." Too many interruptions, many of his own causing; too many projects, too. At least he has gotten reliable Juergens to edit his "Ten Trials" for publication [It never happened].

We talked of Livio Stecchini who is working on ancient measures and geography. His writing may never see the light. Why? "He cannot bring things to fruition," I said. "The idea is hard," said V., "the inception." I added "The conception." "The conception is a pleasure, the birth is painful," said V. and he left it at that. He went to the library. He loves it and works unceasingly and effectively there. The sky in Princeton is low and the air smells of snow. Scholar's weather.

Velikovsky's *Earth in Upheaval* assembles "the testimony of stone and bone." "Wherever we investigate the geological and paleontological records of this earth we find signs of catastrophes and upheavals, old and recent." It gives an old-fashioned sense of the geology of the last century, before jargon swamped its literature. The feeling is deceptive. The plain speech was deliberate, both because little technical language was required to make his case and because his large audience could not be embraced if jargon intervened between the writer and reader. He also avoided exoterrestrialism, so as to show that you do not need to introduce comets in order to prove catastrophes had befallen earth. However, he allowed many implications to be drawn from geological data pointing to astronomical reorientation of the Earth. And in his conclusion, he made the point forcefully that "The earth repeatedly went through cataclysmic events on a global scale, that the cause of these events was an extraterrestrial agent."

He did not deal with electrical phenomena, a strange omission for one who preached an electrified cosmos. (It entered into a supplementary paper that was printed with the book itself.) That much material on electricity could have been considered was shown by William Corliss, who began compiling it during the 1970's, and by V.'s friends, especially Ralph Juergens in the 1960's; then too Eric Crew in England, Milton, and Deg.

Nor did V. take a radical position on geochronometry. He refused close combat with the giant, Time. To defeat macrochronic arguments he carried forward the older catastrophic topics, still valid, with new evidence from biostratigraphy. Although he advanced catastrophic evidence into prehistorical and even historical times, he hardly advanced the theory and methodology of time-determination. He did not attack the long-time conventional view of Earth history. The best work on short-time geology or microchronism was done by Melvin Cook. V. rejected continental drift and his arguments against Darwinism were those well-elaborated by creationists and scientists of "saltationist" persuasion long before.

Nonetheless, the work has solid merits; Harry H. Hess knew it well; he could find no falsehood or factual errors in it, only a theory which he could not accept or announce *ex cathedra;* and he recommended the book to his students in geology at Princeton. There was much to be learned from it that a student could otherwise obtain from no single source. It was controversial; the geologists dismissed not only its style but also its catastrophist ideas. V.'s scheme to make headway among geologists by presenting a "clean" book, without assistance from legend or astronomy, failed. Yet, today, after 27 years, his book can hardly be called controversial. It is advanced, not *avant-garde.*

Still it is more complete, logical, exact, clear, and secular than any other work in geology that considers catastrophism. The comparable next best work, privately published and quite unknown, was completed at the same time by geologists Allan Kelly and Frank Dachille. That is: *Target Earth: The Role of Large Meteors in Earth Science*. Also more daring and provocative, and also highly professional in method, is geophysicist Melvin Cook's *Prehistory and Earth Models,* published obscurely in England a decade later, which employed purely terrestrial forces in explaining Earth's features. Both books are superior in method to Velikovsky's book, more complex and more original. Both books, I hardly need add, are practically unknown and un-cited among geologists and general scientists; indeed, they were not common currency among cosmic heretics because V. would not mention them.

When a true believer is excommunicated or goes apostate from a charismatic cult he is, if let go scot-free, inclined to start his own cult, and in science or art, there is every reason to wish the apostate or excommunicant well. Robert Stephanos left V.'s circle and found a new interest, another cosmic heretic, by then deceased.

William Comyns Beaumont is hardly known today but was a top-ranking English editor and a brilliant catastrophist. His work turned ever more to the — quite mad — idea that the Egyptian dynasties up to the 13th century B.C. ruled in South Wales and that Jerusalem was originally located in Edinburgh; this plunged him into obscurity, even among catastrophists! Stephanos resurrected Beaumont, located what was left of his materials, and formed a committee to promote his work. He prepared a list of his ideas, culled from *Riddle of the Earth* (1925), *The Mysterious Comet* (1932), and *The Riddle of Prehistoric Britain* (1946); he sent them to Deg who verified the list. Beaumont, on evidence not at all execrable, positioned Atlantis on the British platform and accepted what the Egyptian priests told Solon, that their ancestors had been at battle with his Athenian forebears when the great Island sank amidst frightful tumult.

Here were Beaumont's more "reasonable" propositions:
1. The geology of the world's surface is largely catastrophic.
2. The catastrophe was caused by a cometary collision.
3. All geological formations were shifted as a result.
4. Cosmic lightning played a major role.
5. Hydrocarbons were present in cometary tails.
6. Ancient chronology was several hundred years too old.
7. The Ancient calendars had to be revised because of the catastrophe.
8. Many species were extinguished catastrophically.
9. Religion was born in cometary worship and tied to phallic forms because of the shape of comets.
10. Fear of cometary collisions is inherited by mankind.
11. Vermin were deposited by comets which also provoked plagues.
12. Deities from Egypt, Greece, Meso-America, and elsewhere were identified with planets.
13. Pyramids were both astronomical observatories and "air-raid shelters" for nobility and kings.

14. Planet Saturn, as a comet caused the Noachian Deluge.
15. The Atlantis date (ca. 9500 B.C.) given by Plato had to be shortened.
16. Extensive legendary evidence pictures the "hairy," "bearded," "blazing stars" that were comets.
17. Stonehenge, Avebury Circle and similar monuments were astronomical instruments.
18. Central American legends (and cultures) were contemporaneous with those of the Old World.
19. The intercalary "five evil days" were cursed because they coincided with a world disaster and the ending of an age.
20. The serpent, dragon, winged-globe, caduceus, and other ancient symbols are traceable to cometary catastrophes.
21. Religious festival are dated by cometary catastrophes.
22. Cometary conflagrations are the origin of coal deposits.
23. The ancients had a true 360 day year.
24. The planet Venus underwent great changes in color, diameter, figure, and orbit in the time of Ogyges.
25. Quetzalcoatl (Coculkan-Hurakan) commemorated the cometary dragon for the Meso-Americans.

One significant thesis that V. could not have gotten from Beaumont was that the disturbing comet was Venus, although both identified Quetzalcoatl with the comet.

The list appears to be defensible by the criteria of quantavolution. But once one goes into the books behind the list one enters a jungle of brilliant entangled foliage. Beaumont finds innumerable bewildering geographical, geological, theological, and historical analogies between the regions of Great Britain and the Near East, particularly Palestine, such that the history of the two can be merged into one from the time of the Golden Age of Saturn until the Emperor Constantine (312 A.D.) of the Roman Empire. "The history of the Old Testament is the history of Atlantis," he writes. Constantine ("born in York") had definite motives for transferring the arena of Jewish history and that of Christ to another region altogether." *(Britain: Key to World History)* Obviously, to enter Beaumont's world is a pleasure allowed to few.

The reader may have noted that most of the theses occur in Velikovsky's, and also de Grazia's books. It is easy enough to explain the similarities in the case of de Grazia for he drew heavily upon Velikovsky. It is not so easy to explain the parallels between Velikovsky and Beaumont. Velikovksy never mentioned or cited Beaumont. Could Velikvosky have read and forgotten Beaumont's books? His method of proof is entirely different; practically everything — style, format, language, method, and evidence — is different; only the conclusions are the same. And I should stress that when Deg came into possession of the Beaumont materials, he found them mostly unusable for methodological and theoretical reasons; Beaumont's stress upon Thoth, however, helped convince Deg that a catastrophic age ought to be assigned to the god Hermes and the planet Mercury.

Moreover, with regard to both Velikovsky and de Grazia, too many of Beaumont's conclusions are the same to explain them as sheer coincidence. I guess that either in the 1920's or 1930's, when V. was in Palestine, the books, published in England and dealing with matters of interest to the Near East, made an appearance in the bookstores and were seen by V.

A second possibility is that during the 1940's V. met with the books at the Columbia University Library where he spent thousands of hours in research on his own books. The Columbia University Library possessed of Beaumont's relevant works only *The Riddle of Prehistoric Britain* which was published in 1946. By this time *Worlds in Collision* had been written. V.'s library time during which he achieved his major beliefs relating history and geology to exoterrestrialism had been spent in the Columbia University Libraries.

However, a note exists in his archive, mentioning having read Beaumont's 1932 book; the note dismisses the work. Yet V. expresses his wonder whether Beaumont had gotten his (V.'s) ideas by telepathy. V.'s memory was prodigious. Could there have been a 'Bridie Murphy Effect?" This case, it will be recalled, involved a Colorado woman whose accounts of "another life" in Ireland were substantiated by investigations of her "home, family and neighborhood" in Ireland; it developed that she had been unwittingly retailing material conveyed to her by her Irish nurse in early childhood and duly registered in her memory.

V. had an unusual interest in mnemonic phenomena. One time Deg was visited by a nurse from India accompanied by a high official of the Indian Foreign Ministry. She possessed a rare factual and numerological memory. Given any long set of numbers, she could recall them and reorder them. She could also do tricks such as supplying a person's year of birth, knowing the day and month. When younger, she had possessed only an ordinary mind, then had global amnesia following her mother's death, and afterwards had been led slowly by her father to relearn everything. Despite her prodigious abilities, she was a modest person of ordinary intelligence. V. came to meet her and a seance was held. Deg's term for the type was "idiot savant." V. did not use the term, and he was unusually taciturn, leaving Deg wondering whether V.'s mind possessed a similar competency.

***

V. one day confides in Deg that he has discovered in the course of his research certain geographical locations where oil and gas were exuding in ancient times. It might be profitable to explore there. They talk again and again about the information, and Deg draws up an agreement which they both sign. If they can interest an oil company in purchasing their knowledge, they will divide the proceeds. V. chooses a location. It turns out to be in Turkey. Deg buys maps of oil concessions and wells for the area and finds that the spot mentioned stands seemingly outside the boundaries of existing rights to drill, although quite surrounded by concessions. Better Turkey than Syria, certainly, they think. However, Deg knows the problems of Turkey, political and bureaucratic, the tangle of laws, the high costs of concessions. All that they have to sell is a dozen words. Give away the words, and the project explodes. So Deg talks to friends and telephones to experts. He speaks to his friend Robin Farkas, who is Treasurer of Alexander's Department Stores and who has friends engaged in oil speculations. The situation is ridiculous: there is no way to proceed, except by trusting strangers; give them the information and if they can persuade the most appropriate corporation or government agency to spend half-a-million dollars drilling, and if they strike oil, they might be counted on someday to compensate the "owner" of the magic words. V. writes Deg, who is somewhere in the Near East, on August 12, 1968:

Dear Alfred:

Enclosed is the contract [for a book, never signed]... Ralph left on a cross-country trip...

As to oil in Italy, I shall write you separately but I would also like to know how would you like to proceed if we come to an agreement as I hope we will... [Is] the Italian monopoly holding oil company entitled also to off-shore exploration and exploitation?...

And what is new concerning Turkey? ...a concession there?

In the matters of Cosmos and Chronos [etc]... I assume you have received my former letter (or letters), last to Samos.

I wish to think that you have achieved many goals during this trip and also piece of mind and serenity that usually eludes very active minds — though you may be an exception.

I look forward to a letter from you and shall answer speedily.

With warm regards,

Yours,
Immanuel

Deg is nonplussed, and heavily occupied. He cannot figure out an easy way to get in and out of an oil arrangement. He had had the same kind of difficulty once before when he wished to engage the Xerox corporation in a system of information retrieval. There seemed to be no easy and reassuring way of handing over useful knowledge. Perhaps it would be best to publish the information for the benefit of all those interests that might want to scramble to profit from it. Or give it to a friendly government, or to a friendly corporate officer. Or hire someone to run around among the oil companies and venture to the historical locations; he would need funds, must be made a partner, and had to be trustworthy.

Nothing more was done, and the several indications of petroleum rest in their ancient sources. In recent years, oil explorers have come to hire dowsers, several of whom claim to be able to sense oil locations simply from maps. Deg asked an Exxon official whether the company might not profitably set up or contract for an office, which for a million dollars could carefully read every ancient document that exists to discover relevant references. After all, to dig a hole costs half a million dollars. Deg wrote a memo about it. The idea seemed to Exxon rather odd. (They hadn't yet heard about dowsing.) So Deg quit trying to sell information from ancient sources.

By 1970 there are intimations that Deg would be moving into the field of geology. Typically, he notes some striking fact and then reviews his life experiences to weigh its significance. Then he moves out in a number of forays, both intellectual and operational, some of which lead nowhere, others foolish, still others abandoned midway, one or two coming to a conclusion. But meanwhile, like a beaver's dam, the sticks begin to make a frame, the holes are plugged up, the waters are stemmed and a structure manifests itself. Folders begin to collect notes and ideas. Years may pass, during which time little that is directly relevant and purposeful happens in the field, for he is occupied with other writing, or with education, politics, war, and personal concerns. Still, a cluster of opinions begin to form and he is infected by the specific ambition. He has fantasies of a message to be conveyed with fierce logic and compelling force but is already telling himself in a small closet of the mind that he must be respectful and persuasive.

Then he foresees an opening of Time and feels inspired to create a book. He reorders his ideas and notes in a dozen successive outlines; several introductions appear and vanish; meanwhile he writes one after another the chapters. A bad chapter is washed out. A bulky chapter is broken into two, and a section of it is floated into a new position somewhere else. The writing is heavy labor and becomes increasingly furious and fluent. What ends up as *The Lately Tortured Earth*, written in seven months of 1982, began as a note on strange ashes, following a reading of passages from Schliemann's report of his discovery of "Troy."

Deg's journal, Stylida, July 7, 1970

Early in World War II, the Germans air-bombed Rotterdam as a terrible 'object-lesson' to the Dutch to obtain their surrender. Then late in World War II, the British and Americans bombed Hamburg, Dresden, and other cities, using many thousands of incendiary missiles. In no case, despite high buildings, much wood construction, and inflammable objects, did the immense fires leave thick layers of ashes.

How do we explain, then, the heavy compressed layers of ashes that cover so many ancient cities. I cannot go along with the many experts who casually assign these remains to an invasion, the loss of a battle, or accidents. They are really "playing with fire." Schliemann's pretty little story of his discovery of "the treasure of Priam" is a case in point. He implies that somebody carrying a large casket of gold objects and other precious goods had to abandon it suddenly during the final stage of the siege because he or they were pursued hotly. Over a copper shield "lay a stratum of red and calcined ruins, from 4 3/4 to 5 1/4 feet thick, as hard as stone." He nevertheless could extricate the shield and the casket of articles associated with it with 'a large knife.'

He [Schliemann] writes, 'It is probable that some member of the family of Priam hurriedly packed the Treasure into the chest and carried it off without having time to pull out the key [whose wooden handle was gone]; that when he reached the wall, however, the hand of an enemy or the fire overtook him, and he was obliged to abandon the chest, which was immediately covered to a height of from 5 to 6 feet with the red ashes and stones of the adjoining palace." How remarkable that this kind of reading of the ruins has prevailed to this day! And I have noted others from stories of the Near East, Etruria, and Meso-America.

All references to ash layers in ancient times need to be collected. The levels should be recorded, along with the normal data on what is above, below, and the site location. Of course, C. Schaeffer has done something like this in the Middle East and Velikovsky had added some other reports. A special study, however, is lacking. It should also be noted that the original layer must invariably have been much thicker than the final layer as discovered by archaeologists. This was mentioned by Nicola Rilli in his book on Etruria; yet he persisted in speaking of a Ligurian invasion and other mishaps, not associating the ashes to natural catastrophes or the deluge that he believes overcame Tyrrhenian civilization. The Pompeiian, Herculaneum, Krakatoan ashes should also be measured.

Ultimately, we should sample the ashes to determine whether their origins were local or distant, terrestrial or celestial (this may be possible now that we are beginning to know the geological composition of Moon's surface and perhaps soon of Venus and Mars; they must, of course, be dissimilar; if similar, we may be stuck).

In 1973 he goes to work seriously on the case of the Trojan ashes. The literature on what he calls paleocalcinology is nil. He prepares a memorandum and sends it to several experts, asking them for citations and an opinion about the possible sources of the heavy calcinated debris of the "Burnt City" of Schliemann. They gave him other names, until he has a score of informants, practically all of whom are curious and helpful insofar as they have something to offer.

Craig C. Chandler, Director of Forest Fire and Atmospheric Sciences Research for the Federal government, wrote him a letter that might serve as a model of scientific altruism. I quote it at length, for that reason alone, even though its contents are in themselves fascinating:

Dear Dr. Grazia:

Forgive me for taking a whole month to "reflect briefly" on your letter of February 8. The delay is even less excusable since I have come up relatively blank on the citations you requested.

I do however have a contact who I know is quite interested, and deeply involved in archaeological investigations of past natural fire history.

You should contact:
Dr. Edwin V. Komarek, Sr.
Tall Timbers Research Station
Route 1, Box 160
Tallahassee, Florida 32301

All the half dozen references I have been able to unearth that deal directly with prehistoric charcoal and ash deposits stem from Ed Komarek, so you will undoubtedly get them, and more, directly from him.

I found your manuscript fascinating. However, there are some points you should understand before going too far with a theory that credits wood fuels, either forest stands or urban constructions, as a source for 15 to 20 feet of ash fall.

A natural forest can easily meet or exceed the 200 ton biomass figure quoted by Kelly and Dachille. However, in a living forest, only the material less than one-half inch or so in diameter is ever consumed by fire, regardless of the fire's intensity. This practically never exceeds 30 tons per acre unless the fire has been preceded by some other catastrophic event such as massive insect kill, logging, or exceptional weather anomaly.

The "ash" residue from the complete combustion of wood ranges from 0.1 percent for white pine to 2.2 percent for western hemlock. Actual residues from naturally occurring fires are much higher, ranging from about 10 percent in low intensity fires down to the proximate analysis value in firestorms. Thus, there would be less than 3 tons per acre of "ashes" produced by the burning of the densest forest. This is an amount about 10 times as great as the fertilizer you spread on your lawn in the spring.

There is an abundance of practical experience on distribution of ash from large forest fires. The Pestigo Fire of 1871 burned more than 300,000 acres completely surrounding the town of Pestigo, Wisconsin. Contemporary accounts mention "ashes piled nearly an inch deep in the streets." I have been in several forest fires where newspaper accounts played up "ashes falling like rain." In every instance with which I am personally familiar, the resulting deposit could be measured in millimeters.

Cities, of course, have much heavier fuel loadings than do forests. But again, ash residue from the burning of a city is measured in inches, rather than feet. The accounts from the 1906 San Francisco earthquake and fire are good evidence on this point.

In firestorms, forest or city, there are no ashes left. Firestorm winds scour the burned area clean.

Although it is completely out of my field, I would theorize that the only possible way in which a deposit of wood ash many feet thick could be produced in a single event would be to mechanically reduce the wood to rubble (earthquake) cover it with an inert material at high temperature so that the combustion could not occur (volcanic ash fall), and reduce the wood to charcoal and "ash" through distillation. I have never seen *"red* ashes of wood" in natural fires, and the term sounds much more like a distillation residue than a combustion residue.

I hope the above discussion is helpful. Please don't hesitate to write if I can be of further service.

Deg's exchange with Ed Komarek may also be worth quotation:

Dear Dr. Komarek:

In an endeavor to pursue a number of baffling contradictions in ancient and pre-historical times, involving the life and death of ancient settlements and the development of various human traits and customs, I have come upon indications of huge conflagrations involving layers of ash deposits that to my mind could never have originated, as the archaeological community tends to believe, from the ravages inflicted upon the settlements by conquerors with torch in hand. Several strata of the city of Troy (Hisarlik) in ancient Anatolia give evidence of inordinate destruction, sometimes by earthquakes, sometimes by fire and sometimes by both. Yet there appears to be no great volcano that might have exploded or collapsed nearby. Although perhaps noone has done so, it appears to me that a chemical examination of these beds of ashes of the different centers of exploration in Asia Minor and the Middle East might tell us whether hand-set flames, volcanic fall-out or some other less familiar element may have been involved.

May I ask about the nature of your studies and work in this field, and whether you could put me on to some literature in it, and further whether you know others besides ourselves who might be interested in it? I would be most obliged for your advice.

April 29, 1974

Dear Prof. de Grazia:

...I am much interested in some of the comments you make. If the sample of the ash could be examined under an electron scanning microscope we might be able to tell a little bit about where it came from. In fact, if you could ship me a small package of it I will certainly put it under an electron scanning microscope and see what I can determine.

Under separate cover I am sending you several of our publications, particularly one in connection with particulates from forest and grassland fires. With this technique it might be possible to pinpoint what type of ash you have found. Of course, many of these early cities had a tremendous amount of woodwork inside of them and of course, these would burn even inside of stone buildings. We certainly should be able to tell the difference between volcanic particulate matter and that from wood or grass.

[He goes on to describe the work he has been doing on natural fires and the origin of cereals in Anatolia, and expresses interest in the continuation of the Trojan project.]

May 28, 1974

Dear Mr. Komarek:

Thanks for your letter of April 29 and for the many materials that arrived subsequently. I have been having a field day with them.

The enclosed paper on "Calcination in Pre-historic and Ancient Times" carries some of the logic that has led me to my present interest in the testing of ashes (and, I may add, mega-lightning or Jovian lightning, which, I think, may have been almost qualitatively different and/or vastly more frequent and destructive at some periods than during recent times).

I wish that I had samples of ancient settlement ashes to forward to you, so that the testing might begin. But I am afraid that their collection awaits a field expedition of some complexity. I am going to Greece and Turkey this summer, leaving June 23, and may be able to arrange some permissions and even to scrounge some samples. I am seeking support for the research as well, although I fear that the novelty of the approach, its threat to conventional theories, and the fact that my qualifications for the work, whatever the distinction I may hold in other fields, are not specific to the problem, will all handicap my efforts. Apropos of this, may I say, in asking for help, that you will give aid and consultation in the analysis of the obtained material?

Thank you again. Incidentally, I note that we did not miss one another by much at the University of Chicago. I began my studies there in 1935 with $50 that my father borrowed for me and a trumpet that sounded a lot better to people then than it would now....

On Naxos, Deg had met Professor Georg Keller, geologist of the University of Freiburg, and sought his advice as well. Keller knew Aegean geology and assured Deg that there were no volcanoes near Troy, neither now or anciently. He doubted any possible source of ash from Thera or elsewhere. Ash falls are not uniform, even on a small island like Kos, where in one place he found 40 cm of Thera ash but in many other cuts on the island nothing at all was visible.

Deg's journal, June 3, 1973

Everything is understandable when it is simple and it is simple when only one or two things happen to it at a given time — and the longer the time without their changing, the even more simple is the scheme.

Thus the mechanics of the earth seem understandable when 'a presumed history' is said to permit only a couple of motions and even these are under severe constraints.

However, when in fact, the *real* history of earth is shown to have involved *large* changes in not only a couple but in many motions, then an exact explanation of what happened may be impossible, especially so since no reliable observers reported most of the events.

One reason why uniformitarianism evolved rapidly and persisted is that it created a simplistic history, evening out things over time and subjecting them to "normal" changes.

One reason why there are so many theories explaining natural history is that each man can barely cope with the possible effects of his one favorable type of motion and change.

He ruminated about oil, about tectonism, about the Thera explosion of 3000 years ago, about the earthquakes that long ago shook the now seemingly stable earth beneath Athens. Here he is at New York University, noting a meeting with Professor Charmatz of the geology faculty on Oct. 9, 1973:

Deg's Journal

Lunch with Prof. Charmatz of the Geology Department. Nina came along and we ate at the Faculty Club. I worked to minimize threat, arrogance, conviction *re* our subject, the question of how ashes of ancient times are laid down and composed, in relation to Velikovsky's theories. I needed all grace and tact to do so, for young Charmatz was ready to lecture me on my foolish dilettantism, I could see; he was nervous and prepared to give and receive aggression. He had hardly ordered lunch before he blurted out the V. cited sources that could only be found in some exotic library, that one good guess did not make a theory right (he cited the surface heat of Venus), and that V. was an astrologer. I let it all go by with sympathetic murmurs and a soupcon of rebuttal. Then he smoothened out, and began to talk to the point.

As usual, what seems simple is difficult to bring about in experimental science. I did discover that no sure blocks confront a set of distinctions among ash-heaps of varying chemistry, origins, duration, quantity. A crucial test is possible. We need an interdisciplinary team — archaeologists, chemists, geologists, zoologists, geographer, engineer, mythographer, and maybe even a social theorist or methodologist. Then we *need* to find sites around the world where these ancient ashes lay, analyze them, and try to explain their presence in depths varying up to an original 12 feet. Charmatz became quite involved and is willing to go along with me into the possibility of such a project. When he loosened up, he began to release particular information of much value. We talked also of magnetism, of what is to be found in the bottoms of old lakes, and of petroleum. He declared that *all* ('not one exception,' at my prompting) petroleum had been found in sedimentary rocks from ancient seas. 'But not all sedimentary strata have oil?' No. 'And if we found one non-sedimentary pocket of oil, the theory would be blasted?' 'Probably.' 'Tell me: is it possible that only in sedimentary rocks where oil *has* been found *can* oil collect? Or are there other formations that could hold oil over time?' He seemed puzzled by this query. I repeated it twice more, in between answers that were not direct. I still do not know the answer, but it may be important. For if oil can only be held in one kind of rock pouch, then it is indefensible logically to claim that the oil and the rock are generically related. If all my pockets have holes in them except one and my money can be kept only there, it is incorrect to reason that this pocket coined the money or witnessed its coinage.

How helpful it is when scholars of different fields come together on a problem. That is what a university community should be. There is so little of it, however.

P.S.: He began to ponder the fact that oil would decompose everywhere; that ashes would decompose, geology cannot tell.

Now again he is searching for anomalies in archaeological reports of ancient times, and writes in his journal of January 21, 1973:

I am dismayed by the material that I must digest. This morning I scanned *Chronologies in Old World Archaeology,* a fat little encyclopedia edited by Robert W. Ehrich. I search for evidence of clear breaks between cultures. The authors do not give them. They classify but do not explain a multitude of changes in strata and objects. In a couple of instances 'sudden' stoppages are mentioned. Done in 1965, noone mentions Velikovsky, one mentions Schaeffer (he could hardly miss him since Schaeffer appeared in 1948 and the author is specialized in Northern Syria and Northern Mesopotamia).

All are using R-C dating (adjusted) and grumbling about it.

It is difficult to say whether the dates given reflect a sampling of possibilities; e.g.:

If all the dates are put into a frequency table, would gaps show up and would these point to a destruction over part or whole areas? Is this statistically inferable?

Look up possible catalogue of all R-C and P-A dates for the world and make a frequency table from them. If there is
1) any consistency of clusters or gaps?
2) any consistency in parts of the world; i.e. axis tilt or even another disaster would hit certain parts of the world worse than others.

Later, the whole picture could be slid into a true chronological space.

All dates seem to be later than 10,000 B.C.

Then he is in Athens and has looked up Professor G. Marinos of the University of Athens Geology Department:

Dear Professor Marinos:

The Doxiades Organization informed me that you were supervising the analysis of the core drillings being made at a number of sites in Athens in connection with the proposed subway route....

I am interested in any evidences that your drillings may show of levels of calcination in the historical and pre-historical stratigraphy of the area. By calcination I mean burnt debris, ash coverings, and earth subjected to heavy thermal stress. At the same time, I would be interested in concurrent evidences of flooding on a large scale, associated with or independent of the burning.

Professor Marinos is happy to oblige and introduces him to the engineer who is drilling beneath the city. The engineer takes Deg on a tour of the drilling sites, and shows him profiles of many cores. The drilling is too crude to tell him what he wants to know: what comes up is an already infinitely fractured Athens schist; no way of showing thin or scattered ashes. Athens must have shaken a great deal in ancient times, he thinks, but no indications of flooding or ash falls. Could the surface of Attica have been shaken, washed away and blown away? Possibly. The Acropolis was originally part of a larger mass according to Plato and said to have been well-watered.

He sails for Naxos, whence he writes to his old friend, Richard C. Cornuelle, in Manhattan:

...I have nearly concluded that the ocean basins were created about 15,000 years ago, and promptly filled with the waters of heaven. And I bought a beach ball, painted it white, and, with much effort and complication, finally succeeded yesterday in drawing upon it in crayon, a map of the all-land (Pangea) earth, the old poles, the old ice caps, and the fractures that split and drove apart the continents by an expansion of the globe. I had hoped to sketch the book this summer but the problems have come so hot and heavy that I think maybe another six months will be needed just to outline the work so that people like you can look at it and see that I'm not all that crazy.

There's a good little foreign crowd here this summer, writers, artists, sculptors, teachers, drifters, even two (not one) belly dancers (American). Wish you might visit. Can give you the absolutely isolated stone cottage away from town where you can dwell stark naked on the land and in the sea. Or send someone you love.

I meant to go to Turkey to get a sample of Trojan ashes, but the crisis, the out-of-pocket expenses, and other risks of the adventure made me put the trip aside and I may get a friend to do the job in the fall or come back in the spring, hopefully with a small grant in hand, to do it myself....

It is clear that Deg was working to explain global morphology by earth expansion. He had yet to achieve the idea that a lunar eruption from the Earth would cause the oceanic fracturing and rafting of continents, and explain many other mysteries at the same time.

Deg's journal, Naxos, August 15, 1974

"New war crisis. Turks are going too far. People around me disturbed. How do I proceed with my strange far-away thoughts and study?

Met with Gerhardt Rosler for two hours today, three hours yesterday. He wants to talk politics, I geology. We talk mostly geology.

Today we figured out together the parallel faults between Paros and Naxos. May be important. Whole strait between may have collapsed recently. Very 'recent' fault, 'fresh,' according to Gerhard.

Stylida is an everyday sight, by geological standards. The area is not such as to excite the torpid theoretical tempers of geologists. If I can say *something* about recent changes here, it will show that one can go *anywhere* in the world with the aid of catastrophic theory, properly framed, and find 'potential support,' at a minimum.

Gerhardt dug up a note he made on a broadcast in Germany when he was a high school student. It said x m^3 of hydrogen per second struck the earth. Where did it go? Hydrogen is not part of the atmosphere. Does it combine with O to drop into the oceans as H_2O?

He had made some rough calculations. It is enough to account for all the oceans at 2 x 10^{25} grams, we discovered, if E = 4.6 b.y. old. *Cf* this with canopy theory. This held rings derived aboriginally, therefore there is no need for the continuous flow.

But if hydrogen and oxygen met in a different gravitational situation — when Earth was in Uranus-Gigans [later designated by Deg as Super-Uranus] complex and orbit — they could compose the rings. Then, relieved from Uranus-Gigans, the rings fell and the stored H_2O deposits with them.

Now, since then, waters would be building up with them *directly!* Is this so? Continental shelves — have they been filling and dropping?

Back in America to teach for the Fall Semester, on November 11, 1974 he telephones Dorothy Vitaliano who with her husband Charles worked as a geological team. Indiana University Press had recently published her *Legends of the Earth,* the aim of which was to establish uniformitarian interpretations of both catastrophic folklore and of geological sites assertedly catastrophic. Her book's sales were disappointing. It is not so easy to sell anti-quantavolution books; although well-received by editors and professors, they lack an enthusiastic audience.

As an example of her method, she presents an Arancanian Indian legend according to which in ancestral times two serpents made the sea rise. Earthquake and volcanism were followed by a universal flood. The survivors took refuge on a mountain top which floated up close to the sun. Ever thereafter, the Indians repeated their climb up the mountains, carrying bowls to protect their heads from the sun, they say, whenever an earthquake occurs. There must have been numerous similar earthquakes and tsunamis, claims Vitaliano, to perpetuate the legend and its associated behavior.

The myth and associated actions are, in fact, rather clear examples of universal responses to a universal flood, preceded by violent quakes and volcanism. The "sun" was probably Saturn gone nova (the infant Horus and Jupiter). The twin serpents were twin comets either from a second confused catastrophe or debris from the nova. The bowls are means of protection from fall-out of all kinds. The continual repetition of the behavior is a form of compulsion, whether it occurs during "normal disasters" or in celebration of the anniversaries of the primordial disaster. The concept of *illud tempus* (the First Great Day, so to speak) that Mircea Eliade, the famed comparative ethnologist of the University of Chicago, employs, explains the psychic nature of such events. Deg's *Homo Schizo I* transfers the concept from a solely psychic complex to a complex based upon primeval experience.

Now, at this point in time, Deg and the Vitalianos' should have gotten together to discuss their findings and differences. Not at all. Scientific development seems at times to proceed as a series of missed encounters and perpetuated misunderstandings. A small problem in business — say a sentence in an annual report — as Deg could observe among his friends in government and corporations, will arouse a rich system of conference telephoning, airplane rides, xerox fireworks, and overnight express mail. Not that the scientists need to have agreed but that they might have erased 50% of the differences and retire both enlightened.

Often impatient of delays, and often pushing things to a conclusion — not always qualities either pleasant or helpful — Deg was poignantly conscious of the defects in scientific and intellectual business:

Talk about Pop and Mom grocery stores! The intelligentsia is driven to work at the lowest support level of technology and economy. And is brainwashed besides to accept its lowly status. There is a mythical complex of incompetence and insufficiency which are inextricably rationalized and justified as a single process usually called creative or scientific, and worshipped as a whole. Yet how can you be sure that they would not waste the technology if you gave it to them. Every other occupation does, the military, the bureaucracy, the corporations, everybody except Mom and Pop. There's the paradox: the least efficient is the most efficient, the least costly is the most effective. We can't all be Mom and Pop, but everything else is worse in its own way!

The Vitalianos were part of the Thera volcano study group, a combined geological-archaeological effort at understanding the explosion that tore apart a thriving island in the Aegean. The peculiar shape of the remaining land excited suspicions as to its history but no historical reference to it occurs. At first, therefore, modern volcanologists assigned it an old age. Then Spiridon Marinatos excavated cultural remains of the Bronze Ages; finally a town of Late Minoan Age was uncovered, Akrotiri.

The geologists followed Marinatos' lead in assigning the destruction to about 1500 B.C. and tying it into both the Exodus and the sinking of Atlantis. Eddie Schorr, a graduate student of the University of Cincinnati, working for Velikovsky, showed (contra-Velikovsky and all concerned) that the event could not be of 1500 B.C., but rather must have occurred around 1100 B.C. or later, and also that it could not be Atlantis. Deg adopted Schorr's view, even though he would have liked to see it dated at 1500 B.C., when there was a felt need to discover universal destruction surrounding the major Venus disaster. The others went merrily along writing books and articles to profit from the glamorous Atlantis and Exodus connections, which I think shows how readily 'hard' scientists will buy meretricious goods. V. was quiet, though his voice, correcting his error and endorsing Schorr would have carried weight. Schorr should have been granted his doctorate promptly upon the publication of this brief piece and his two articles disposing of the Greek Dark Ages (hence 500 years of supposed time) that appeared at the same time.

Such was not to be. Indeed, he published the articles under the pseudonym of Isaac Isaacson, so fearful was he of being evicted from the PhD program of his University. V. was disposed to support his fear; movements are made of martyrs.

Deg could not figure out how justified was their fear, but was concerned with the self-destructive aspects of it. V. had paranoiac tendencies which fueled even stronger and similar suspicions on Schorr's part. Good for one another intellectually, they were bad for each other emotionally. Schorr was highly regarded at Cincinnati. Yet he finally left the University and retired to his family's business in Houston. His research continued privately, and he remained in touch with several other heretics if only through letters that are extremely long, brilliantly correct on Aegean history, and malevolently critical of practically everyone, including his correspondents.

In one of these letters to Greenberg, he attacked Deg's articles on Troy first for not crediting him enough for his advice and counsel (in what name he should have received credit was not made clear), secondly for small errors that could and should have been corrected in a letter to Deg or to the publishing magazine, *Kronos*. Greenberg passed the letter to Deg saying, you see, here is what I have to deal with (for the rest of the letter was furious on other matters as well), or perhaps he was saying, see here, I am not the worst of the Furies. Efforts were made by Elisheva and others, following V.'s death, to consolidate Schorr's unpublished work on the Dark Ages into V.'s lean manuscript on the subject, to no avail.

Deg offered to speak to the Cincinnati authorities on Schorr's behalf, but he was warned against doing so; the prophecy went on to fulfill itself. I cannot say, however, that word of the pseudonymous scholar did not leak to the Cincinnati network, for Deg told his daughter, Dr. Catherine Vanderpool, who dwelled in association with the Athens terminus of the network, of Eddie's predicament; and when Eddie put Deg in touch with Professor Cadogan of the University of Cincinnati surely he must have been tempting, or even admitting, self-disclosure.

Deg, we recall, was on the trail of Trojan ashes. One day he was working at the Library of the American School of Classical Studies in Athens, and found in one of the volumes a remarkable sentence to the effect that samples from numerous levels of Trojan debris had been collected by Blegen's team in the 1930's. Yes — Jerry Sperling, a visiting scholar from Cincinnati told him, who had worked on Troy and was at the Library at the same moment — this showed the thoroughness of Blegen; no, he said, I do not know what they are or where they are.

Deg had friendly access to James Caskey, head of the archaeology department at Cincinnati, through Cathy's father-in-law, Professor Eugene Vanderpool, a friend, and highly reputed as the "Grand Old Man" of the School of Athens. Yes, the samples were in bags still, and were about to be analyzed by a geologist, Professor Bullard. So said Caskey. And Deg spoke to Caskey of his interest in the calcinology of the debris.

On September 18, 1974, Deg called Reuben G. Bullard who, it developed, had left the University to join the faculty of the Cincinnati Bible Seminary. Deg found him well-disposed and even willing to undertake the work from his new position. The samples were contained in about 400 cloth bags in the attic of McMicken Hall. Deg wrote to Caskey and meanwhile reported to his friend Bruce Mainwaring, another cosmic heretic who also on occasion dug into his purse to help move along a publication, a radiocarbon test, or research expenses. Mainwaring responded, "very enthusiastic about your idea for an 'ash' project... and hoping "to try to organize a program which embodies some of Eddie's ideas as well..."

Then Caskey decides the same action should be taken; he writes Deg:

3 Nov. 74

Dear Professor de Grazia,

Thank you for your letter of October 22. I am interested in the project, but must ask for a bit of time to inform myself further, it was a shock to me to hear that Bullard is no longer at the university. I shall be leaving Greece soon but shall be in Cincinnati only shortly before the Christmas holidays. Therefore I'll take up the question - as soon as possible - after the opening of the winter quarter in January. It is important. My colleagues and I shall give it careful and serious consideration.

With apologies for the delay and, again, thanks, I am

Yours sincerely,
John L. Caskey

There is no recognition, here or otherwise, that Deg might render theoretical or operational assistance. Deg sent a copy of his manuscript on paleocalcinology and Trojan ashes to George Rapp, whom Dorothy Vitaliano had recommended as having had an interest in Trojan geology. Deg now applies to the National Science Foundation and is turned down. Time passes. On May 12, 1976, Deg called George Rapp, who is at the University of Minnesota in Duluth, and notes down the substance of their discussion:

Conversation with Prof. George Rapp
Department of Geology
University of Minnesota at Duluth

1200 hrs. May 12, 1976

Has rec'd NEH and NSF grants to study the 350 sample bags from Troy. Is applying a range of chemical analyses to all bags. Has found some pollen and wood that can be 14C analysed. No reports yet and possibly for another year or two. (Student asst is going away for summer on job.) He is expecting to look at the terrain himself in December. No signs of vitrification in the samples. Visual inspection cannot often reveal ashes, but he will know whether there has been fall-out from volcanism or local incineration from torch or accident.

I asked him about the Scottish vitrified forts. He never heard of them. I described the findings of a century ago and said that the theory called for brush or log fires set outside the walls to harden them. He questioned the temperatures, as did I. 1000 degrees needed well-focused, as is done in ceramic baking (with help of venting.) When I told him that the fusing had entered a couple of feet into the crevices, he dismissed any brush fire. So one more important detail is cleared away. The vitrified towers are definitely of unusual origin. I asked him whether the soil of Hisarlik contained the same kind of ferruginous clay that we were talking about and he said he did not know but would look see when he visited the site. (He had been there before but had not noticed.) He said that the vitrification would be noticed by the archaeologists at Troy but noone mentioned it. I am not so sure they didn't. What was the calcination if not vitrification? But the copper and lead deposits would have performed the same lightning attractive function as the ferruginous clay. Hislarlik is a lonely tell and promontory, also attractive.

I told Rapp that I would rap with him come fall to see if anything new had happened. He said he doubts if anything new will have happened. He said he doubts that he will ever have final answers.

On June 15, 1977 Eugene Vanderpool writes to Deg:

Dear Al,

Here is Caskey's reply about the Troy samples, written from Kea.

About the Thera conference sponsored by Galanopoulos and scheduled for July, I am told by Jerry Sperling that it has been postponed until next year. He heard this from George Rapp.

All well in Pikermi,

Yours,
Gene

J.L. Caskey to E. Vanderpool June 14, 1977

Work on the Troy samples is proceeding, very thorough, under George Rapp of University of Minnesota at Duluth. Progress satisfactory, I am told. The results are to be put together in 1978, with the plan that they be submitted then as a supplementary Monograph in the Cincinnati TROY-Series (Princeton U. Press) [actually the results were published in 1982] Slow, but I trust worth the time and effort (and money).

If you are in touch, tell Prof. de G. I'll try to write to him one day but am not sure just when. I haven't got the facts, and probably could not understand them if I had. Nothing definite has been reported yet, in any case.

In 1982 the report finally appears, dedicated to Caskey who had deceased, extravagantly published by the Princeton University Press, and offered at a price of $52.00. Deg who has been following closely its production calls his friend Jerry Sherwood of the Press. She invites him to sit down in their offices and go through the book. He is disappointed. There are no findings of consequence from tests of the debris. The only organic elements of significance are from the straw used in making bricks. There is no indication that any of Deg's hypotheses was considered, even if to refute them.

What could be concluded from this study that occupied several years and cost a hundred thousand dollars? Either nothing unusual had occurred beyond the man-caused or accidental burning and earthquakes, or the proper tests were not employed, or the samples were defective to begin with. Schliemann's Burnt City remained a mystery, so far as Deg was concerned.

Only some of the samples were used. He argues that the remainder stand for future investigation. Regardless of the sinister hypotheses of strange fall-outs or electrical-thermal emanations from underground, there are other more conventional hypotheses that would be worth further study. An outside team, say, such as Blumer of Woods Hole Oceanographic Center led when he was alive, might be asked to evaluate the samples on a much wider range of tests, seeking gases, polycyclic hydrocarbons, lightning residues, and volcanic tephra.

On the one hand this may seem to be the suggestion of a crank who is never satisfied by proofs against his pet theory; on the other hand this may be one of those cases (so well-known in the record of the U.S. Food and Drug Administration, for instance) where decades of one-sided proof turn out to be bad and new theories and tests bring about retraction of the "proofs" and significant new discoveries.

* * *

At Chapel Hill, North Carolina, Deg was visiting fire-dance expert and archaeo-astronomer Elizabeth Chesley-Baity, and paid a courtesy call to the Political Science department. Professor Andrew Scott was cued in to Deg's quantavolution and suggested he get in touch with his relative, John William Firor by name, who was Director of the National Center for Atmospheric Research. An exchange of letters followed. One notes that the inquiry strikes into two lines of study: the possibly catastrophic origins of mankind and geophysical catastrophism. Firor's letter stuck in Deg's mind as he wrote the chapters on exoterrestrialism and the atmosphere in *Lately Tortured Earth.*

June 3, 1976

Dear Dr. Firor:

As I was explaining my present studies in the origins of human nature to Andy Scott recently, he came up with the suggestion that I address you on one type of problem which I've encountered. In my scenario of practically instant creation of the psychocultural human from a closely similar *homo sapiens* anatomy, I have had to set up models of genetic change, cultural traumas, and atmospheric change (plus combinations). In the atmospheric context, one major question is whether there occurred a radical change in some atmospheric constant, which then assumed a uniformitarian guise and which is not observable presently therefore, but yet is producing distinctively human behavior.

For instance, what are the limitations (low-high) of the gases and particles or combinations thereof that an essentially human physical type can absorb or endure without expiring and secondly what mental and anatomical operations would be continuously altered by the different possible mixes?

High altitude deoxygenation, nitrogen bends, oxygen poisoning, carbon monoxide poisoning, x-ray and ultra-violet effects are some cases of relevance. I wonder whether certain gases can affect the endocrines continuously; I postulate this because a constant heightening of endocrinal output will result in pathological exaggerations of typical behavior.

Among the hypothetical constructs for abrupt change in atmospheric constants might be included increase or decrease in oxygen; CO_2; ambiant ionization; x-rays; solar particles; heavy volcanism and gases over centuries. I have not mentioned changes in barometric or in atmospheric mass weight, nor of the effects of high, heavy ice-water rings or canopies that were removed in a series of cataclysms. The chain of causation may be complex, e.g., a life span increase (decrease) brought on by changed gas mixture promotes longer training and group memory and skills.

Perhaps I haven't provided enough detail even to permit considering the subject. If so, please tell me. If this suggests to you some ideas of studies that you would care to relate to me, I would be most grateful. I call my field revolutionary primevalogy; the atmosphere which may be the most delicate of all ecological factors, is part of it.

8 July 1976

Dear Professor de Grazia:

I have given considerable thought to your June 3 letter asking whether there has occurred any radical change in some atmospheric constant. There are three areas that I can comment on: atmospheric composition, climate, and ultraviolet radiation.

The present notions concerning atmospheric composition do not suggest that there have been sudden changes. Those who have thought about the history of the atmosphere take as a starting point a gradually cooling earth which has exhaled a good deal of carbon dioxide. In this situation, some sort of primitive plant life begins and the plants themselves begin to produce oxygen. When the oxygen content reaches some particular level, then animal life becomes possible and it too begins its long evolutionary chain. I am not an authority in this area, but my reading tells me that no one has yet proposed any cataclysmic changes in composition. There is some notion that we have reached an oxygen content which is self-regulating, that if plants produce enough oxygen that the atmospheric content tends to increase, the likelihood of lightning-starting forest fires and other events would increase enough to burn up the extra oxygen and bring it back up to its regulated level. I do not know how accepted this notion is, but if anything, it works against what you are looking for, that is a sudden change.

There are sudden changes known in the dust content of the atmosphere as a result of major volcanic eruptions. When the Agung Volcano erupted in the early 60s, it's well established that the dust in the stratosphere went all over the world and stratospheric temperatures changed for a year or two afterwards as the dust only gradually washed out. However, no ground-level effects of this process were measured and, hence, nothing that might easily fit into impacting a Homo sapiens anatomy.

The climate does change. The northern hemisphere warmed up between 1890 and 1950 and has cooled off since that time by a similar amount. The changes are larger in some parts of the northern hemisphere than in others. This particular change is not particularly large and perhaps not cataclysmic enough for what you are looking for. There are suggestions, however, in the paleoclimate record that larger changes have occurred more rapidly. Around 500 B.C., evidently, in the space of a day, or a month, or a year (after this long a time, it's hard to tell the difference) the climate of Europe cooled strikingly, clogging certain well-known mountain passes with snow, changing the dates of which harbors were free of ice, and producing dramatic effects on the trade arangements, travel patterns and so forth of the time. There are other tantalizing bits of evidence of sudden changes in climate—a rodent in Canada found frozen in thousands-of-year-old ice-covered terrain. Climate change and climate theory is a very active area of study just now and I would suspect a rapid accumulation of new information in this area in the next few years.

Finally, ultraviolet light. Recently, we have found that a sudden stream of fast particles from the sun on one occasion struck the high atmosphere of the earth, produced nitrogen compounds that in turn destroyed some of the ozone and suddenly admitted more ultraviolet light to the surface than before. The effect went away fairly quickly as the ozone layer healed itself and indeed the effect was rather small. But it suggests that if during the changing patterns of the earth's magnetic field there occurred a moment when there was no general field of the earth, hence, no magnetosphere to protect us from solar particles, we might have an era in which the atmosphere would have much less ozone and, hence, the ultraviolet radiation at the surface would be considerably larger than today. It is hard to say how rapidly such a situation might begin. I suppose one could also not rule out the possibility of a major and sustained emission of particles from the sun which would begin essentially instantaneously and diminish the ozone layer for weeks or months, but we have never observed that much solar activity. Very recently you may have seen an article in *Science* magazine written by a scientist here at NCAR in which he pulled together many lines of evidence to indicate that during a 70-year period in the late 17th century, the sun seemed to be free of sun spots and the character of solar activity was very different from anything we have known in modern times. This fact at least holds out the possibility that sustained changes in solar activity can occur and I would suppose if they can occur negatively, that is the vanishing of sun spots of solar activity, one might have eras of higher than normal solar activity. The carbon-14 record, which was used in the *Science* article as corroborating evidence, suggests that the changes in cosmic rays producing carbon-14 and controlled by the sun were of the same relative size of that occurring during the sun-spot-free period in the 17th century.

I hope these rather crude thoughts are some help to you in thinking about revolutionary primevalogy.

Sincerely yours,
John W. Firor

* * *

The ancient Roman Encyclopedist Pliny mentions that the Etruscan city of Volsinium had been destroyed long before him by a thunderbolt from the sky. Noone paid serious attention to the remark, except the cosmic heretics. Deg, who had campaigned during the War in the region, would have liked to investigate Pliny's claim, a pleasant location for a critical test of the veracity of legend and the activity of Zeus the Thunderbolter or another god.

After he had become aquainted with an authoritative figure of Italian geology, Professor Piero Leonardi of the University of Ferrara and the Academia Nazionale dei Lincei, he wrote Leonardi about Bolsena and received a disappointingly assured reply:

10 March 1977

...I read with interest what you said in your letter about the Lake of Bolsena and the publications of your friend Juergens on the possible attribution of the craters and 'sinuous rilles' of the Moon and Mars to enormous electrical discharges, but I must confess to you that the arguments of your friend do not convince me, for a complex of considerations shared by almost all planetologists. I am sending you separately a work of mine on the origin of the 'sinuous rilles' in which you can discern my opinion on the matter...

He voices, too, his opinion that meteoroid impacts and volcanism can account for the craters.

> So far as concerns the Lake of Bolsena, one is dealing undoubtedly with a normal volcanic structure, and I do not believe at all that its origin can be attributed to extratellurian phenomena.

He goes on to address himself to a query of Deg concerning a nineteenth century report of human bones and pottery found in Pliocene deposits and deposited at the Museum in Florence, and says that the report was probably made before proper stratigraphy was carried on, thus permitting a mixture of materials of different epochs.

Naturally Deg was not satisfied. Comyns Beaumont had written many years earlier of the erratic nature of volcanic eruptions and suspected that meteors and volcanos transacted electromagnetically. Stephanos found a striking instance of this reported by the noted oceanographer Beebe on the ship "Arcturus" approaching a volcano at Albermarle Sound. In one day, two brilliant meteors came out of the sky and shot into the crater of the volcano. Noting that Flaugergue's Comet preceded the frightful New Madrid, Missouri, earthquake in 1811-1812, Deg figured that a correlation between comets and meteors on the one side and volcanos and earthquakes on the other side should present few difficulties.

Deg is also corresponding with Professor Ernst Wreschner at this time, inquiring whether he has news of the discoveries at Ebla. Wreschner on March 30, 1977 responds:

> ...On the Italian digs and tablets. There are two possibilities for the destruction of the town, 1) A natural catastrophe, 2) A man-made one. The time: ca 2200 B.C. I do not think that a natural catastrophe destroyed the town and left the tablets intact. The shortlived Semitic (Jewish?) Kingdom of Eber had powerful neighbors in what is now Iraq. The time is also known as the beginning of the Hittite expansion..."

Other cosmic heretics are also alert to the fate of Ebla. Its destruction occurs in Deg's Mercurian period, a highly electrical period. The nations are in turmoil; the natural forces of the Earth — volcanic, seismic, aquatic, atmospheric — respond to exoterrestrial forces, attributed often to the planet Mercury and his identities as Thoth, Hermes, *et al.* Deg laid down the challenge: that no exceptions will be found to the catastrophic destruction of settlements of this period. Concurrently, radar engineer M.M. Mandelkehr published his first study, this "An Integrated Model for an Earthwide Event at 2300 B.C." that extended Schaeffer's Near East investigations to demonstrate on all continents "a global catastrophe caused by an extraterrestrial body." Philip Clapham made his debut as a cosmic heretic in 1983 with two articles in *Catastrophism and Ancient History* on

Ebla, fitting it into the catastrophic chronology of the Near East.

* * *

One of the most promising ventures of the mid-seventies was the little magazine that Hans Kloosterman, a Dutch geologist, put out from Rio de Janeiro. *The Catastrophist Geologist* went on for two years and subsided, but not before it had brought to light materials of German and Russian catastrophists quite unknown to the English-speaking heretics, and of a high degree of sophistication. Noteworthy especially was Otto Schindewolf, a paleontologist who had begun his publications in 1950. He favored the hypothesis that fluctuations in high-energy cosmic radiation caused the periodic extermination of most species. He contributed the essential concept of anastrophism, the positive side of catastrophism, attributing the birth as well as the death of species to radiation disasters.

Deg heard first from Kloosterman in May of 1977 and replied to congratulate him. He absorbed material from at least half of the contents of the journal into *Lately Tortured Earth*.

Kloosterman removed himself *a priori* from an association with Velikovsky, a step sincerely taken which would perhaps help to bring a new line of contributors to the field; however, it also put him out of touch with devotees of Velikovsky and actually incited antagonism to his work. He knew that catastrophists were few without realizing perhaps how very few. He and Deg never met and Deg would get snippets of news about him from Dutch heretics. The journal, which could have matched *Kronos* and *SISR* had it continued, brought in professional geologists, an element conspicuously absent in quantavolutionary circles.

* * *

What Deg meant by ideological features of geology and science generally was amply explained in a note later on:

> As I moved from the theory of human behavior into the study of Nature, my intellectual baggage included the concept of a "scientific fiction" which had given me good use for many years and which may be hypothesized when encountering phenomena that are unproven or lead too far afield to explain, yet are needed to move ahead with an exposition.
>
> I discovered surprisingly that most natural scientists are not sceptical about some major guiding concepts, conceding to them the 'hardness' of reality (reality itself being a fiction of undeniable universal utility). Several scientific fictions can be named, however, that may be losing some of their utility and therefore should when employed be watched for what they are doing to one's mind and the facts being ordered.
>
> Practical fictions of Science:

a) the Ice Ages
b) Natural Selection
c) Continental Drift
d) "In the Beginning," "primordial melt," "the primitive solar system," "as the Earth was being formed," "illud tempus."
Such a fiction includes:
a) the indexing function
b) the classifying of material
c) an explanation of phenomena
d) defense mechanism phenomena
e) license to work (freedom)
f) acceptance (reward)
g) allows one to conjecture freely
All may have in common defense mechanisms vs. catastrophism.

Maybe analyze with similar concepts articles in *Nature* before 1970 and several Sci. encyclopedias' usages of these terms.

Cf. Hans Vaihinger *Philosophy of 'As If*

When no longer functional, these may and should be reviewed to pass muster.

* * *

All the while the cosmic heretics were sure that the planets and the Moon would display catastrophic effects along with the Earth. Planetary and satellite geology was carried on actively in the pages of *Pensee* and subsequent media of the heretics. The high heat of Venus was the central topic of the debate, but V. kept extending his list of claims to other planets and the Moon.

For instance, in a letter to H.H. Hess, July 2, 1969, he wrote:

Some nine thousand years ago water was showered on Earth and Moon alike (deluge). But on the Moon all of it dissociated, hydrogen escaping; the rocks will be found rich in oxygen, chlorine, sulfur and iron.

Velikovsky had not then or later a fixed idea of when the Noachian Flood, which he is talking about, occurred. Here it was 9000 B.P. Sometimes he said 4000 B.P., at other times 6000 B.P., and it was this last date that Deg also chose when the time came to postulate a catastrophic calendar.

Unlike V. and other heretics, Deg accepted the theory of "continental drift" that triumphed in geology during the postwar generation. He went far beyond it, pulling the Moon from the Earth at the beginning of the continental movements, in proposing that then the drift was a rapid "trot," assigning the total quantavolution to a large passing sky body which he called Uranus Minor.

NEW FASHIONS IN CATASROPHISM

24 December 1981

A Merry Christmas and Happy New Year to SIS and yourself!

The Editor, SISR
Dear Sir:

Dr. Peter Smith's "Open Earth" (V SISR 1 1980-1, 30-2) is not open enough to some tastes. If, as he rightly says, "The only certainties are that our sphere of ignorance is huge..," then he should let some quantavolutionary theory squeeze through along with the gang of speculations about continental drift. I do not call if "drift" but "rafting." (See *Chaos and Creation*, 155) In fact, I considered calling it a "trot." Its course has followed a negative exponential curve since its catastrophic beginning. The simplest explanation of the mosaic of jostling crustal pieces is an initial set of heavy shocks from a passing body that wrenched away half of the crust, cracking the remainder and sending it sliding hither and yon toward the great basin exposed by the lost material.

For the moment, geophysicists are enchanted by the shivers of movement and the designation of the creeping pieces as major and minor plates. I have seen the most marvellous reconstructions of the Earth going back "half a billion" years; one is published by a University of Chicago paleographic project under Alfred Ziegler. In my view, the original plate until a few millennia ago was the whole earth covering the globe. What we can chart now are the millimeters of creep of the long uniformitarian tail of the exponential curve of decline from the original precipitous outburst of crust.

To accomplish their uniformitarian infinitesimalism, most geophysicists have taken refuge in billions of years; thus can the curve be smoothened out. This imaginary flat curve they then prove by elaborating geological and radiometric tests of time, the very foundations of which were destroyed by quantavolutions. But, too, tests of time aside, if Dr. Smith would provide us with a single study proving subduction of frozen mantle back into the molten depths — carrying with it light crystal material or, worse, where is all the stuff dumped along the shores? — or if he can supply any other type of hard proof that the continental plates move under an Earth power that is *sui generis* and not originally extra-terrestrial, we should be most obliged.

On the other hand, I do not intend to support Dr. Velikovsky's view of continental drift, which was always to my mind a non-view, "fence-straddling" (to allow an American political expression). As he says, "My position on continental drift was (and is) intermediary between ..." Between what — an orange and a banana? Maybe he did not want to hurt Harry Hess' feelings, Hess having fathered the plate theory, for Hess was one of the few establishment leaders who treated him with a full hearing. Had Wegener's life not been cut short, he might finally have come upon the best explanation of continental drift, for he already had unblinded himself of major geological theses and had the basic components of continental rafting mechanisms in mind.

I hope that Dr. Smith's youthful journal, which you advertise, will open up to articles employing condensed time scales and depicting external forces playing upon the terrestrial globe.

        Sincerely yours,
        Alfred de Grazia

Deg's theory of recent lunar fission began in long fits of staring at the physiography of the globe. He was attracted by Carey's advocacy of a considerable global expansion as the basis for the globe-girdling fractures, but then put off by M. Cook's comments that the heat of such an expansion would have dissolved the Earth. Still, invoking exoterrestrial help, he worked up first an expansion model, as is related in his letter to Cornuelle of August 1, 1974; then, after a year of worrying that expansion great or small could not explain the actual disposition of the continents, he decided upon an explosion-expansion model. Only Milton actively endorsed the concept. The cosmic heretics, who could visualize Venus flying by the Earth 3500 years ago, balked at picturing the crust of the Earth exploding into space to form the Moon a few thousand years earlier. But Deg found that the model, proposed in *Chaos and Creation,* of a binary solar system, recently disintegrating, could accommodate lunar fission along with every major feature and dynamic of the natural and biological sciences, together with the earliest grand legendary themes of mankind.

When he finally got down to writing at length about geology in *The Lately Tortured Earth,* the work came easily. It was simply a matter of taking up in turn the elements of the biosphere, atmosphere, lithosphere and hydrosphere and applying to them all the material that he could gather about exoterrestrial forces playing upon the Earth. The more he wrote, the better he felt about the possibility of adapting conventional gradualism to quantavolution.

It seemed to him that the scientific fields were still far behind, needlessly so, even when they were boldly led. After he had completed the book and sent it off to India for production, he became aware that a striking conference had been held at the resort town of Snowbird, Utah on October 19-22, 1981. Sponsored by the National Academy of Sciences and the Lunar and Planetary Institute, and funded liberally by several foundations and institutions, scores of experts gathered to report upon their separately supported and conducted researches in "Geological Implications of Impacts of Large Asteroids and Comets on the Earth." Deg was of course unknown and uninvited; he recognized having met personally only one of the participants! Their papers were published a year later by the Geological Society of America.

The conference would have been a practical impossibility a generation earlier. It displayed contemporary geology doing what it could do best, technical variations on a theme: given unmistakeable traces of the occurrence of certain meteoritic falls, how might these be distinguished and measured, what excavations could they have caused, what chemicals could have been scattered about, what animals and plants would have died — all of this tightly bound up with uniformitarian experience and highly mathematicized. One searches hopelessly in the volume for an enlarged philosophical and cosmogonical inquiry.

Many topics went unaddressed, among them the possibility that important exoterrestrial transactions of the Earth involved pass-bys of large bodies without impacting; that planets might have played a role in cosmic disasters; that the measures of time employed might not be infallible; that the Earth's tortured crustal morphology might in its most general features be an exoterrestrial effect; and that heavy fall-outs of non-exotic material such as water and gravel might have occurred. When Deg examined the papers, he felt keenly the ambivalence and loneliness of a front-runner in the course of thought. The elation of being far ahead was countered by the fear of being disoriented and by the longing to be moving forward amidst a body of kindred spirits.

CHAPTER TEN

ABC'S OF ASTROPHYSICS

In his journal of January 12, 1968, Deg writes of a conversation with Professor Lloyd Motz of Columbia University, the same who had called the attention of scientists to Velikovsky's successful predictions of Jupiter's radio noises and Venus' high heat:

> Motz turned out to be a cheerful sort, full of admiration for Velikovsky, but of course entirely convinced that the laws of gravitation and thermodynamics are much more positive proof against Velikovsky than are some historical events of which Velikovsky may have proof positive. (...)
> Motz is going, obviously, by deduction from laws that he regards as immutable. He feels simply that, whatever the historical evidence may be, it would be impossible for enough energy to be generated on Jupiter to launch Venus by eruption into the heavens. He wonders whether there might not be some third body that had appeared in space and constituted a counter force that would have drawn off or helped draw off Venus from Jupiter or whether Venus had come from somewhere else in space. I pointed out that Velikovsky is firm at this time that Venus must have come out of Jupiter by eruption (But not volcanic eruption — rather from disequilibrium owing to Saturn) and that we have no knowledge of a strange third body that may have been in space at that time within the planetary system, else we might have heard the name given this body in the records of the times. Still it is worth keeping an eye out for such an intruder. Motz says the same problem besets those who think of quasars as a high-intensity explosion, an eruption from larger bodies. Where can the energy come from, he says, and how could it gather together?

With Director of Antiquities Spiridon Marinatos in 1968, Deg met astronomer Constantinos Chassapis who had studied the Orphic Hymns and derived certain conclusions about Greek astronomy in the second millennium B.C. The Hymns, he asserted, had originated between -1841 and -1382, but probably in the 17th century. They showed the Greeks to understand heliocentricity and the sphericity and rotation of the Earth, and spoke of the attraction of the Sun as the source of orbital movement, and named the planets, the seasons, the atmosphere, and the ether beyond. Their calendar was of twelve lunar months; they identified Saturn with time; and they referred to a universal law that regulated the universe and stabilized the Earth.

Stecchini, Santillana, and Von Dechend, among historians of science known to Deg, were quite persuaded of the advanced state of the most ancient known science, so Deg was rather more impressed by the indications of modernity in Orphism, which Chassapis was exhibiting at the same time. If the hymns had originated so early, though, they went to prove a uniformitarian history of the heavens. Incompetent to challenge Chassapis' readings, Deg could but question the definitiveness of the poetic lines, which seemed indeed vague, and the technique of retrojecting the present celestial motions unjustifiably.

The Orphic Hymns, Chassapis also maintained, evidenced an early knowledge of lenses. This, too, rankled with Deg. He had worried over a mention of a lens-like object found in Nineveh's earliest levels, and had discussed the general question with Stecchini. If the Bronze Age peoples had been able to magnify the stars, meteors, planets, sun and moon, they might also have derived proportions and distances among the planets, this making Jupiter the King and Saturn the retired king. Too, they might thus have perceived the rings of Saturn and bands of Jupiter. They might then for religious reasons, and because humans are anxious animals, have created a body of legends ascribing to the heavenly bodies the various adventures, including approaches to the Earth, that the revolutionaries said were historical occurrences.

Stecchini believed that the ancients had lenses, or at least would have built concave disks of copper alloys polished to a high reflectivity. He wavered often in his basic position about cosmic encounters. Always quite happy to play the game of catastrophic models, he might still be readily influenced by Santillana or another colleague to believe that other solutions might be found in the messages sent down through the ages by the earliest voices.

Deg, on the other hand, even when he postulated ancient telescopes, could not explain away the concordance among ancient voices; did they have telescopes everywhere? Moreover the explosive speech of the modern skies and terrestrial crust were seeming to make a point. Not until 1980 did a space vehicle confirm the great and incessant electrical discharges of Jupiter, but by then he had for fifteen years been persuaded that the legendary electrical behavior was real, and on a much larger scale than anything that might be observed today. The same concordance on many other matters was consistent, too, with ancient legend. If the ancients had telescopes, they would have previewed the catastrophes but could only have modestly exaggerated them in their mythology.

A possibility existed, he thought, that the theocratic elites, here and there, using telescopes, would purvey to the masses distorted history, where legends survive and where are perpetuated some happenings and forecasts; but there would be no compelling reason for widely divergent cultures to achieve consensus on these. Why, let us ask, would the priests of the Jupiter (Yahweh, Zeus) age, using telescopes upon calm heavens, invent catastrophic heavens of the time of the birth of Jupiter, and of the earlier times of Saturn?

For that matter, the great telescopes of the past century have not induced uniformitarian astronomers to alter their dogma of a calm celestial history. On the contrary, they have made an increasing number of observers proto-catastrophists. So telescopes, even if the ancients possessed them, could not impress catastrophes upon men who had not experienced such. If Venus simply seemed big and beautiful enlarged 50 times, why would men go berserk, catatonic, orgiastic at her regular, safe distant approach? Fossil telescopes could not affect quantavolutionary theory. They might even support the notion of cultural hologenesis that Deg espoused.

The great Book of Venus was of course Velikovsky's *Worlds in Collision*. In Deg's long acquaintanceship with the book there developed practically no significant errors of astronomy or geology, errors or omission of sources, or misreporting of legends. There is some exaggeration and "purple prose", as in the title that suggests explosive impacts between the planets Venus, Mars, Earth, and Moon, which he does not claim in the book itself. The style is less timid, hesitant, than might be deemed appropriate. There are hints of arrogance as he warns of the dire fate awaiting the theses of Darwin and Newton (less unseemly today than in 1950, however). There are no appeals to religion, only rare confusions of "ought" and "must" with the factual "is". A certain repetitiveness occurs that may be impossible to avoid but which nevertheless tends to overstress and amplify some catastrophic occurrences. He avoids scientific and pseudo-scientific jargon and the coinage of terms.

I cannot here defend all of this, of which the first statement is already shocking: that "there are practically no errors of astronomy." How can a book that enraged many astronomers commit no errors of astronomy? Apart from the main reasons, which are sociological and psychological, there occur two substantive reasons: Velikovsky establishes his natural history by assertions of fact; certain events either happened or did not happen and we weigh the evidence tending to the one and the other to arrive at a judgement about planetary behavior. Second, after this is done, Velikovsky asks how can the laws of astronomy permit such happenings. He understands the laws. But when the behavior of the heavens does not conform to the demands of the laws, he offers briefly some ideas as to what may improve the laws, such as the introduction of a larger measure of electrical transactions into solar system behavior. He reasons the same in respect to geology.

In legendary matters, he follows Euphemeris the Sicilian (fl. 300 B.C.) who established the scientific canon that a myth is to be explained by natural causes. And when Dorothy Vitaliano years later attacked Velikovsky while espousing euphemerism herself, she failed to realize that she was merely reducing Velikovsky, not supplanting his method, which was the same as her own.

By the standards that Cook, Bruce, Juergens, Milton and Deg came to set for sky-body conduct, Velikovsky was actually conservative and conciliatory to the establishment. He was heretical but not a full-scale quantavolutionary. Deg came to feel almost perfunctory when he argued for the middle-road quantavolutionaries like Velikovsky.

If a mini-microphone had been implanted in one of Deg's large ears, we would be entertained by a litany of quantavolution over the years, emerging from an analysis of his stream of discourse whenever the subject occurred, whether it would be in Greece, Manhattan, or Washington, Princeton, London, Thailand, or India. What happens is this: most educated people are unaware of the case for quantavolution; the subject is perennially interesting; it is impossible to state or argue a full case; certain sloganized propositions are proven over time to have an enlightening and convincing effect; these slogans are packaged and delivered in personal and group conversations, with a couple left out where unnecessary or deemed inappropriate.

I have not had the advantage of an elaborate study, but I notice the frequency of these statements, prefaced by something like: "more has happened to change the world by catastrophe than by gradual evolution."

"Religions are obsessed with primeval disasters."
"Mankind has always been fearful of the skies, such that terrible events must have happened there."
"Venus is hellishly hot and locked to the Earth."
"Mercury now is believed to have been recently relocated."
"Cosmic disasters destroy time measurements."
"Big changes in the biosphere are connected with general catastrophes."
"Ancient legends from around the world confirm each other."
"The surfaces of Earth and its neighbors have been torn up recently."
"The world is electrified from universe to atom with potentials that can overwhelm gravitational forces when exercised."
"You can't determine what happened in natural history by natural processes nowadays."
"Science is as non-rational as any other kind of behavior."

And other such simplicities occur more or less frequently. Whether tossed out in defense or in exposition, the expressions collide with a variety of phrases with which the well-educated person is equipped, such as:

"Gravitation accounts for the solar system."
"All methods of chronology give very old ages."
"The solar system has been functioning as it is for billions of years."
"You can't trust legends: they say everything and nothing."
"Evolution is a fact: it took millions of years to change the horse's foot to a hoof."
"The oldest features on Earth are hundreds of millions of years old."
"No imaginable force can move the Earth without exploding it."
"Venus' thick clouds work to make it like a greenhouse."
"First came myths, then religion, then magic, then rational science."
"Any local disaster can be exaggerated to huge proportions."

After the clash of these sets of slogans is amplified somewhat, the discussion is usually turned off or diverted. Book reviews and scientific table-talk infrequently go even as far. Once in a while a foray in strength is launched by one or the other side. Even so, rational discussion or exposition does not ensue, but rather an elaboration of one of these slogans with the citation of authorities, or with dogmas more elegantly stated.

Rarely does the exposition break out of the brush into the clearing. It would not be an exaggeration to state that in the two decades about which this book talks, no more than a dozen public presentations have occurred in which a systematic attempt has been made by a practiced and specialized scientist in the face of opposition to destroy and bury one or another facet of quantavolution, such as the capacity of moving the Earth without destroying it.

If this condition appears incredible, it is because so few people understand the sociology of scientific communication, or human discourse of any kind. Scientists can answer questions that they pose for themselves, and spend most of their time doing so, and encourage their " stooges" to ask these questions; but they cannot well answer questions that are asked by others, true others, who come out of a different mentality and have different purposes in mind.

Take an example from Deg's experience in these years from a quite distant field, political science, where in parts of three different books he proposed a single equal tax on every living soul: that the annual budget be divided by the population to figure the tax of each one. The shocks, reverberations, incomprehension, suspicions, reservations, indignation and flustered unmediated ejaculations assailing the idea make it practically impossible to present or discuss, even to the point of starting up research in the subject. Yet when he captured an honors seminar at New York University and forced the students to expel all their preconceptions and prejudices, and to dig up fresh facts, the single equal tax was not only understood by the small group, but was also preferred by them, as one after another of the terms were defined, the data researched, a sample of people interrogated, and the idea drafted into the common and understandable form of a legislative bill.

On the proposition: "Venus is a young planet," first reactions tend to be equally obstreperous and incredulous. The attack builds up rapidly:

"The solar system is very old and stable, Venus included."
"The heat of Venus is an effect of its great cloud banks."
"A planet cannot be moved by any force without exploding."
"No force capable of moving a planet exists actively or potentially."
"Existing records reveal at least 4000 years of Venus observations."
"Bode's law of planetary spacing forbids its moving from elsewhere or being elsewhere."
"Planets cannot move from ellipses to circles, and to move they must take up elliptical orbits for a time."

Against these, the quantavolutionary argument, as it was developed by Velikovsky and his friends, asserts:

"The arrangement of the solar system is only stable by our recent historical observations."
"Venus is an exceptional planet in its dense atmosphere and with its great heat of 900 degrees F."
"The heat of Venus is an interior heat moving upwards to the surface and into the clouds."
"The hot planet Jupiter could have contained Venus, expelled it by fission (nova), and given it its great heat."
"Venus rotates retrogradely unlike the other planets."
"Venus is locked to the Earth (not to the 10^4 times larger Sun's tidal force) in two ways: each inferior conjunction (243.16 days) finds it presenting the same hemisphere to Earth; and its axis of rotation is perpendicular (within one degree) to the Earth's orbital plane (even while 3 degrees off its own orbital plane)."
"The postulation of historically active electrical forces allows a planet-sized body to move orbitally, axially, and rotationally without destruction, as an effect of the distribution of charges throughout the solar system and of the near passage of a large body."
"Sacred and secular legends from around the world allude to the deviant behavior of Venus in vicinity of Earth."
"The Venusian atmosphere, compared with the Earth's, contains 300 to 500 times more Argon-36, a gas thought to have been dissipated from the planets shortly after they were formed."
"Venus practically lacks a magnetic field, it being 10^{-4} of Earth's."
"Venus possesses a comet-like tail blowing away from the Sun that is much

longer than the Earth's relative to their respective magnetosphere radii."
"The Venusian surface is heavily featured, despite its great eroding heat and eroding wind turbulence, but has no ocean basins."
"Fires seem to be burning on the surface of Venus, which may be caused by burning methane or hydrocarbons."
"Chemical composition of the clouds indicates no hydrocarbons (or components) yet, but the question is not closed."
"Slight indications are present that Venus may be cooling off."

* * *

The idea of a double sun, the system of Solaria Binaria, as Deg named it, came with shocking suddenness. It was a monster that came leaping at him even before he had a name for it, and before he conceived of a dynamic for it. On April 28, 1963, shortly after becoming concerned with cosmogony, his journal reads:

> Discussions with Velikovsky and Livio have not cleared up the phenomenon of the similar planes of the planets in solar revolution (maximum of 7% off) or even of why they rotate. Velikovsky and Stecchini are not very concerned, since Velikovsky's theories hold anyway. But I wonder whether the nebular hypothesis that has the sun throwing off the planets in an initial series of explosions is true and ask:
> Could the Sun have cast off the planets at different times, or more importantly, could the planets be created on their common plane by the pull between the Sun and a second sun or planet revolving around and near (a twin). Then from time to time a planet would be released from one or the other...

While the people of his camp were arguing with conventional scientists over the origins of the heat of Venus and the chronology of Egypt, he took the time to wander about the cosmogonical fields and ponder what his friends might have known better than he, that is, that changed motions of large celestial bodies signified not aberrations but somewhere back in time a basically different order.

The old order must have functioned on some basic principle, probably a simple principle. What could it have been? He knew next to nothing about formal astronomy or paleontology or chemistry. What he was picking up might be scornfully and legitimately called static, a buzzing of voices, weak signals from many directions, from alleys and haunted houses of science, disreputable astrologies, occult references, stern and orgiastic religious cults and sects, ancient poetry, restless cemeteries of legends, the rage for science fiction, anomalies, contradictions overlooked and brushed aside.

Probably if he had not experienced the hubbub of politics and warfare, where all is said and done and almost nothing is true, he would have avoided all of this, shut his eyes, clapped his hands over his ears. Even earlier, the presumptuous liberal education at the University of Chicago, which combined in a nettlesome but hardfast marriage with skeptical sociological pragmatism, had irrevocably attuned him to ideological quarrels.

Perhaps, too, had he not been pummeled by contradictory and obstreperous personalities among his friends and family, his neighborhood and his schools, he would have been quick to settle upon a regular line of thought. And, to be sure, the din was pierced by his immoderate ambition, which clamored louder than all else for solutions. He did not wait upon his betters.

He asked himself what he could contribute, and in line with his character it had to be "the bigger, the better." It had to avoid competition with superior heretics, not to mention superior conventional scholars, whenever there appeared a well-worn path — solar chemistry, celestial mechanics, the fossil record, and so on. His head contained a large quantity of whispers and scratches telling him what to avoid and what might be chosen. He disagreed with most of Teilhard de Chardin's work, for instance, but in reading *The Appearance of Man*, he caught a fine phrase that would describe his own mental set: "On the cosmic scale (as all modern physics teaches us) only the fantastic has a chance of being true." Chardin followed this course by continuing as a Catholic priest; Deg followed it more specifically.

It was strange that an old, different order of the heavens did not suggest itself much earlier. However, going through the hundreds of titles that Earl Milton and he had compiled for the research on *Solaria Binaria*, Deg could find no statement that the solar system had been anything but a great sun which had cast off its planets in its early history. The history had been stretched greatly over the past century, from some millions of years to several billion years. A rotating hot ball of gases, interrupted by its own violence, perhaps, had operated as a centrifuge. An alternative theory had predicated a passing body which by gravitational attraction had pulled off the planets and gone its own way.

Perhaps somewhere in the literature, as there always seems to be precedents, an obscure passage or writing would suggest that the Sun had a companion that had withered away, or, who knows, even Jupiter may have somewhere been called such a companion. If so, it remained hidden to contemporary discussion.

How did it happen that a few minds adventured in new directions? Let us extract some of the ideas that seem to have influenced the turning of thought.

Legends were gaining respect. After two centuries of general neglect, the idea of Giambattista Vico that behind legends stood a substantial truth began once more to pick up support. It is not without significance that Giorgio Tagliacozzo, an economist and employee of the Voice of America conceived a lush Tree of Knowledge whose fruit was of all the sciences and schools of philosophy and brought it to Deg for publication in the 1950's. Then Tagliacozzo went on a one-man crusade to resurrect the figure of Vico and Deg became the recipient of a continuous flow of material, which, however irrelevant to *Solaria Binaria,* carried a message of the validity of ancient materials. There were others to come, the historians of science, Stecchini, de Santillana, von Dechend, and of course V.

But, going back, too, some twenty-five years, there were the anthropologists and sociologists whom Deg knew at Chicago who respected the customs and ideas of so-called primitive peoples. By his simple and radical logic, it seemed always that if these people were so smart about the present, what they said about the past could not be more stupid than what the great religions said. And, if the two — the "civilized" and "primitive" — agreed that a great god blew a great wind over the Earth, burned it and flooded it, here might be the beginning of a historical truth.

Perhaps this was not all so easy. The anthropologists hardly went farther. Nor did the historians of religion: Mircea Eliade went a great distance to establish the obsession of peoples everywhere with their traumatic beginnings, and the beginnings generally correlated; Eliade just failed to take the step, enveloped as he was in the uniformitarian song of science, to say that these earliest peoples spoke some universal truths.

Nor was it a simple matter to detour around Sigmund Freud, Carl Jung, and other psycho-historians. Freud had his own basis for reality, a primeval cultural event establishing the oedipal complex, guilt, obsession, recapitulation and, for the cosmogonies and catastrophes, nothing but uniformitarian principles. Jung had archetypes, primeval to be sure, cosmic also, but purely psychic in origin.

Velikovsky's was a different story. He generated a formidable sometimes caricatured obsession out of ancient catastrophes, and, further, had attached to the beliefs-cum-faith of mankind an original series of skies that carried two explosive bodies besides the Sun, Jupiter and Saturn, then later Venus that looked like a *Sun* in its approaches to Earth.

To my mind, there is but little doubt that if Velikovsky had been able to focus upon the general cosmological problem of the solar system, in the last decades of his life, he would have provided an ingenious explanation of the behavior of Saturn and Jupiter within a dynamic system. He understood that Jupiter's behavior was akin to a "dark star" it being "cold" (i.e. non-luminous) but with turbulent gases, and suggested that it sends out radio noises; his unpublished talk on the subject preceded by less than a year the actual announcement of the detection of the radio signals by Burke and Franklin (1955). In the same paper containing the bold surmise, he had been arguing on the solar system and, just before mentioning Jupiter's radio noises, he had used the analogy of a close binary or double star to illustrate the presence of electromagnetic effects between stars. He had also brought forward late studies demonstrating a correlation between the positions of the planets and electrical effects detected upon Earth.

He had argued in *Pensee* and in conversations that Saturn must have gone nova to eject immense waters some of which flooded the Earth during the Noachian Deluge. Then X-ray emissions were discovered to emanate from Saturn, a possible sign of recent nova. On 4 November 1976, Milton was asking Deg's advice about mentioning this in a Foreword to *Recollections of a Fallen Sky*. "Ransom suggests that I not draw attention to this claim until Sagan *et al* make some claims about Saturn's heat, magnetosphere, and X-ray emission. The point is relevant to Velikovsky's talks, but Ransom may be right, 'don't give them any points to avoid, let them commit themselves first.'"

In no case, however, did Velikovsky venture the concept of the solar system having a full binary history. In several passages here and there he broaches the idea that Jupiter and Saturn may have encountered the solar system and wreaked havoc from a distance, and he appears to have favored the idea that collisions between Jupiter and Saturn may have caused the Deluge and later on made Venus erupt from Jupiter. It was difficult to try to discuss such matters with him, and when, in his last years, Deg mentioned to him working upon a theory of Solaria Binaria he let the subject pass like a report on the local weather.

Meanwhile, most cosmic heretics who followed Velikovsky were devising schemes by which the major encounters among the planets occurred incidental to their clustering as satellites around the two giant planets, a kind of independent Olympian system interacting at a great distance from the Sun. They believed that the present solar system was occasioned by the forcible ejection of the planets into their present positions in consequence of disruptive encounters of Saturn and Jupiter, after which these large planets spaced out. What may exist in the way of specific scenarios for these occurrences rests still in private files unpublished. When Deg and then Deg and Milton came out with the model of Solaria Binaria in detail, they met with an initial refusal within V.'s circle to consider it; it was lamented that these two had "made up their minds;" the existence of Ouranos as a sky god was denied and other key assertions were denigrated.

* * *

The respect and patience of Ralph Juergens towards Velikovsky assumed proverbial proportions. Juergens devoted most of his professional life to establishing a fully electrical theory of the solar system, including especially the explanation of solar radiance as the reflection of an accumulation and dissipation of electric charge from the galaxies. When Deg asked Velikovsky, more than once, whether he could accept Juergens' theory, he would reply with a definite negative. He adhered to internal thermonuclear fusion as the secret of the Sun's radiation. Because Deg respected Juergens, and then came upon Melvin Cook and then Bruce and Milton, he was never of this opinion. And now, looking backwards, one must wonder whether Velikovsky should have spent with Juergens the many hours that he spent instead, and wrote a book about, with Einstein.

In introducing a posthumous paper of Juergens, a "pioneer in the study of electric stars," in 1982, Milton comments that Juergens perceived the astronomical bodies as inherently charged objects immersed in a universe which could be described as an electrified fabric.

"The Sun," writes Juergens, "is the anode end of a cathodeless discharge extending from the perimeter of the solar system." The solar photosphere is comparable to the "tufted anode glow" in an electric discharge tube. The Sun gathers electrons from galactic bodies and plasma, and sends out an ion current, the solar wind, to the galaxy.

Juergens dismissed the thermonuclear explanation of the Sun's heat in favor of a galaxy-solar electric exchange. The thermonuclear theory, recently developed, sought to explain the Sun's properties of luminosity, temperature and stability by its essential chemical composition, mass and size, assuming that the Sun and its behavior are effects of the conditions in galactic space, not in its interior. So, much of his time went into seeking ways of detecting and measuring the suspected inflow, capable of reflecting a continuous output of electrical power amounting to 4×10^{26} watts, or 6.5×10^7 watts per square meter; this, it happens, registers 0.137 watts per square centimeter at the Earth's position in space. The searched-for input must amount to 4×10^{26} watts as well.

Now whereas scientists have for a long time accepted the invisible source of power known as gravitation, they have largely ignored and disdained the possibility of an invisible source known as electrical discharge in a gas. "Electric discharge is a known and observable phenomenon, yet we might live immersed in a cosmic discharge and know nothing of its existence."

V.A. Bailey of Australia published in *Nature* (1961) his calculations, based on the data of Pioneer space probes, that the Sun must possess a net negative charge with the potential of the order of 10^{19} volts. Bailey visited Princeton to meet V. and there Juergens and Deg became acquainted with him as well.

V. was always excited by indications of unforeseen electrical forces playing about the universe. Still he never accepted Juergens' theory, possibly, as he told Deg, because the thermonuclear theory seemed solid to him, and it is indeed regarded as fact by physicists, astronomers, science publicists, and of course the educated public. Since V. never read or discussed Deg's theory of Solaria Binaria, which accepted Juergens' theory and satisfied so many requirements of V.'s own reading of natural and astronomical history, it can be surmised that Juergens' theory was not working for him, V., and should be tolerated because of the usefulness of Juergens' ideas and work, whether as an ever-respectful historian of the V. Affair or as the indefatigable discoverer of electrical forces and effects on

Earth, Moon, Mars, Venus, and in planetary encounters. Long after Juergens pulled up stakes from the Princeton area to find a new life in Flagstaff, Arizona, partly to be "his own man," V. tried to coax him into returning to collaborate on one or another of his books.

Juergens persisted in developing his theory, while repeatedly coming to V.'s aid in the astrophysical exchanges in which V. engaged. Never was the issue of the origins and prior shape of the solar system introduced to systematic discussion. V. generally reacted negatively, even harshly, when material which he objected to or deemed irrelevant sought its way into the magazine *Pensee*. Ultimately the magazine was discontinued in part because of a disagreement between V. and the Talbott brothers on the question of broadening the magazine's scope. However, he behaved gently towards Juergens' material, and Juergens' ideas did receive their initial publication in *Pensee,* where Deg could study them, along with the rebuttal of them by Princeton physicist Martin Kruskal, to learn something about the Sun. The date was 1972. Juergens had already moved from Hightstown, New Jersey, to Flagstaff, Arizona.

Deg was by now knocking the planets around like billiard balls, looking for the right pockets. He came to realize in the legendary succession of Greek gods, which might be afforded backup from divine successions in other parts of the world, a possible sequence of real cosmic events. His basic god became Ouranos (Uranus), generally ignored by V. and the other heretics. And, reading in the century-old esoteric papers of Isaac Vail, and elsewhere, he found an original divine Heaven which eventually produced a Sun-like figure which was still called by the name of Heaven. Thence the succession of events took shape; Ouranos-Heaven, Ouranos-Sun, Kronos (Saturn) Sun, Zeus (Jupiter) Sun, and the antics of the Olympian family of planets — Earth, Ares (Mars), Hermes (Mercury), Apollo, Poseidon (Neptune), Uranus-Minor and Venus. Each and every one of these had been a principal in catastrophes upon Earth, and victim of catastrophes itself.

Deg thought that these might be interacting meaningfully and in a series or succession, ending at the beginning of the present historical period, when Greek philosophy was born, which could be regarded as the Solarian Age. From that time onwards, the Sun (and Moon) seem to have been the dominating bodies of the sky and no intruder — planetary, cometary, or meteoroidal — appears to have played a major role in the sight of mankind, excepting always in the beliefs of astrologers that carry down to us their fossil memories.

Deg speculated as follows: there were three legendary Fathers — Ouranos, Kronos, and Zeus. Hence only these three major bodies had to be accounted for as the basis of the earlier solar system. But, since Zeus was the son of Kronos, and Kronos was the son of Ouranos, only one body had to be accounted for, that is, Ouranos. Now, since Ouranos was originally a thick cloud enveloping the Earth when mankind's legends began and was the first subject of creation legends, this canopy-sky must have been an atmosphere thicker than any in historical experience, thicker even than those provoked by known catastrophes such as the temporary darknesses of Exodus and other legendary or pre-historic episodes and the recent volcanic explosion of Krakatoa. But finally Ouranos emerged and exposed himself, enveloped in clouds. To some, he was the Cosmic Egg.

The birth of Kronos and his revolt against his father were readily pictured as successive explosions of a super-Uranus and the establishment of the new body, Kronos. The birth of Zeus out of Saturn was analogous. The planetary children of Zeus, of different mothers, remained under his nudging regime until the settled skies eroded his rule and, indeed, all planetary rulership, except in myth and astrology.

Deg imagined that electricity might do what seemed impossible for gravitation, although he clung to both powers until Earl Milton persuaded him that all the problems could be solved without gravitation, letting Deg cling only to the inertia which he had cherished all along as the vital element in "gravitational" behavior. In 1976, he was in touch with Milton, who was coaxing a key paper from V. for his book, *Recollections of a Fallen Sky*. He was also in correspondence with Juergens, and he told both of them what he was up to in *Chaos and Creation*. Both were sympathetic.

On April 22, 1976 he wrote to Milton a memorandum of "Alternate scenarios for the shift of planets, including Earth, from a proposed binary system to the unitary solar system." He conceived the planetary system strung out between Sun and Super-Uranus and rotating around the common electrical axis while the axis, carrying the whole set, wheeled in revolution around the Sun. He is becoming enthusiastic:

> I am beginning to feel my oats, Earl. I can visualize as neat and elegant a model as anyone might wish, replete with formulas. What great blooper have I made, cher colleague? Are you still holding to your generous offer to collaborate? Is scenario II our preferred kick-off? We are having a thunderstorm with lightning.
>
> Perhaps Jupiter knows!

Further exchanges took place: then came a week's discussions in New York in 1977, ten days together in Washington, D.C. in early 1978, the same in Princeton in early Fall of 1978, some days in London and two months on the lonely promontory at Stylida, Naxos, by the Aegean Sea in the Spring of 1980, where most of *Solaria Binaria* was written in its final form. On May 26 1980, Deg notes in his journal 'Finished 1st draft of chaps II and III of *Solaria Binaria* with Earl Milton 12:30 hours." He tells how they would discuss heatedly from early morning until early afternoon, sometimes arguing stridently, their voices echoing over the rocks of Stylida, putting their only competitors, the crows and seagulls, to flight. Afternoons and evenings they would write in their separate rooms. In the early summer of 1981 they met again in Princeton and New York, and again in late 1981, spending a strenuous ten days at Edward de Grazia's beach house at Rehoboth, Delaware to complete a manuscript of the full work. Leroy Ellenberger, not far away, called repeatedly but was not invited to come, for a visitor would have disrupted the relentless pace through the manuscript. (This incident may have triggered Leroy's animosity, who before had been deferential and complaisant.) Pages of notes and reprints lay in piles about the large room, on the floor, the chairs, the tables. Upstairs Ami worked quietly at her novel. Outside the low sun beat weakly upon the great beach and roaring waves. They drove to Annapolis to visit St. John's College where Bill Mullen and Joe de Grazia were now teaching. Deg and Ami dropped Milton off at the Washington Airport amidst a howling blizzard for his long flight back to Alberta.

The notes and manuscripts had traversed the continent and the Atlantic Ocean several times, punctuated by messages and phone calls, and by "Did you receive...?" letters, with chapters and cassettes chasing the men like heat-homing missiles. By the Spring of 1982 the book was completed and stood in line for publication.

So ambitious a work should have been created under ideal conditions, with at least a solid year of side-by-side collaboration and next to a giant library. If they had waited for this setting, the book would never have been written. Milton had been troubled by asthma most of his life. He was placed under great pressure in the writing of *Solaria Binaria*. The discussions were heated, the environment often strange, yet he was less troubled by poor health when they were exerting themselves upon their creation to the point of exhaustion.

Milton worked steadily over the years to make a respected place for V. and quantavolution in Canadian thought. He was a popular teacher and, at some risk to his career, he systematically introduced the new ideas into his courses. Canadian higher education employs outside evaluators whose word goes far on matters of curriculum and promotion. He was able successfully to fight off professional criticism of his innovations in teaching and writing, and ultimately achieved an influential role as spokesman for quantavolution.

He was a principal agent in persuading his faculty to offer an honorary doctorate to V., the only one ever given him, and within a decade he was once more agitating at the University for the same honor on behalf of Deg. He held meetings, journeyed to contact potential supporters, wrote reviews, spoke on the radio, and was an organizer of the Canadian Society for Interdisciplinary Studies. He was the principal Canadian representative in England and the United States. Only Irving Wolfe, at the University of Montreal, and Dwardu Cardona, living in Vancouver, approached him in effectiveness and productivity. Two papers of Milton, written at the turn of the decade, one erasing gravitation as a necessary concept in celestial mechanics, the second dealing with Earth-Venus close transactions, are among the classic expositions of astronomical quantavolution.

Ralph Juergens was struck down by a heart attack in 1979, a few weeks after Stecchini and a few weeks before V. died. He was gearing up to participate in the writing of *Solaria Binaria*. I doubt that the final manuscript would have been much changed if Juergens had taken an active hand. Milton thinks not. He had gone over the general theory with him, and Juergens had received in 1976 and 1977 Deg's skeleton of the book and chapters from *Chaos and Creation*. In Juergens' home, Deg's accumulated manuscripts were used as a raised seating facility for Milton's little son Davin, when they were visiting.

Afterwards Milton examined Juergens' rigorously organized archive of materials and manuscripts; Solaria Binaria would have been improved, but no contradiction would have ensued, given Juergens' outlook. Deg and Milton dedicated the work to Juergens, for his electromagnetic theory was deeply implicated in it. To the dedication the ancient fragment 64 of Heraclitus was appended : "Lightning steers the universe." Deg wrote a poem to his memory and sent it to his widow. It was printed in *The Burning of Troy,* along with an *Oratorio* to Stecchini and a memorial to V.

On December 8, 1980, Deg writes to Milton:

> My *Chaos and Creation* is due for March 1 publication, already outdated in certain respects by what you and I are doing in *Solaria Binaria*. It makes me uncomfortable to know this, but then it helps to recall that Galileo had already committed worse "crimes" in science and philosophy by the time he was brought to trial for heliocentrism. It will bring pleasure to admit errors in *Chaos and Creation* if the truth is measured by what appears in *Solaria Binaria.*
>
> I don't think that we need to fear competent appraisal and criticism. Apathy is a more real problem. Physicists and astronomers are ordinarily paid to go about their work without making waves. They are not philosophers, or even interested in philosophy. Nor are they competent in more than their specialized areas; it doesn't pay them to be so. That is why remarks like, "It isn't physics," or "If that's astronomy, then I'm King Tut," often carry weight. Phrases like these are the shock troops of reaction in science. If they fail, then somebody — hopefully someone else — is awaited, to bring up the heavy artillery. But then maybe the heavy artillery is not there; maybe it is rusted from disease; or maybe there is mutiny among the cannoneers. We shall see.

In 1979 he was beginning a friendship with geology Professor Frank Dachille at Pennsylvania State University to whom he sent *Chaos and Creation,* and who engaged himself in the new astrophysics. Dachille wrote to Deg:

> ...In the earlier letter I indicated that (have browsed through your mss;

since then I have read it completely through, but not with hypercritical attention. I expect to read it again, but I doubt this will be done before we leave for Africa. Frankly, I am quite shaken and taken by the intensive physical processes described, generally fitting well the human recordings of the time. However, I still feel that I would have to understand the processes analytically before I could accept them without reservation. Shaken, too, I was by the views that the Moon was not always up there; also Venus. So, I went back to Velikovsky, am now reading *Worlds in Collision*— really the first time. My first contact with V. was in a magazine article about 1950, when I browsed through *Worlds in Collision,* but was turned away by what I felt was his cavalier treatment of I. Donnelly, and the too easy flip-flopping of planets. Kelly and I were already working on *Target: Earth*— that is, I was going over his original manuscript, started by him about 1947 or so. I was deeply involved trying to quantify the mechanics of the collision process, including axis change, orbit change, figure of rotation, inertial response of water, slippage of shells, atmospherics... My contributions were just intended as suggestions to Kelly, but he asked me to come aboard as co-author. I think you can identify my work by the diagrams, calculations, chemistry, white bills, dry points, epilogue. In all this time, while I was, or *we* were aware of V., his work did not contribute to ours in any way. I did feel however that his work strongly supported Kelly's historical presentation, that is, the ancient records were, in fact, describing horrendous events touched off by what Kelly called Cosmic Collisions. As I said before, I quantified the collisions, based on impact processes, and found that sub-planetary, or small asteroidal bodies would be necessary agents. I did not consider electric fields between bodies at a distance. To me the very clear evidence of impacts on the moon provided the simplest, continuous, mechanically sufficient process or mechanism — collisions involving objects up to 600 miles in diameter. Combining the size-frequency distribution of collisions with the erratic records in the geologic and evolutionary columns, I found support for the impact processes; it was not necessary to involve *planetary* approaches.

However, after reading your book, and going into V., I think that occasional close passages of large (but not quite planetary) bodies will have left their marks on the Earth. So, it appears to me now, massive collisions by the hundreds of thousands have forged the earth in its *ca* 4 1/2 BY history; by the tens or hundreds close passes by generally larger bodies will also have left their marks. As you know, Kelly has been suggesting close passes as a process operative on the geology of Mars, perhaps even Venus. It seems that Bob Stephanos has a fly-by process. Beaumont too. And, of course, Donnelly. It was Donnelly's work *(Ragnarok, Atlantis)* that got me thinking in this area, plus my activity as an amateur astronomer...thinking about electrical charging of the "spheres." I do not know enough EM theory at this time to quantify the mutual interactions of two oppositely or identically charged planetary bodies. Then there is the problem of conservation of momentum and the scale of energies involved. The energy in the earth's magnetic field is many, many orders of magnitude less than that of its rotation and orbiting. How a flip-flop can be affected by magnetic or electric coupling I cannot understand at this time.

Well, you can see that I am thinking along with you. The Cosmic Collision, in all its variants, must be of utmost importance in the history of the earth and life. Last winter term I introduced the subject to my students in the Geology of the Solar System. The coming winter term I intend to intensify my presentation...

On August 3, Deg replied from Naxos:

'Dear Frank,
Thanks for the excerpts and clippings. Io is full of surprises. Purely sulphur volcanoes, some one writes now. But note the pulsing electric arc between Jupiter and Io. It compares with my postulated arc between the Sun and its binary partner, Super-Uranus.

Your work on collision-electricity interests me. Also sphere-charging, and passby-electricity. Regarding the last, you should certainly know Ralph Juergens. Eric Crew has done some thinking, and an article on the funneling effect in meteoroid and lighting strikes. I hope to get a chance to read your full articles when they are available. I can give you the Juergens and Crew stuff when I return. Juergens, you know, would say, in reply to your query as to how a million craters could strike the moon in a few thousand years, that a great many of these are the marks of lightning bolts, not of meteoroid falls. Further I imagine that after the major passbys, and a couple of collisions ("Apollo") and fissions (novas) as conceived in *Chaos and Creation,* the space would be jammed with a great many millions of pieces of debris. Ovenden sees the asteroid belt as remnants of an exploded planet many times the size of Earth, not too many millions of years ago. I call it Apollo, set it in human times, and can readily imagine the debris of Apollo and its Destroyer. We have a big gap to close between our solar system time scales; if you grant the conceivability of what I say in my chapter on the subject, I'd like very much to discuss with you the seemingly impossible obstacles to it. I guess you won't see Olduvai Gorge; there's a fine place (the African Rift) to test the theories of chronology given the hominid and hominid finds on various levels..."

* * *

It is depressing to many minds to think that the planets may have once undergone displacement; it is much more depressing to think that they may have changed motions recently. Of course we must admit that displacements must have occurred to bring the planets into existence, and to place them where they are now. But very few astronomers and philosophers have let the planets shift thereafter, and practically none allowed this within the time span allotted to mankind.

Malcolm Lowery, in a letter to the *London Times Literary Supplement* August 27, 1976, named several latter-day movers.

In 1960 W.H. MacCrea — then president of the Royal Astronomical Society — calculated that no planet could have formed inside the orbit of Jupiter. In 1965 T. Gold concluded that the planet Mercury could not have been in its present orbit for more than 400,000 years, as it is still rotating with respect to the sun. J.G. Hill's 1969 model indicated that Jupiter and Saturn were originally the outermost planets to form, and that Uranus, Neptune and Pluto were displaced into their present orbits by planetary encounters.

Robert Bass in 1974 exposed the prevailing common misunderstanding of the mathematics describing planetary stability, even when based upon present recorded behaviors, such that planetary orbits could not be proven stable for more than a few centuries or millennia. W.M. Smart, wrote Bass, "demonstrated unquestionably that the interval of assured reliability of the La Place-Lagrange perturbation equations is *at most* some interval 'small' relative to 3×10^2 years; Prof. W.M. Smart's exact words are 'one or two centuries.'"

Bass went on to apply to astronomers the kind of pragmatic critique that impresses experts in propaganda analysis: "...Whenever these authoritative statements about time intervals of validity have been made, they are without exception accompanied by words like 'supposed,' 'appeared,' 'hope,' 'seems,' 'might,' and 'think,' revealing clearly that the writer was relying on his personal intuition rather than quantitative evidence."

Bass repeated his findings at a Glasgow (Scotland) conference held by the S.I.S. in April 1978, where there appeared to speak also Astronomy Professor A.E. Roy. Roy agreed with Bass, saying that "even under Newton's law of gravitation, we have no celestial mechanics proofs that enable us to prove that the orbits have not changed by more than 1 or 2 per cent over a period of more than, say, 50,000 years." This figure allows humanly witnessed perturbations, but is not enough for the wilder of the cosmic heretics, who want to bring changing planetary orbits within memory of myth-making man and even historical mankind.

Thus it occurred that when Melvin Cook, Ralph Juergens, Earl Milton, Eric Crew, Deg and others — and V., in principle — wanted to move the planets more certainly and recently, they turned to electromagnetics, and Bass once more, now in 1978, applauded their heretical stance, affirming that "if planets approached closely, there *would* be electrostatic and electromagnetic interactions not predicted on the basis of orthodox theory."

This was not enough. The solar system had to operate as a electromagnetic system, and, though Bass produced an awareness of the sources of such theory, in Juergens and Cook, it was Milton who, with Deg cheering from the sidelines, took the fatal leap onto the plane of non-gravitational fully electromagnetic operation of the solar system.

In a paper circulated in 1979, called "$10^{-36} = 0$" to connote the vastly superior forces at the disposal of electricity by contrast with gravitation, Milton wrote that the phenomenon of gravitation implies "an interaction of slightly unequal strong electrical repulsions between distantly separable objects (or centers) that yield a weak net attraction." Thus masses vary when determined gravitationally insofar as they represent an electrical transaction between two bodies of unequal negative charges. In close enounter masses undergo polarization and transact strongly as dipolar bodies. Rapid and forceful exchange of charge then occurs which can modify motions significantly and suddenly. Hence the absolute level of electric charge on a body is indeterminate, as is, for example, absolute motion under relativity theory.

Deg's image of the whole solar system as consisting of bodies lined up between Super-Uranus and Sun within a tube of gases and rotating with the gases around a discharging electrical current, with the whole system falling apart recently into its present configuration, proved to be just the mechanism to display a non-gravitational system, and Deg, who had never quite understood gravitational mechanics in the first place, was happy to observe his model work nicely within the systems of permissions and restraints belonging to electomagnetic theory. He was doubly pleased because he had been so fond of Juergens and found Milton so congenial: one should not dismiss compatibility in scientific achievement; any scientific (or social group) manager will be glad to elaborate the proposition: compatibility is as important as computability. An eloquent instance of this proposition suffuses James Watson's autobiographical account of the construction of the DNA molecule in his book, *The Double Helix* (1968).

V. was the Great Hostess, in the earlier time, of this whole business; he took no active part at all in it, and the heretics dutifully thanked him at every opportunity in their writings. It will be remembered that Juergens left the Princeton area in flight from the domineering proximity of V. Milton was too far away to be captured intellectually, though he was continually active in defending V.'s views. What Deg received from V. in the theory of Solaria Binaria was nil; all he got from V. was the useful dogma that electricity had been neglected by scientists and was an essential factor in cosmic encounters. Whether V. discussed much of importance with Einstein will not be known until the manuscript devoted to this subject is made available. My hunch is that Einstein retarded V.'s growth in electromagnetics just as V. retarded the growth of some heretics in this regard.

V. made no attempt to relate his work to that of Charles E.R. Bruce of the Electrical Research Association of England, whose seminal work of 1944 on electrical discharges in astrophysics had been the basis for correspondence initiated by Juergens in 1965, and whose work was introduced by Juergens in *Pensee* in 1973. Bruce was a cosmic heretic whose ideas made little or no impression upon British astronomy. They were carried into the British quantavolutionary circle by Eric Crew when it was organized. To this day his one hundred and more articles and notes have hot been published in assembled form. Milton caught on to Bruce in the early seventies, Deg after his meeting with Crew in London in 1976.

Bruce observed the first identity between the velocity of propagation of a solar prominence and an electrical discharge in 1941, when at a lecture he heard of Evershed's photograph of a solar prominence that had reached a height of a million miles in an hour. He writes, "I thought, 'If that isn't about 3×10^7cm sec^{-1}, I'll eat my hat." It was, as a little mental arithmetic, confirmed on an envelope when the lights went up, established — and I was in business as an astrophysicist." He thereupon published privately *A New Approach in Astrophysics and Cosmogony*, copies of which several cosmic heretics came ultimately to possess.

Galaxies were seen by him to be structurally determined as electrical fields. Magnetic fields spring up around cosmic flares and bolts. In cosmic discharges, matter aggregates along the discharge channel, and in this process of electrical breakdown "one can forget about the force of gravitation, as every arc welder knows." This discovery Bruce attributed to Bellaschi of the American Westinghouse Company in 1937. Jets and balls of hot gases are formed in the process. Bruce also applied the notion of pinched-off discharges under extreme pressures to the extinction of novas. Juergens and Milton pushed Bruce's electrical interactions between stars and atmospheres into stellar interiors, the greatest step in obviating the need for gravitational theory.

V. lacked the capacity to give and take; he would disrupt any on-going thought processes to call all hands to shoo the chickens out of his backyard. Those heretics, like Rose and Vaughan, who opted to exercise their intellects in his garden, found themselves becoming over-specialized in certain crops, interpreting Venus tablets and calculating conceivable orbits under conventional restraints. This is only to say that such heretics became unfortunately limited despite their eminent suitability for larger tasks; they were also diligently occupied, as was the solaria binaria trio, in developing the larger network of heretics and playing firemen for V.'s fires (some of which were arson).

The progress of quantavolution in the astrosphere required an electrical model. Fortunately it could profit from a considerable advance along the whole front of electomagnetic studies which was occurring in conventional science, as well as from the work of the heretics themselves. But one ought not forget that the theory of quantavolution in the astrosphere was sustained too by heavy inputs from faraway fields: myth analysis, paleontology, and critical geochronology.

Deg's assurances that the fossil voices of myth and legend were speaking truths of the skies kept the theory from flying off to join the conventional dogma that change could only happen hundreds of millions of years ago. They also blocked the hopeful theory that comets and meteors could take the place of the planets.

In paleontology we have this remarkable logical position, perhaps exposed for the first time by Professor Roy in explaining why astronomers should prefer a longer rather than a shorter period of celestial stability:

> Most celestial mechanics — orthodox *and* informed — would say that we *suspect* (it's probably no more than a hunch) that the solar system is stable over hundreds if not thousands of millions of years, but we cannot prove it by the methods of celestial mechanics that are available to us today. We have to go to geophysical, astrophysical and selenological evidence — and there, of course, we are again on ground which has been disputed by those who advocate the very short time scale. The fossil record would appear to have been laid down in the rocks over the past two thousand million years, and in those fossils we have very complicated animals. If the orbit of the Earth had changed drastically in that time, then conditions on the EARTH would, it seems to me, have been such that those creatures could not have existed. In addition, one could say that, even if the orbit of the Earth had not changed in that time, but the Sun's output of radiation had changed dramatically, then again the fossil record as we know it could not have existed. The ages of the rocks by radioactive methods would appear to be 4 1/2 thousand million years; similar methods appear to make the oldest lunar samples of that order of magnitude in age. Theories of the energy output of the Sun make it appear, from a consideration of the helium/hydrogen ratio, that the Sun has been operating with much the same output as it does today for something like five thousand million years. And so on...

What Roy is saying here is that, for no other reason, a long term stability of the solar system is acceptable because it has taken so long, according to the fossil record, to evolve life and its peculiar, complex structures. Further the rocks are datable by radiochronometry and the Sun is datable by its self-burnup rate. This is nice: here we have the queen of sciences, to which the other sciences had looked for *their* assurance, abandoning its throne and asking for refuge among the fossils of the rocks and the furnaces of the Sun.

Effectively, however, the quantavolutionists had spotted this cross-disciplinary mutual rescue society, and had begun to launch assaults against the positions of the other disciplines as well. Juergens had fully disestablished the thermonuclear theory of the Sun, so far as some heretics were concerned, and substituted (with Cook) a galactic electric-collecting model.

So far as the fossil record is concerned, Bass in 1978 accords Cook the honor of having achieved the main victory over radiochronometry. (The old catastrophists, such as Price and V., had done the job on conventional stratigraphy and erosional gradualism in geology.)

In a footnote that should be a placard, Bass writes:

> ...If I believed those long-term radioactive dates in the fossil record and elsewhere, I probably would also believe that the Earth has not changed its position for thousands of millions of years. However, in another book, *Prehistory and Earth Models* (London, Max Parrish, 1966), Dr. Cook has had the audacity and temerity to take on the entire historical, geological and geophysical establishments, and has reviewed in great depth and detail every radioactive dating method, short-term and long-term. After several years making up my mind, I have come to the conclusion that Melvin Cook is right and has established that there are enormous and inescapable fallacies in the uranium, thorium and lead dating methods; and I don't think it can be maintained that the surface features of the Earth have been in their present form for more than 30,000 years.

Deg had supported Juergens in several works, and had relied heavily upon Cook in attacking the full range of dating tests offered in support of great ages of time. I have not yet introduced the several other contributors to the demolition of time measures. They appeared in the pages *of Pensee,* the *Creation Research Society Quarterly* and the *SISR* for the most part. The attack requires hundreds, not a dozen, writers, however.

But still there must be an elite, leaders of the republic of science, like Robert Bass. Everyone got a lift in spirits with his appearance upon the scene, a stocky dark man, bespectacled, a convert to Mormonism it appeared, with a weakness for women which, Deg reflected, was in keeping with history and not incompatible with his experiences of Mormon friends who came out of the West to the University of Chicago in the 1930's. Bass was associated with Brigham Young University, where, paradoxically, catastrophists were unwelcome in the sciences; a story goes that Bass forgot to sign and return his contract, lost his tenure, and, in order to retrieve it, was asked to agree to submit to pre-censorship of his publications, which he refused. Bass was covered with the medals of scholarships and degrees and when he showed up, it was like a troop pinned down by continuous fire greeting a marksman with just the right gun.

Bass took aim at the brain center of the opposition, the reliability of planetary motions, and fired. The shot was on target. Blasted was the astrophysics of orderliness. His troops cheered. The opposing line continued firm; hardly a surrender or desertion. It seemed that the facing army lacked a brain center. It operated just as well by rote.

CHAPTER ELEVEN

CLOCKWORK

Deg's Journal, Naxos, July 3, 1973

The animation of the night skies is both poetic and heuristic. Each meaning enhances the other and creates a third set of meanings that are beliefs. These beliefs join the stream of myth, color, and shape it, change its direction somewhat, make its fundamentals more difficult to understand. Cosmopoeia is the imagined form of stars, a guide for students and navigators by sea and land, the astrologer's subject of story, the marking of the passage of bodies and the occasion for anniversaries of related events, be they births, deaths, or disasters. All of these functions are important to humanity. But that they flourish should not be pretext for diminishing or denying the occurrence and greater importance of erratic and special heavenly changes.

Similarly, the world as we see it in the "normal" processes of constancy and incremental change is a true and real world. The tides flow, the sea suddenly beats the shore, the rains wash down soil and the winds abrade rocks. This everyday vision lulls us into somnolence about natural forces, or when aroused, to a discrete excitement about tornadoes, volcanoes and earthquakes. Like the animation of the skies, the ordinary experience of nature is a reality that is also a screen and a censor, concealing and prohibiting the colossal, historical and potential behavior of nature.

As it is with the skies and earth, so it is with life. The recent fixation of species, based ultimately upon an operational definition involving interreproducibility, gives a truth that must always have been real: gradual changes occur; species can develop in isolation, by occasional mutation. But all the time that biology can beg, borrow, or steal is not nearly enough to present us with the fantastically organized and behaving conglomeration of animals and plants of 1973. The validity of received evolutionary theory must become minor, while the heavier reality of catastrophic change and origin of species by potentiation comes forward.

It was inevitable that Deg should end up in defiance of billions of years of time. He could hardly lie on a beach unless he was exhausted from swimming and diving. He knew and disliked the stereotype of the American as restless and impatient, so he cultivated various devices and appearances that would let him seem to be casual and unconcerned with waiting upon the world. Since he was raised without the time-consuming liturgies of religion, religious routines were not a common means for stopping his time or feeling it. Sports, smoking, drinking, eating much and long, running — these and many other tricks were used against time. More than all of this, he played games against time, sometimes excruciating acrobatic games pitting motion against time. He wanted quick results in everything he did; but the world is not constructed to provide results, much less to provide them quickly.

The same urge to quick results inclines one toward intellectualism, because so much can be solved in the mind and the world of the imagination can be rich and malleable; fat gobs of time can be reduced to frizzled specks, and one can leap over far spaces and epochs. However, intellectualism is also opposed to both physiological and mental time-control in that it forces one to be physically inactive over long stretches of time; research and writing are termite mounds of time and a single footnote, a single bad line, can drive one to despair.

Sometimes I think that Deg was one of Alfred Adler's pure compensatory characters, who set himself very often to do precisely what he was unfit to do because of his unfitness. If under such circumstances he was not destroyed by the contradiction, it was because he often escaped into the activities already noted but also into sex, travel, brief adventures, commitments to things extraneous. Most of all, and too important to call an escape, was his taking on two or more large tasks at the same time, so that while to the outside world he appeared to be proceeding carefully along one line, at a measured pace, he was in fact speeding along other lines and then doubling back to the first line of engagement.

Paradoxically, the intellectual who is so fretful of time's arrow hastens but to sit and stare upon dead written pages, to pitch his nervous system and organs upon his several moving digits, gaze at the stars, watch the rats run, listen, observe, and discuss only that world that his mind will accept for consideration — all of this consuming such enormous amounts of time that those who in turn observe the intellectual cannot be blamed for thinking him mad for his dissociation and hatred of reality, his obsession, his wrestling with details, his fear and guarding of his own thoughts, his ruthless hunting down of words and meanings, amounting in the end to the squandering of the very object of his anxiety, time itself, time in the thousands of hours of which every minute, he insists, counts dear, and if this lunacy is not sufficiently oxymoronic, the time-saving time-waster can dedicate himself to time-studies.

Perhaps one-fourth of all Deg's work on quantavolution over the years dealt with time. Perhaps a quarter of the three thousand pages that he wrote were concerned with or governed by calculations of time. Before he had entered the field he had been possessed by problems of time and had written but not finished what was supposed to be a lengthy philosophical and psychological poem on the subject. By virtue of the tricks I have already alluded to, he would escape the psychiatrist's verdict of obsession, but in fact he was obsessed and his impatient and striving character often led to pitched battles against time; it was the most uncontrollable element in life.

He beat time as a child by being precocious, stripping off three years of schooling, and he became the youngest member of his graduating class at the University. But then time reacted smartly at war and he felt the full poignant irony of "Hurry up and wait," the life of the soldier. He nosed his jeep into many destroyed towns where clocks were stopped; hanging crazily, sober and still, or startled faces staring from the rubble — they were all wrong. Are all clocks wrong? Madness about time was a disease of the poets, literati and humanists; turn to scientists, and 99 out of 100 are perfectly satisfied that they are measuring an absolute, an ever-so-old process; they are like the bureaucrat who is content to keep the entrepreneur waiting, because his check comes in regularly no matter what, while for the businessman time is money. For these scientists, there was something called the relativity of time, which was reserved for their Sunday outings.

All of this joins in with Deg's anti-authoritarianism and republicanism (which goes back to sibling rivalry) and gave him his ideological stance confronting time. If authorities would say time was long, well then he would be pleased to discover time to be short, and thus more containable and controllable. There was a contradiction here, however, but it can be explained away. Deg had always been a darwinian, but might this not have been because Darwin was anti-authoritarian, anti-theologian, too, while trying to be nice to the traditional believers? Deg was exactly like this, against the scriptures as authority, against church authority as such, but then respectful and even loving towards the many "nice" and "gentle" believers he met. How could he join the theologians, the short-time creationists? Well, he didn't really. He found them to be the most active critics of macrochronism. They were experienced microchronists, who knew the history of the defeat of microchronism well because it was their history.

The problems of time came in two batches. First there was the historical batch, epitomized in V.'s *Ages in Chaos*. Second, there was the geological batch, which could also be epitomized in V.'s *Earth in Upheaval*. Let us see what V. did with time in both regards.

V. aligned and connected Jewish and Egyptian history which had hitherto gone along on separate tracks. The alignment settled upon the Exodus at about 1450 B.C., the Biblical date, and tied it into the end of the 13th Dynasty of the Middle Kingdom of Egypt with the Hyksos invaders as the Amalekite enemies of the Biblical Hebrews. He begins the splendid 18th Dynasty of Egypt at the time of Saul and David. King Solomon he places alongside Queen Hatshepsut of Egypt, and has her, as Queen of Sheba, visiting his court. And so on.

The reconstruction attempted in his volumes on later time, as I have already indicated, fell victim to the scholars of the "British Connection."

Dropping by 500 years the accepted chronology of Egypt after the Exodus, and holding the Exodus at -1450 meant that all dates elsewhere, whether of the Near East, Greece, or finally Italy, which had been set by coordination with Egyptian artefacts and occurrences, required resetting by 500 years as well. In Greece, a gap which had been closed only by creating a barbaric "five hundred years of the Dark Ages," was promptly nominated for elimination. A grateful rush of scholars to profit from the new chronology did not occur; the Greek scholars were frozen to their positions until the Egyptologists (all 30 of them) would admit the loss of the five centuries. Then they would follow suit. Similar scientific lags continued in the other ages affected by V.'s reconstruction of Egyptian chronology.

When did the mistaken chronology begin? V. traced the major error to Manetho of the third century B.C. as reported and adopted later by scholars. Manetho was eager to prove to the Greeks and Asians the superior antiquity of Egyptian civilization. Berosus followed suit, exaggerating for his Assyro-Babylonian country by tens of thousands of years. Eratosthenes, soon afterwards, took up the cudgels for his Greek compatriots and moved Greek dates backwards by approximately the length of the "Dark Age." The motive of ethnocentrism thus played a large part in the beginnings of modern chronology, as it did in V.'s stupendous reconstruction itself. But it was not at all clear that the ancient chronographers following Manetho were wrong for their errors were covered up by a heavy burden of refinements and rationalizations up to the present time. If V. had written nothing else in his life he would have deserved the highest accolades for his essay on "Astronomy and Chronology."

Soon after his first major attack upon Egyptian chronology was published, V. sent a copy to Etienne Drioton, Director General of the Service for Antiquities of Egypt and received shortly one of the most nearly perfect replies an author could wish for, and, for that reason alone, as a model for my readers, I reprint it here. (My translation is from the French original.)

Cairo, May 29, 1952

Dear Doctor,
    You were kind to have had me sent your beautiful book, *Ages in*

Chaos, which I received this morning, and which I have read nearly in entirety, so exciting and interesting is it.

You have certainly jostled — and with what vigor! — many historical tenets of ours which we regarded as firmly established. But you do it with a total absence of prejudice and with an impartial and complete documentation, which is most sympathetic. Your conclusions might be argued at every step: whether they are allowed or not, they will have posed anew the problems and compelled a fundamental discussion of them in the light of your new hypotheses. Your beautiful book will have been, in every way, very useful to science.

I thank you warmly for having sent it to me and I pray you accept, Dear Doctor, the assurance of my sentiments of cordial devotion.

Etienne Drioton

V. received few such letters concerning *Age in Chaos*. Actually, a number of archaeological discoveries were made in the years following *Ages in Chaos* which tended to corroborate V.'s reconstruction of time. One of the most important of his priorities for testing was at the town of El-Arish, between Egypt and Israel, where he believed might be uncovered the capital of the Hyksos, Auaris, and, if so, then there might be demonstrated the further correspondance of Biblical and Egyptian history in revealing that the city fell to a joint Egyptian-Judean army, one led by an Egyptian Prince (Ahmose?) and the other by King Saul. This excavation has not been accomplished

V. paid attention closely to developments in carbon-dating, for here was one of the places which he thought might give him a quick and decisive victory. He corresponded with experts, beginning with Libby, founder of the C14 system of dating. His pathetic and persistent efforts to achieve a dating of 18th dynasty objects were put into a manuscript called "Ash," selections from which were published in 1974. To Libby he writes (October 7, 1953), "I also assume that if analyses of organic objects dating from the time of Hatshepsut, Thutmose III, Amenhotep II, III, or Akhnaton were made, the results will indicate a *reduction* by as much as 500 years from the conventional figures; and over 650 years for objects of Seti or Ramses II or Merneptah." At the same time, he suggests the dating of Pleistocene fossil beds and petroleum deposits, predicting a late date. Libby was unhelpful, but said petroleum datings by C14 had shown "great antiquity."

Now V. begins a circuit of frustration. Finally a German admirer, Ilse Fuhr, who was later to publish a fine work dealing with comets in early times, with courage and persistence obtained 25 grams of three different bits of wood from the tomb of Tutankhamen. V. was delighted and expected the results to show -820, not the conventional -1350. In another letter he did worry over the effects of original atmospheric contamination of the samples owing to a catastrophe. The University of Pennsylvania laboratory performed the tests and came up on the middle, between the conventional and the heretical dating. Bruce Mainwaring had used his strong ties with the University to help arrange the tests.

Seven years later the British Museum tested reed and palm nut kernels of Tutankhamen's tomb and emerged with dates of about 846 and 899 B.C., both of which dates were never published, and then seemingly lost or misplaced them. Other dates of the 18th Dynasty appeared in time, not so definite and reliable as to dismiss V.'s claims, but not such as to please him.

By this time, Deg had read Melvin Cook's article of 1970 in which, retrocalculating the C14 in the atmosphere, using the rates of Libby, Cook figured that the atmosphere would have had to have been constituted (or reconstituted) some 13,000 years ago. Deg's deduction was that a series of catastrophes would have created the same effect. Further, Deg observed increasingly wild fluctuations as well as a secular swing of the C14 dates from "known" dating and bristlecone pine dates as time marched backwards, and, without straining the discrepancies overly much, he could conclude that carbonating would be both invalid and unreliable before 3000 years ago, which ushered in the Venusian Age (in his terminology).

Deg was further impressed by the studies of John Lynde Anderson and George Spangler, which he read in 1974, not long after their publication, that challenged the very constancy of the radiocarbon component of the atmosphere. Thenceforth he paid small heed to earlier radiocarbon readings, whether they seemed to support or oppose his theories. On the other hand, V. who had expected salvation in C14, could not readily denounce the system afterwards, and played on occasion the game of using C14 dates when convenient to do so, nor did he ever renounce C14 in principle.

To this day, Deg has not been able to understand how V., having succeeded in restructuring the chronology of Egypt to the end of the 18th Dynasty, could then have made further drastic changes needlessly, displacing forwards the great kings Ramses II and III. Deg had so much confidence in V.'s ability and so little knowledge of later Egyptian history that he accepted the new chronology *in toto* as it came to him by word of mouth, by hasty readings of manuscript pages, and by the published volume of *Peoples of the Sea* when it appeared in 1977, after many years in manuscript and printers' proofs. Very soon thereafter doubts were heard coming out of the "British Connection," from persons whom Deg had come to respect. None of the Americans around V., nor V. himself, had met any of the British and they were inclined to put on airs or to rant against them.

Deg did not try to follow the controversy, which was based upon close historical analysis. He thought to wait until the dust would settle. He was made uneasy by a lurking contradiction in V.'s position. The great catastrophist seemed to be putting aside catastrophism in ordering the centuries. In early 1972, William Mullen had written in *Pensee* that

Two assumptions from *Worlds in Collision* are taken as fundamental: first that no chronology using retrograde calculation of the positions of heavenly bodies is reliable earlier than -687; second, that the principle clue for synchronizing histories of ancient nations should be the break caused in all of them by the catastrophic events.

The second point is at issue here. Deg agreed with Mullen. For example, he made the following note:

It is interesting that in one of his articles Isaacson, doubtful perhaps of the strong basis for celestial connection, ventures that V.'s reconstruction of chronology can be separated from catastrophism. This I think not to be so. First, V. would never had revised chronology so boldly if he had not discovered the key to chronology in two parallel accounts of the same disaster — one in the Papyrus Ipuwer at the end of the Middle Bronze Age of Egypt, the others in Exodus. Second, the evidence of catastrophe is what explains the end of the Mycenaean civilization and ties it directly into the Archaic Greek culture that succeeds it, both in the 8th-7th centuries, and then ties both of *these* into the Biblical accounts and many other accounts of the same disasters at the same time. In short, it is catastrophic theory that sired the revised chronology of V. and if the genius of that reconstruction is extraordinary, it is the effects of hereditary genius, a "fall-out" of genius from a single elemental key idea, as Juergens has written. I say this while reminding myself that the Exodus disaster was the key, but the motive came in the desire to reverse the order of Moses and Akhnaton: to recapture Moses and monotheism for Israel. Not that V. cared for monotheism in itself. But since the world regarded it as an invention of paramount importance, he was ready to fight for it.

Not until 22 December 1981, do we find Deg at the denouement of his doubts; writing to Derek Shelley-Pearce (S.I.S.) in England, Deg says:

The Glasgow Chronology is in full swing, it appears, with John Bimson (SISR 5:l) and Martin Sieff (Workshop 4:2) pushing it mightily. And the readers, no doubt, a bit giddy.(...)

I am glad to see that Claude Schaeffer's work has come into its own with Geoffrey Gammon's article in SISR 4:4. It is one of only several general studies of value in cultural quantavolution. Gammon approached two points that he might have developed more fully. First, the best benchmarks of past ages are catastrophes: cultural quantavolutions coincide with natural quantavolutions. For a century scholars have been playing at quantavolutionary theory unwittingly by using catastrophic age-breakers. It reminds me of how some early geologists tried to dismiss the word "strata" because that implied discontinuities, and discontinuities implied you know what...

The other point to stress is that the end of so many settlements around

-1200 (conventional dating) indicates that this date actually falls between -780 and -680, that is, the Martian period. Gammon seems to shunt aside this evidence when, with his mind perhaps upon Egypt, he says, regarding the destructions that ended the Late Bronze Age, "the evidence that these may have been due to natural causes rather than the agency of man remains scanty."(p. 107)

Perhaps Velikovsky did the same, in order to progress with his idea of further shortening Egyptian chronology; that is, he abandoned his fix on the Martian episodes. To me, the term "Peoples of the Sea" is a euphemism for the Martian-Moon-Venus disturbances, a kind of reductionism. Wars, movements of people, and social turmoil are expectable in natural disasters and are a concomittant and effect of them. To show that they happened certainly does not prove that extraterrestrial events and general catastrophes did not happen, but the contrary. Applying the term "Peoples of the Sea" to a construction of a fourth century Ramesses III is already a warning sign of trouble ahead; one cannot move Martian events to the fourth century; one may not give Ramesses III a special "Peoples of the Sea" of his own. The Glasgow chronology may find its clincher by research of Martian period disasters in Egypt, possibly finding the evidence around the time of Merneptah or Ramesses III (...).

He goes on to write:

As Sieff says, "By placing the 19th Dynasty so late, Velikovsky ironically obscured the cause for these destructions which he himself had found." The reasons why he did so are also obscure. Granted that my offhand remarks should carry little weight, surely some scholar who understood the catastrophe-culture-history interfaces must have read and disputed this part of the reconstruction of history. When Velikovsky was writing this book with the others still to appear, was he by-passing his own catastrophic benchmarks to complete a descriptive history postulated on different grounds? When the Glasgow Chronology began to surface after his relevant book, soon two books, were in print, I heard recriminations and ducked out. I should have given more attention to this breakup of the consensus around him, but there were too many intimations of the "Love me, love my dog," kind, for which science has no place. I am going to have trouble with this matter when I come to it in the course of writing "The Cosmic Heretics."

There were to be four volumes of *Ages in Chaos*. The first scored a large success with a group of competent heretics. The second and third volumes, not treating of catastrophe, but of chronology and archaeology, failed to persuade most of the heretics and their dates were soon replaced by a new reconstruction that tied into the first volume very well.

The reviews in the orthodox media were bad, usually attacking V. for the wrong reasons. The fourth volume was held up indefinitely by Elisheva and her daughters. Deg advised that it be printed, even if it held a basic flaw, because V., though increasingly doubtful, intended that it be ultimately published, and because V., even when he was wrong, was more instructive than most people when right.

None, among the anti-heretics, seemed to notice that V.'s supporters, supposedly so slavish, had quickly and thoroughly analyzed and rejected two thirds of his general theory of Egyptian chronology. Indeed the opponents would still proceed as before, talking of his cult and his claque. There was restraint among the heretics in attacking V.'s newer books, and *Kronos* hardly attended to them at all. Evidently, the heretics could also ignore books that they didn't like. Or is this what one ought to do with books that are neither catastrophic nor correct?

For a catastrophist to limit his concerns is difficult. Once you have the planets misbehaving, you must acknowledge that it may have been their wont in earlier times as well. V. decided that he had better investigate the earthly effects of prior cosmic disasters; if prehistoric catastrophes could be demonstrated to have occurred, then historical ones might become more believable. So he wrote *Earth in Upheaval*. V. did not set up a timetable of catastrophes. However, he adduced more evidence that the -1450 to -687 periods suffered grand natural disasters, and he introduced doubts ranging backwards. He paid little attention to the burgeoning science of radiochronometry aside from carbonating, nor did he ever exert his powers in this area. To strengthen the case for late catastrophism, he brought forward instead the studies of others on glacial melting rates, sudden ocean level drops, very recent alpine orogeny, rapidly drying lakes, waterfall cutbacks, late fossil assemblages, surprisingly recent C14 datings, the simultaneous devastations of civilization (using Schaeffer), excavations of warm-weather life forms and human settlements in impossibly cold zones of today, Indian traditions of orogeny and other quantavoluntary events, changes in magnetic orientations, and the large-scale ash levels on ocean bottoms.

He did not know Otto Schindewolf's work, then appearing, which tied the great periods of biosphere destruction to cosmic events and consequent radiation storms. He followed Dunbar's *Historical Geology* in examples of very early disastrous effects. He advanced the idea that coal was formed from biosphere masses propelled and dumped by huge tidal waves, without specifying which waves and when, and used Heribert Nilsson's studies of German coals to prove his case. He relied heavily, too, upon the early English catastrophists. He used also the work of American creationists.

In a few lines, he expressed his feeling that the uneven lengths given to the ages were "basically wrong;" the remark is strange, cryptic, confused.

He "does not suggest either a lengthening or a shortening of the estimated age of the earth or the universe," and then adds irrelevantly and naively that a religious mind should not be upset by great ages. It was all rather humanistic and old-fashioned.

Deg found that the accretion of evidence of catastrophes was much easier than the application of a time scale to them. V. had not set himself to demolishing the new techniques of radiochronometry, possibly because he believed them valid, possibly, too, because he felt that he could obtain the right to his catastrophes down to Noah (about 6000-9000 years ago) without contending with radiochronometry, which does not begin to operate, except for C14 and certain tests still in the realm of the exotic, until 100,000 years back.

Also V. had done practically all of his writing before the issues of radiochronometry came forward, before several of his supporters engaged in its study on their own accord, and before the creationists had worked to discredit it.

Deg set himself two tasks. One was to set up a model of past catastrophes, hence of the ages. The second was to classify and survey all existing techniques of measuring geological time, and to state the grounds for believing then invalid. He had always to bear in mind that one of them — he ultimately included over fifty measures — might be valid, even if grossly valid, and thereupon would seriously damage his model of natural history and at the worst render the model only an intriguing metaphor. He was surprised repeatedly as he went from one test to another to discover that none existed without a flaw or a question, either of which might be fatal to its validity or reliability.

His major teacher was a man he had not met, Melvin Cook, who went on a rampage among the uranium-lead, potassium-argon and other tests, pointing out inconsistencies, contradictions, incompatibilities, and arbitrary assumptions. Cook was not an exoterrestrialist. His attacks are almost all from the materials of geology and chemistry. His exoterrestrialism, such as it is, comes in estimating intakes and outputs of gaseous elements from the earth's atmosphere.

Perhaps the valuable critics of radiochronometry number no more than a score. Deg could name a half-dozen besides Cook whose work he regarded as heroic and essential to establishing and maintaining his perilous stance. I mentioned Anderson and Spangler on C14. There was reliable Juergens who showed theoretically that the electrical environment could effect enormous changes in radiation rates, such as to annihilate time. There was N.J.G. Sykes who, in a simple test published in the *S.I.S.R.*, gave grounds for believing that a changing magnetic field would augment or diminish radioactive decay rates. Then, too, there came Roy McKinnon, also writing in the *S.I.S.R.* and Thomas G. Barnes, writing in 1977 on the recent origin and decay of the earth's magnetic field.

R.V. Gentry and his team repeatedly showed, to everyone's

astonishment, that extremely short-lived polonium halos occur in the absence of parent uranium, evidencing that the host rock was formed very quickly. Coal was examined that seemed to have formed in days instead of millions of years.

Deg began to treat the longer-range radioclocks as he did radiocarbon dating, an indicator at best of relative time, and vulnerable to the kind of electro-chemical turbulence that is inherent in natural catastrophes that begin with disorders in the sky. Essentially this freed him to consider together all factors that could have left some indicator of time upon or around a specimen rock or site. Since no technique appeared by itself to be a tamper-proof, independently set, and auto-operative clock, every technique or test had to take its place in the group of indicators of time, some of which were carried into the setting to measure its time and others of which were inherent in the geology and circumstances of the setting. All too often, geophysicists came to believe that there is scientific validity in what is a purely administrative and industrial axiom — that tools and products should be standardized in as few forms as possible — and therefore they assumed that there must be some true superiority in a tool like potassium 40-argon 40 radiochronometry because it can physically be applied to any strange igneous (and now metamorphic) rock, that is carried into the laboratory.

Deg came to rely, too, upon some very general ideas in concluding that the time of the world and of the ages may have been very short. These had an air of philosophy or, worse, homespun reasoning about them that is infuriating to technicians intercepted on their way to their laboratories and machines. For example, Woodmorappe's painstaking survey, published in the *Creation Research Quarterly*, of the successive occurrences of the earth's several eras, as denoted by its surface rocks, shows a preponderance of discontinuities through the series of eras. Also, the macro-geography of the Earth seems to call for a giant micro-chronic integrated episode.

Inevitably, then, the mind was jostled to close up time radically in the period between hominid and man in the face of evidence that the hominids were human-like, and very little time was required to achieve a culture. Thus, microchronism lent itself to Deg's theory of *Homo Schizo*.

Then, upon arriving at the notion that the earth had been recently ravaged, Deg began to wonder how the earth could have survived for very long if it had begun to suffer one after another disaster through four billion years; this led two ways: first, to shorten time in order to admit the fact that the earth still exists and has a biosphere even if, like the old grey mare of the song, "she ain't what she used to be," and, second, to postulate, even then, some backward limit in earth history to a beginning of the period of disasters, and thereupon he asked himself what might have been the first great catastrophe to threaten the world, and what started it — giving him Super-Uranus, and a binary system in throes of disintegration, a baseline of perhaps 14,000 years for the first great destruction, and an initial electrical explosion arising naturally from a pre-existing electromagnetic system.

When Milton and he sat down to discuss the system before the age of catastrophes (now compressed into the Holocene of 14,000 years), they found no need in their binary system, with its highly productive, enormous, magnetic tube, for more than a million years to accomplish all that was new under the sun. Their model of the solar system probably included errors of great magnitude; it might have major system failures; and it might even be basically wrong: both he and Milton freely acknowledged this; but they were ready to race it against any other model in the field.

Having spent much of his life in building (not inheriting) a science, that of the study of political behavior, Deg did not take kindly to inferences or statements that he did not know what science was all about. He replied sarcastically on occasion that indeed he did know what science was about and it was up to no good.

When *Chaos and Creation* appeared, he sent a copy of it to the University of California physicist, Walter Alvarez, in appreciation of the study his team had published, exhibiting the existence of an iridium layer that might have fallen out from a meteoroid explosion, contributing to the demise of the dinosaurs. He took the occasion to ask "whether you remain convinced of the validity of radiometric dating, granted the possibility of catastrophic radiation and heavy subterranean heating."

Alvarez replied, "In answer to your question: I consider radiometric dating to be an excellent tool that gives reliable dates. The systematics are well understood in all except the current frontier areas, and serious practitioners are well aware of the possible sources of problems and how to avoid them."

From which answer, we may all take heart. In accepting kindly the book, Alvarez wrote "It helped me appreciate clearly the difference between the basically anti-scientific, Velikovskian approach and the way a scientist would seek to understand nature." Need I say more?

PART FOUR

CHAPTER TWELVE

THE THIRD WORLD OF SCIENCE

For a decade from the appearance of *Worlds in Collision,* no quantavolutionary circle existed in the world. V.'s correspondence with his readers was voluminous. Immanuel and Elisheva were socially active for several years, but no scholar who could be said to be of catastrophist persuasion was a frequent correspondent or friend. In July 1956, Claude Schaeffer, author of the monumental comparative study of archaeological levels of destruction wrote Velikovsky his appreciation of receiving from him a copy of *Earth in Upheaval.* V. had used Schaeffer's work in preparing the book. In 1957, Immanuel and Elisheva visited with the Schaeffers for a week at Lake Lucerne, in Switzerland. Schaeffer did not agree with any part of Velikovsky's ideas except what Schaeffer himself had printed before V.'s work had appeared, that periods of sudden destruction had befallen Bronze Age civilizations.

Two decades later, Deg and Anne-Marie Hueber visited Schaeffer at his home near Paris. Deg wanted to update Schaeffer's inventory of sites, and they had corresponded briefly on the matter. Schaeffer had offered Deg the materials of his files about which he had written to V. many years before. Then he had spoken of "new confirmations of the reality of these crises on a continental scale which I have tried to analyze. I would be glad if I could write now immediately the contemplated second edition of *Stratigraphie Comparee* in two volumes, for with the new confirmations these crises could no longer be questioned... so striking are proofs and so accurate the dates established by the new discoveries..." V. had not told Deg of his correspondance or of Schaeffer's intention of moving forward. V. had passed up a rare chance at statistically demonstrating his theses. Nor had he exhorted others to undertake work with Schaeffer. Deg had to suggest the idea to Schaeffer as if Schaeffer had never been aware of the possibility.

Schaeffer was ready to collaborate. It was clear to both men that V.'s reconstructed chronology was not to be at issue. Their aim was to confirm the ubiquity and internal cohesion of Schaeffer's set of catastrophes. Deg was made aware of Schaeffer's doubts of V.'s chronology, especially that coming after the 18th Dynasty of Egypt, doubts that were even stronger with Madame Schaeffer, who at one moment was with the group and at the next was out of the room tending to her visiting family. Deg conveyed his belief that the catastrophic sequence of Schaeffer could slip forward nicely, using the same intervals, to fit the scale that he had drawn back to the neolithic age, which included V.'s fifteenth and eighth century disasters. Thus Schaeffer's sequence could serve both the conventional and the quantavolutionary calendar.

Deg sought funds for the research from the American Geographical Society, without success. [The proposal is carried in *The Burning of Troy.*] He tried to reach Schaeffer in Paris in 1983. Schaeffer had just died.

With the appearance of *Stargazers and Gravediggers* in 1983, a reader might see how barren was Velikovsky's personal and scholarly life during the 1950's of the very people who were capable of, or were independently pursuing, studies in quantavolution. The characters in the book are mostly his opponents; few friends and supporters appear. The only persons of catastrophist persuasion mentioned were Alan Kelly (but on nothing to do with his catastrophism) and Claude Schaeffer. Alan Kelly, and Frank Dachille who was his collaborator in *Target Earth (1953),* lived far apart and they worked alone.

In American biology, Goldschmidt and Simpson knew there had been quantum jumps in paleontology and presumably their students acquired some inkling of the anomalies. In circles espousing Biblical literalism, the work of Price and others was discussed. There must have been other catastrophist scientists of the 1950's in America and England, but to this day Deg has not been able to name any. The existence of perhaps half a million readers of V.'s books meant little so far as research and writing were concerned. Some bootleg teaching of catastrophism was occurring, especially among fundamentalist Christians. In Germany there were Schindewolf and Nilssen in paleontology, as I noted elsewhere in these pages.

Significant differences came with the sixties. The civil engineer Ralph Juergens left his business in the Midwest and moved to Hightstown, near Princeton, so as to be near Velikovsky and to use the libraries of the University. Warner Sizemore, a minister and graduate student of philosophy appeared on the scene at the same time. Stecchini, historian of science and unemployed professor, was already there, indulged by his wife Catherine, a star teacher of young writers at Princeton High School. While teaching at the University of Chicago in 1950, Stecchini had signed a letter of protest to Macmillan against the treatment given Velikovsky's book.

When Deg met V. and decided to publish his story, there was noone else in sight. They thought of Eric Larrabee, but noone would be paid to write, and Larrabee was busy with unrelated affairs. Since Deg could not do the whole job himself, Velikovsky recommended Juergens, then working for McGraw-Hill as a scientific editor, and Deg and V. persuaded Stecchini to do an historical portion. Thus, all the effective resources of V. amounted to three men who could and would write about his case in depth. This was the first time any cooperative group had engaged itself in the study of V.'s problems. It was also the first time that V. realized the values and capacities of voluntarism in America. He was, however, cunning about the media. For instance, as soon as the *American Behavioral Scientist* was in the mill, V. could persuade Larrabee to write an article for *Harper's Magazine*. Larrabee was spurred into action and the article came out two months before the ABS issue appeared.

V. was inspired and a new outlook, that of a movement, of helpers, even of collaborators, dawned upon him. Before then he had been a lone wolf in his field of study. Now he had friends who talked his language. Sizemore began to organize locally and to suggest that others organize in other places clubs or study circles under the name of "Cosmos and Chronos." V. referred often to these ghost legions. Sometimes they sprang to life to extend invitations to V. to speak at various places, or they were used as a letterhead denomination when rebuking critics. It was, for example, on "Cosmos and Chronos" stationery that the Philadelphia disciple and high school teacher of psychology, Robert Stephanos, addressed the Franklin Society in seeking to arrange a lecture invitation to Velikovsky. When the Society reconsidered and hastily closed its gates to V., it brought a certain public disgrace upon itself.

Inspired though he was by his association with new and competent men, V. himself could not be organized by them; he could seek only to determine all of their activity, without becoming controlled by them. Time and time again, spurts of organization occurred, with excellent initial results, but thereafter the efforts would slump and expire. The most successful organizing and activity was done out of his reach, in Canada, England, and in Oregon. He was too immense to allow himself even to be the leader; for a leader implies followers who are assigned responsibilities, are allowed judgement, employ initiative, and can be trusted. V. allowed none of these. There was to be no control over this leader; he was superman, distinct from the following, distinct even from a field of science for he refused to call it by a name, such as catastrophism. He would deny such allegations and not even perceive the distinctions. Nor would others, because it was unbelievable. It was nonetheless true of him.

Among the types of activists of a movement there may be distinguished: the theorist, the researcher, the publicist, the agitator, the organizer, and the fundraiser. A movement is oligarchic to the degree that the functions are concentrated in a few hands; it is bureaucratic to the degree to which the oligarchy assigns and restricts these tasks to specialists; it is democratic to the degree to which tasks are shared and distributed by common consent; and anarchic to the degree to which anyone can do whatever one pleases.

Pensee was an oligarchy, *Kronos* developed beyond oligarchy into autocracy. The S.I.S. was an oligarchy with high turnover and open access. The cosmic heretics as a total aggregate were anarchic, and formed and transformed plastically, so that one could perceive the aforesaid stable organizations, then glimpse pairs, trios, bands, circles, and groups in process of becoming (such as C. Marx's small Basel group that embraced Professor Gunnar Heinsohn of the University of Bremen, and Milton Zysman's Toronto band, and Luckerman's small Los Angeles operation). The attentive public shape itself over the period into *ad hoc* opponents and task forces (such as the AAAS panel), into members, supportive audiences, subscribers, book buyers, gossipers, fund-donors, materials-copiers-and-circulators, — reflections indeed of the several functions, anarchically undertaken.

An instance of the highest type of voluntarism came with Alice Miller, a San Francisco librarian, who put to herself uninvited and uncompensated the task of indexing intensively the works of V., and V. made the necessary arrangements to publish the book. The few scholars who obtained this work could now search to their heart's content for the fullest play and nuances of ideas (where such fullness existed) and for contradictions and errors. The first operation to be performed in serious criticism is an index; the memory of a reading or two rarely sets up written material adequately for analysis. Would that every high school student who today is being hastily introduced to a computer would be instructed in the philosophical logic underlying the indexing of content. Deg longed for an Alice Miller for his Q Series; his indexes were inadequate, even more than V.'s, because his work contained a larger proportion of abstract materials, which are harder to index. He found, for instance, that searching for "monotheism" in V.'s own indexes was useless; in Alice Miller's the idea came forth nicely, even beyond what V. might have wished to expose.

We return to Deg's favorite pastime of counting, listing, and categorizing, and to his figures of the numbers involved. They are impressive for they may be exponential. Despite the casualties, the deaths, the desertions, the languishing, and the waywardness, and counting parallel little groupings and isolated active scholars, by the end of the decade of the sixties there were perhaps thirty true scientific catastrophists who had come up by the non-establishment route into the field of quantavolution, and by the end of another decade, there were fifty more creative workers in the field. Shadowing these, watching intently, and supporting them were several hundreds of others, close in.

Shadowing the cosmic heretics, too, were a new group, union-card holders of the establishment, who are distinguished most readily by their denial that they are or ever were sympathetic to Velikovsky or any other quantavolutionist, or that they have ever sought or do now seek any ties with cosmic heretics. And these were equal and greater in numbers, carrying out the revolution by partial incorporation, the process whereby a revolutionary movement, as it advances, meets an opposition that has already been infected by and has adopted in part the principles of the revolution. It is at this point that most successful movements subside or are destroyed; their heirs are their enemies.

As one can see, if workers number, say, 15 in 1 decade, 30 in another, and 80 in the next, a doubling process may be occurring, against all predictions that might be based upon resources available, unchanged state of the opposition, and so on. At this rate, with 150 to 200 in the 80's and 400 in the 90's, taken with the activists who lend support to their views, the quantavolution viewpoint should enter the 22nd century primed for a large role in scientific thought. At the same time, it should be borne in mind, there will be attrition and desertions, doubling, and trebling the numbers of quantavolutionists outside of (but beginning to merge with) the establishment. But the threat of nuclear warfare to all civilization overshadows projections of science. One is tempted, in all of this speculation, to recite Keynes' ironic words, not about short-term economic policy but about short-sighted world politics: "In the long term, we'll all be dead."

* * *

Be it admitted that Deg, publishing a special issue of the *American Behavioral Scientist,* had a perfect subject and extraordinary materials in the Velikovsky affair. But why should he stick with Velikovsky? Let Velikovsky say his piece and then be done with it. What of next month's issue of the magazine, and the month after? The journal needed continuous attention. What of the state of political science, and of higher education, of which he had always been so critical? What of the state of the nation, *ibid?* What of his family staggering into adolescence in the disturbed and unruly Princeton atmosphere? What of his meagre fortune, skating on a thin monthly bank balance and a home mortgage? And his friends, the women and men who had been no more conversant with Velikovsky than he himself? And his book contracts: especially the *American Way of Government,* a good textbook in need of revision, whose care would lift his finances from year to year and carry his name around to hundreds of college communities. And the radical book on behalf of congressional supremacy that he was writing?

What of his reputation, that, in line with the customary in academic careers, should now begin to rise to a peak, abetted by the constant "mending of fences" and "nursing of the constituency" ordinarily pursued among scholars in his circumstances? Or should he not now throw in his fortunes with a political party, Democrat or Republican, it mattered not, for in both he had "friends in high places." Close friends welcomed his participation in Barry Goldwater's camp and in Hubert Humphrey's; this would appear strange unless one understood that subjectively Deg was confident that he was his own man, and that he could find equal opportunities in both camps to exercise his skills and ideals, which, to put them in several words, were: decentralization, basic income guarantees, voluntarism, legislative rule at home, and representative government for the world. The American party system, however, no wise shared his bent for change.

In all of this and through it all, why did Deg continue to involve himself with Velikovsky's problems? Did he not have enough problems of his own — larger and more serious and worse? Did be not have as grand and earth-shaking ideas himself? Most of all, if he was to spend a great deal of time in promoting somebody, and it was not to be "the next President of the United States," then why didn't he build up his own reputation?

He had had mean reviewers, scornful ones, too. His books had not sold very well. He had not yet won any considerable prize, no Pulitzer, no National Book Award. Still he could drum up audiences at colleges around the world. Bill Baroody wished that he might tour the country on behalf of the reconceptualized *American Enterprise Institute,* addressing public issues and garnering funds in the end. He was in mind as a political campaign manager here and there in the nation. He was offered the job of heading the social sciences division of UNESCO in Paris (and refused).

Why should he waste his time on a political campaign in science, especially one that had already been victorious *in principle* (Jastrow, Polanyi, Sagan, Motz, Neugebauer, Kurtz, Hadas, and dozens of other personages had sooner or later pronounced themselves against the ill treatment of Velikovsky). Did not Elisheva insist to the end that he had opened up the final phase of Velikovsky's public appreciation? Was the establishment of the motions of Venus so important? Or the evidence of ancient catastrophes on Earth? Or the likelihood of collective amnesia, a common enough idea of wise men of all ages? Must the world of science sign line by line in agreement with Velikovsky's books — the ultimate wish of a cult? No, none of this was so important. Well, what then? Was he sexually deprived? Did he identify Velikovsky with his own father? Many more motives offer themselves. Can one ever know? Why bother to ask, too? Yet it is a question that was asked at scores of lectures, receptions, meetings, and in personal discussions, a question that came out of the interest that people felt in their own motives, out of curiosity about what might be construed as altruism or some other form of abnormal behavior. It's Alfred's *halva,* Nina would say, meaning the joke about the man who loved sweet "Turkish Delight" and would turn the conversation to it at the slightest cue.

Deg behaved as he did partly because he had enjoyed enough successes in other matters and success bored him. Deg did not attend to promoting his academic career because he was already a tenured professor, "heavily published" as they say, and where was there anything further to be gained; universities and colleges seemed ready to succumb to stupidity or insane revolts, but not to total self-evaluation and reform. They were, with governmental help, becoming ever more bureaucratized and inane.

Besides he found self-promotion an embarrassment, all the more as he watched his acquaintances climb the rows of ladders inclined against decrepit edifices where committees and trustees held sway, and important research was kept in a corner like a bastard. He was not adverse to fame. To the contrary, he expected it to be "handed to him on a silver platter," to use one of his mother's expressions. Subjectively, he desired glory; objectively, externally, he had to scorn it.

He was having his last words on Congress and the executive force, an appeal for the preservation of republican government that went against every major political and economic interest in America (and that communists and socialists when in power also and even more rampantly suppressed). He was, as I said, uninspired by the political movements of the moment, and even more so as they developed through the sixties and seventies of the century. The kindling problems of his family would burst into flame but he had no intention of becoming party to a decade of adolescent rebellion of the kind that ruins the best years of many Americans' lives. Besides, did he not have such splendid plans for going *en masse* to Europe for a year to teach the children foreign languages and escape the menacing youth and drug culture of Princeton?

But look particularly to the controversy surrounding the Velikovsky matter: was it not exciting? The ideas at stake were of the highest order. Not only in sociology: for what sociology is more important than the sociology of knowledge *(Wissenschaftsoziologie)* that he had cut his eyes teeth on with Mannheim, Wirth, Shils, and Leites, and which was really the theme underlying his first book, *Public and Republic,* where ideas of representation were shown to be unconsciously operative and externally effective over hundreds of years and many different political generations?

Also there was excitement in the substance of this strange new kind of science. Scattered about but eager to stay in touch were dozens of intelligent people interested in one or more of the hundred fields upon which quantavolution impinged. More exciting and elevating than yachting, the horseraces, gambling, cocktail parties, tourist travel, religious routines, better than the eviscerated or wrongheaded politics of the times. In the final analysis it was the unlimited firing of sky rockets in all directions that held Deg to the course of quantavolution and bound him to his friend Velikovsky.

There was the intransigent personality of Velikovsky. Even some opponents, Robert Jastrow, Walter Sullivan and Motz, for instance, found him fascinating. He was always there, like a smoking Mt. Vesuvius, whenever Deg came back from anywhere, the tallest mountain in Princeton and anywhere else, so far as Deg could observe.

A series of entries from Deg's journal, most of them from the year 1968, show what I mean. But first a letter from Velikovsky to Deg, before the ABS issue of September 1963 had made its impact, to show that V. had no intention of letting his new friend escape his camp by crossing the ocean:

August 16, 1963

Dear Professor de Grazia:

It was very good to have a letter from you in Paris. I like to hear that you may come to the States in October. No old castles here, no ancient arenas, but you will be most certainly engaged in some skirmishes in the tournament for which the scene is being set. Larrabee's article produced certain effect (I assume it was mailed to you) and the foundations of the establishment are being loosened, (...) A few papers started to comment on the issue, one or two colleges invited me to speak before their students, much discussions going on without reaching the printed page, and I am emerging from the 'shadow of darkness.' (...)

I wish I could bring to our side a few prominent scholars and scientists. I write to de Madariaga about Lord Russell whom he knows. You may say again, 'Cabot', but visualize the effect on the closed scientific ring of one such renegade.

I wish to think that Mrs. de Grazia and your children are enjoying their many new impressions, and the old villa makes them feel that theirs is part of an old heritage. Turgeniev wrote someplace that two urges live in a human soul — a striving for far away lands and a longing for the homeland and home. Mrs. Velikovsky joins me in wishing all of you good health and animated months ahead.

    Cordially yours,
    Immanuel Velikovsky

PS The mail brings an envelope with copies of letters received by Harper's. Menzel of Harvard Observatory writes a 17 pages letter, unfair, emotional: he exposes himself to embarrassing statements of fact. A battle of letters started. At the present, the response runs 50% against 50%. Therefore any articulate supporter — or opponent — should enter the fracas, the earlier the better. Mobilize your friends!

    I.V.

A year later, Deg was not only still in the camp, no matter where he was, but he was suffering privately the annoyances of the camp. His journal of September 1st, 1964 from London is relevant. He is on his way to the University of Gothenburg, Sweden, to lecture on American politics and will from there go to Marina di Massa where his daughter Catherine will be wedded to the best-looking boy on the beach, Dante Matelli (future star journalist of *La Repubblica*).

Left for London at 10 AM. On way to airport penciled a crude note to Velikovsky, finally telling him bluntly of my feelings towards him. I said, "Dear Immanuel, I am writing this on the bus to the plane.

"Last night I went again over the letters and material for Rabinovitch, to the detriment of many pressing affairs. I finally decided to send out nothing at the moment.

"You will receive the page proofs on the Margolis critique. Please make only absolutely necessary corrections (I do not care if you offer to pay for them.) Issue is already late.

Please do not call my office or the printers. Your inability to let go of anything will be the ruin of our friendship and of the magazine. Sincerely, Alfred".

I handed the letter to a passenger agent just before stepping aboard the PanAm Clipper. It culminated a day of annoyance and desperation that began when I courteously called Velikovsky to say goodbye. To those who know him well, the history of the next 24 hours was to be clear. He wanted to rewrite letters, call lawyers, discuss imbroglios, in short, utterly and without conscience disrupt my carefully measured out and urgent last hours before departure. And worse, he succeeded.

This hardly matters. The friendship, the campaign, continues, and V. is still the mastermind. When Deg goes abroad in 1966, V. has ideas of how he should spend his time in Israel and Egypt:

Feb. 14, 1966

Dear Mrs. de Grazia: Please *do not* send this letter to Alfred if he already left Italy. Im. *Velikovsky.*

Dear Alfred:

I received your note written before leaving for airport. Should you visit Jerusalem you may wish to give personal regards to President Zaluccan Shazar — our friend, especially of Elisheva, of many years. He will be glad to hear that Elsheva is active as sculptor and as a chamber-musician (as good as ever); and Elisheva wishes him to know of the change in the attitude of the scientific world to my book with many discoveries of the Space Age; the fact that I am invited to speak at Yale, Princeton, Duke, Pittsburgh, Wisconsin, Oberlin, Brandeis, etc. is an indication.

I wish you good weather (pleasant driving, good new friends, and many invigorating experiences).

Regards from Elisheva and my regards for Paul and John.

Yours,
I*mmanuel.*

[P.S.] It would be good if at the Cairo Museum you could obtain some organic object of the time of Ramses II or Ramses II (or of both) for radiocarbon test (better seed, mummy swathing, leather, papyrus, linen - and not wood, if possible) at the lab of the University of Pennsylvania (Dr. Elizabeth Ralph.) To apply to Dr. Isnander Hanna (Director at the Lab at the Museum). The material needs to be sent from museum to museum with all the precautions. By far better not to mention my name.

If any difficulty, I shall try to obtain the samples by asking Dr. Ralph to write to Dr. Hanna.

Deg's journal, January 18, 1967

Phoned Velikovsky tonight. Elisheva came on the wire too, at his request. I told them what I was doing to institute a Foundation. He was quite subdued. He is not used to having anything taken out of his hands. Both were happy, I could tell, at the thought of something they had talked so much about moving so quickly to a climax.

Anti-Velikovskianism's first line of defense is the impossibility of his theories. Then, I suppose, if proved right, it will be said that he was a simple scribe: he read an inscription which told what happened. That position will not endure, either, for he worked in a superhuman way to piece together the shattered mosaic.

Deg's journal, November 15, 1967 9 pm

Immanuel called me at twilight to tell me Stephanos had called his attention to the Nov. 3 issue of *Science* magazine wherein Professor R. Eshleman of Stanford University, Electrical Engineer and Co-Director of the Stanford Center for Radar Astronomy had raised briefly the question whether the baffling puzzle of Venus being 'locked-in' to Earth might be answered by the Velikovskian hypothesis of an historical collision of the two bodies. A year ago *Science* refused to accept an advertisement for one of his books. "Who knows, Alfred, whether the Nobel prize, which has had a poor record very often, might not come." I said, "Immanuel, your biography is your triumph. You do not need these foolish prizes."

Deg's journal, 1/4/68 [Providence]

At 2:30 I left the ribald company of Mike N., N., Jim Kane, Al Saglio, Tom Yatman, and Edwin Safford at the Spaghetti House to visit Prof. Otto Neugebauer at Brown Univeristy. His office is in an old red brick house next to the new Library and has an entrancing scholarly air to it, closed into the basement, holding several tables, everything with a century old appearance that I too should find a perfect atmosphere for quiet study and work. O.N. was somewhat suspicious of me, as well he might be, knowing that I sponsored a special defense of Velikovsky's work. However, like most true intellectuals, once engaged, his defenses were down and he spoke vociferously, indignantly, said he couldn't waste time on the foolishness and trickery of V. but proceeded to amplify at great length, his little blue eyes peering directly into mine and his slight but determined German voice carrying effectively, even colloquially, his arguments. He disputed hotly the idea that there had been or was any conspiracy against V., (I stated that I too disagreed with V. on this point), and he felt that V. was employing the tools of propaganda and sophistry against him and others. Who can deny this, too? But there seemed to be little reason to go into the political aspects of the controversy, inasmuch as O.N. could not know, more than V., the dynamics of this process, and I essayed questioning him upon several critical issues concerning Babylonian tablets.

He declared twice that he had "no investment" in the words of the tablets and could take or refuse any interpretation, depending only upon its truth. They were only a minor interest with him, not even "minor," less than minor.

He said he had not read Stecchini's interpretations of Kugler's work (and declared offhandedly but vigorously that much had been learned since Kugler's time anyhow). He declared that the observations in the Venusian tablets of Ammizaduga came from erroneous reportings of lunar movements that, in turn, had been used by the Babylonians to measure the movement of Venus. An amateur, he said, would transfer his ignorance of the ancient reports into a wrong interpretation that it was Venus, not the Moon, that was moving erratically. He declared emphatically that from their beginnings around 700 B.C. there were no unexplainable irregularities. (He kept reasserting, and I had to stave off as not relevant to the argument, which was the empirical facts re the tablets, that the whole V. thesis was mechanically impossible, that any 10-year old schoolboy would know how the Earth would be destroyed by anything approaching a collision with Venus, and so forth). He said further that there was little or no reporting of any planetary behavior in a scientific way prior to about 700 B.C. (I didn't press for the exact date) that, for instance, there was no reporting of Saturn before 400 B.C. Earlier records are largely the oracles which deal with sun, moon, and a bright star (which *could* have been Venus, since it is the brightest and hence would oppose V.'s theories of the non-existence of Venus before ca. 1500 B.C.) He asserted further that Egyptian chronology was perfectly established, on the basis of the Egyptian lunar calendar (based on a thirty-year cycle) that carried back to the very earliest times. He claimed that the whole V. affair showed the basically anti-intellectual atmosphere of the population.

I asked whether it did not show also the failing of the establishment of science to perceive its "public problems," and offered the opinion that if he, and others such as Harrison Brown, had dealt with V.'s work more seriously, there would have been no prolonged vicious aftermath, to which he grudgingly acceded.

Then he added that there should not be such an accent on "going to the moon" so that billions were being largely wasted, for which sums the whole of Mesopotamia could be dug up down its virgin soil. Then, said he, we should have all of these problems solved. To which I agreed.

I asked whether someone should not set forth the thirty or sixty principal factual theses of V. and find specialists on each topic to criticize V. He had mixed feelings about the idea (first taking it personally, of course: "I don't have time for that!") holding that V.'s ideas were too vague to discuss, that this would prove that the "conspiracy" actually did exist; that there would be too few to undertake the job in certain areas (such as his own of Assyriology and Babylonia); but that it might be a proper way to get to the heart of the matter. He was, on the whole, quite negative *re* the general problem and hostile to V. As I was leaving, he said "I just received a letter from Chandrasekhar of the University of Chicago. He is the physicist. He asks whether we shouldn't do something about the *Yale Scientific Magazine* issue of V. I replied that there was no use to if."

I walked out into the winter snow-threatening afternoon and down the streets of exquisite old structures of Providence's East Side, to Mike's house, thinking of what I had learned and of the beauties of this old part of town.

1. N[eugebauer] is convinced V. plays a tricky game: "He couldn't answer my colleague's questions at a Brown University meeting, but said he would reply to them the next day. Then he didn't appear."
2. He believes V. to be a foolish and wicked amateur.
3. His direct assertions concerning the Venusian tablets should be worked into a direct encounter with V.'s words. (...)
4. N. appeared uncertain about Kugler, and unconvincingly dismissed him.
5. N. is persuaded that V. is arguing in a great circle, using established theories as grounds for criticizing deviations and unknowns and for proving the deviations accord with his theories, then destroying the established framework without perceiving that his interpretation of the deviations is itself dependent upon and sponsored by the established theories. N. did not say so, but this kind of problem is fundamental to all theoretical change: man is dependent for what he sees upon what he has been taught to perceive, so how can he prove wrong what he has been taught if his new vision is wholly dependent upon being preceded by the old one?
6. I feel the need to organize an 'Anti-Velikovsky' symposium where highly reputed scholars are asked to address themselves to a meaningful segment of a carefully prepared set of questions that test the whole fabric of V.'s theories. Logically V. cannot dispute this procedure. It would, I think, cause him to be angry with me. So be it.

Deg's journal, January 20, 1968

I have been visiting with Velikovsky once or twice a week since November, and have reread *Earth in Upheaval* and *Ages in Chaos*. Since I have been heavily occupied with the theory of activities of the federal government, the American Government text revision, a plan for a business company should I decide to leave the academic world, and so forth, I indicated to V. ten days ago that I could not organize the magazine that we had always talked of publishing. Then, for some reason, a week ago, I thought "We must start a foundation for V. and his work." I asked Richard Kramer to initiate the papers for organization of a corporation not-for-profit in N. J., settled on PO Box 294 and my home as the address, and decided to ask Juergens, Stecchini, Kramer, and Herb Neuman to join me in the first Board of Directors. I called each man to invite them aboard and received their prompt acceptances.

Deg's Journal, March 2, 1968

This morning, I am resolving to withdraw myself as much as possible from Immanuel's campaign for honors and recognition. A full eight hours went to him yesterday; it is too much, considering what I must do for my own work. In its way, it deserves the same kind of attention V. gives to his and I give to his. My intellectual children may be scrawnier but I cannot turn them out to starve in the cold. I give up lectures that, just like his, might explain my ideas and bring me income, as for example one that I turned down today for $100 and expenses before an audience of civil service officials in Washington. My ideas go undefended, many aspects of them go unexpressed. I do not give them the tender, fierce, loving care that every man's respectable notions deserve. Let's see whether I can behave by this resolve.

Deg's Journal, March 3, 1968

March is come cold and blustering. Jill and I rode our bikes to Mom's where Ed and his young friend, Margaret C., were visiting. We arrived frozen. M.C. has just returned from 2 weeks in Boston, under the tutelage of a Yoga guru. I say to Ed, in greeting, 'Ah, here is the "slim, elegant Sicilian !"', quoting Norman Mailer's autobiographical novella of the March on the Pentagon that is printed in the current Harper's Magazine. [Edward organized the legal defense of the arrested protesters.]

Jill says, of Margaret, 'Girls who have had trouble with their fathers work it off well. Girls who have had difficulties with their mothers do not.' She cites Jung on the point. And we string out many examples. It is probably true, even as an unrefined statement. I ruminate: so important, so simple are basic truths. What conceals it and them? Great truths and discoveries are not hidden by their complexity but by the jamming of our ideological, cognitive, and perceptive machinery.

Velikovsky, the other night, quoted me Butterfield's comment that the very young can understand principles of science and nature that have baffled the greatest minds of history. I think V., who is in essence a philosophical realist, uses this idea in only a limited way. He means that the young haven't had their tender minds distorted by unfact. It is more importantly to be understood that the mind is structured in each generation to receive some truths and reject others, or, better, some half-truths. Both V. and perhaps Butterfield unjustifiably abstract the mind from its context. It has, for instance, been pointed out by numerous defenders of classicism, such as neo-Thomists, that we believe the ancients foolish or unperceptive of truth because of our partial and current truth-idolatry; freed from contemporary ideology, we can understand truth as the ancients discovered it and agree with them.

Deg's journal, April 30, 1968 AM, en route to NYC

Half of this past warm flowering week-end in Princeton has been spent with Velikovsky or on matters related to him. We spent Saturday afternoon going over materials that might be suited for the proposed book "V. and his Critics" that I am discussing with Kluger of Simon and Schuster. We spoke also of the Foundation for Studies in Modern Science, which I have organized. He named eight major problems that are critical to his theories, and I am taking them into consideration in the memorandum which I am preparing on the program of the Foundation. Bob Stephanos called me on Friday night upon my return from NY to tell me that Mr. Mainwaring of Philadelphia, an admirer of V., intended to help financially. Both V. and I had written letters to M., who runs a family manufacturing firm and is, I hear, a person of some intellectual stature. V. was naturally pleased. He talked on and on, I edging him back to a subject from time to time.

Sunday evening, V. seized the initiative and called Prof. Philip Hammond of Brandeis U. to ask about his possible interest in excavating at El Arish for signs of the siege of the Hyksos fortress by the allied armies of Saul and Thutmose, about 1050 B.C. in V.'s chronology. The digging would be a crucial test of the V. theory of ancient history. Hammond, who had given indications of sympathy years ago, appeared enthusiastic. He offered to go to El Arish with two assistants if we could organize the expedition.

After tearing this from V., I called David Dietz to ask whether he would still be interested in taking part in the expedition. He was. Yesterday, Monday, I asked Harry Hess of Princeton University Geological Department to serve on the Board of Trustees of the Foundation. After some demurral (later, V. would be mystified by his hesitation since 'Hess definitely agreed to join,' but I was not mystified, poor Hess who is one of the busiest men alive with his Space Board, Mohole and other activities, couldn't take the leap into the cold water without encouragement. So I purred, gently, sympathetically, and finally he said with a hopeless smile "Aw, hell, OK, put me on!" (...)

Deg's journal, May 2, 1968

N[ina] and I met at the Museum of Modern Art at six yesterday after my discussion with Kluger, of Simon and Schuster. A surrealist exhibition was on. Max Ernst, Nadelman, Matisse, Ram bear up very well. Picasso rarely becomes human enough to excite me. His lines are cold and cruel. De Chirico's colors seem shabby now. It was a brave moment and said a lot. We drank beer and ate cheese and crackers in the garden of the Museum, which filled with grey rosy lights as the sun set. Rodin's Balzac, seen from above, is stern and emotionally stirring. A Picasso She-goat is my great love.

Back at Washington Square, N. prepared a light supper at her place and accompanied me to my work. I talked to Velikovsky at length, recounting my conversation with Richard Kluger and explaining my plans and hopes for the expedition. As usual, he was difficult to converse with but excited more than I've ever felt him to be before. I told him that I thought we should film the El Arish episode from beginning to end, and he was fully agreed. I wonder, of course, continuously, whether we shall find what we are after beneath the town — the siege evidence and artifacts of Saul's army, the Egyptians and the Hyksos.

I hung up the phone and went to work sorting out materials to be used in my Reader on American Government. N. said "Velikovsky can never finish his work." "Nor can I!" I replied. "He has thirteen books to go, when we last counted them. I am as badly off." She asked me what I had to finish. "you have done so much." "Not at all," I said, impatiently. "We do not measure ourselves by other men but by an absolute criterion of what we might conceivably do." And then I ticked off what I imagined I might yet do:

    the publication of my collected papers of the past
    the American Government books
    another book of poetry
    several novels, mostly autobiographical

a philosophy of science
"the new political order"

and whatever would intervene, such as the El Arish story and the government operations study, and who knows what else: editing the *Velikovsky and His Critics* book, for example. (...)

I spoke to Sebastian about other matters on the telephone during the day. We are concerned about the troubles that Eddie is having over the custody of the children in divorcing Ellen. (...)

Bus told me of a quarrel between Renzo Sereno and his wife one time over a lady, possibly a mistress, of Renzo. 'The only thing you like about her is that she thinks you're great,' declared the wife. Bus and I breathed reverently over this gem for a minute of ATT long-distance time and charges. What has come over womankind? What do they imagine to be the foundation for a man's love and devotion, even charm, even presence?

After a day of labor selecting readings for my American Government Reader in the company of Eric Weise and John Appel, I entrained for Princeton, snoozing aboard, and arriving happily into the fresh air of the countryside. John, Carl, and Chris were all in excellent moods, the one fixing things on the old Cadillac, Carl playing his Beethoven pieces, and Chris shooting baskets. Mom came to dinner, bringing some freshly picked and cooked wild cardoons.

At nine I biked to Velikovsky's home, Francie loping alongside and for two hours, while she stretched comfortably in the middle of his parlor, we talked and argued over who should do what about books, magazines, and the ever-growing prospect of the expedition to El Arish. Prof. Philip Hammond caught me by telephone soon after I arrived from N.Y.C. to reaffirm his interest. I asked him whether he would, in addition to his usual excavation reports, accept co-authoring of a popular book on El Arish that I was proposing to Simon and Schuster and he accepted promptly. I like the sound of him, though we have not yet met.

V. was difficult. He holds out things and then pulls them back. He wants to do too much himself I try to take responsibilities off of his shoulders and he fights to keep them and even to take new ones. He wishes to discuss every small decision, to control every document, he is elated over our plans but becomes more demanding and even a little more paranoid as events speed up. He has a poor sense of organization and scheduling where other human beings are involved. His own immense mental world can grab and hold everything and shake it out in marvelous patterns, but the world of affairs has its own ruthless laws that treat all men equally and make their own patterns.

Now came time for the Foundation to form and the incorporators met to elect themselves and additional members to the Board of Trustees, and to transact business. R.P. Kramer, L. Stecchini, R. Juergens and Deg coopted Horace Kallen, Harry H. Hess, A. Bruce Mainwaring, John Holbrook Jr., Robert C. Stephanos, and Warner Sizemore. The date was June 2, 1968, a day that would not go down in history.

Deg was chosen President and other preliminaries were disposed of. Then the ill-fated excursion to El Arish, where the capital of the Hyksos supposedly lay buried, was taken up. Everyone knew already that Mainwaring and Holbrook had put up some funds, that a Dr. Hammond had been approached to lead the group, and a contract had been drawn up. Deg set forth a budget, even the minimal costs of which were well beyond the pledged resources of the group. Besides the preliminary soundings at El Arish, papers on the "hydrocarbons" of Venus and its temperature changes were to be commissioned, a publication was to be prepared, preparations to receive and use V.'s archives were in order, a magazine was to be inaugurated and besides, there was provisions for work on collective amnesia, dating systems, magnetic polarity, evolutionary theory, the psychology of catastrophe, an electromagnetic cosmic model, and the reception system of science. A happy set of prospects indeed, every one of which the foundation was to fail to inaugurate, much less carry on to any extent. The case of El Arish will suffice as an *exemplum horribilis*.

In June, A. Biran of the Israeli Department of Antiquities wrote to Deg saying:

> Indeed there is much interest in the archaeology and history of the area but unfortunately it is not always possible to satisfy this curiousity. Even I with all my interest and curiousity have not yet been either to Kadesh Barbea, Mons Cassius, or Qantara...

July found Deg in Naxos, ready to go to Israel if needed, and John Holbrook had gone to Israel to seek permission to begin a site survey at El Arish. Deg is getting a variety of inputs from his assistant:

July 10, 1968

> ...I spoke with Velikovsky today. He told me that Holbrook had arrived here yesterday. A copy of all the correspondence is on its way to us. The gist of it is that Holbrook saw Biran and Dotan, the chief archaeologist, and that the Israelis would like to see more solid support from Americans, Biran said that FOSMOS seems a bit fly-by-night to them. Another problem is that they don't want to grant foreigners the right to dig in occupied territory. But apparently they have softened a little, and if they could see something more established in support of the dig, well then... So Holbrook is going to ask somebody at Yale about it, a Professor Popo.
>
> I read your report of the Natural Museum with interest. I will probably get to the Met sometime this week. The figures you described on the one

vase are usually interpreted as Amazons, and I am going to compare the costumes with those of the Busiris vase, out of curiousity. I think there is also a book on Greek arms, which should have something in it about helmets.

I am sure you are enjoying Greece — it's so wild, beautiful, clean and clear...

Meanwhile John Holbrook is grinding his gears in Israel and is addressing a set of marvelously detailed letters to V., a copy of which he then sent to Deg.

Holbrook writes to V. on July 10, 1968:

Now I am in a bit of a quandary. First, I have no reason to doubt Bihran's word that the military situation in the Sinai area prohibits any extended work at El Arish at this time. Second, although I shall certainly see Dothan when he returns from the field at the end of the week, I cannot pledge the support of the foundation to the extent of $50,000. Although we have great hopes for it, the treasury of the foundation is still a bit empty, that being the case, I can only explore the possibility of organizing an expedition to El Arish at some indefinite time in the future (when military situation permits) on the most tentative basis. Much will depend upon what I learn from Dothan. At the very least, I hope that I shall be able to get a look at the site before I leave.

One other matter deserves mention. There is no way of telling the extent to which opposition to your work played a role in the rejection of our proposal, there were too many other reasons for rejecting it.

Later Holbrook ventures an opinion on the actual site:

Quite frankly, although I am sure that a complete archaeological survey of the Wadi El Arish and its vicinity might be extremely useful, I am willing to bet that the first trench which is dug in the area which I have described above, the northern quarter of town, will not be found empty or unrewarding.

Little could be done with the El Arish party, upon which V. had set the highest priority (and did for the rest of his life and rightly so, says Deg). The failure was bad enough, but to Deg the most disagreeable part of the episode was the way in which V. began to find grounds for opposing Hammond after he had agreed on his competence and leadership qualities, and had invited him to lead the operation. V. soon convinced himself, and then Holbrook, that Hammond was pro-Arab and would be *persona non grata* to the Israeli authorities, until they were actually approaching the Israeli saying in effect "we know how you must feel about Hammond, but we are aware of this situation and are taking care of it," whereupon the Israeli, in the case of President Shazar, said what are you talking about, who is Hammond?

Deg's journal, October 20, 1968

Velikovsky and I talked for the first time in a week yesterday afternoon and again last night. He leaves for a grand lecture tour of Texas today. We have counseled him not to go to California to talk, a little later on, because he would become tired and he absolutely should finish *Peoples of the Sea*. He continues to add new data to the work, which is slender still though, like a stick of dynamite.

We argued over the final contract details of *Velikovsky and His Critics*, which I am not keen to do anyway, given my poor financial state and other projects of greater personal importance, he wanted us to guarantee mutually that we would not submit the final manuscript without his approval, in effect. It is of course a perilous idea, for he hangs onto everything and cannot suffer any criticism. I drew up an appropriate missive but added words to the effect that we would also be jointly responsible if Simon & Schuster publishers sought damages from us for non-delivery of the manuscript. As I suspected, he balked, and talked of legal formalism. I laughed and expostulated: "But you want everything, complete authority and no responsibility!" It is the same with the Foundation we are creating: he wants it to follow his every wish, but does not think that he should be identified with it.

He then said, "All right, Alfred, we will agree just among ourselves, without a paper. You will not submit it without my approval."

"O.K."

And then we went on to argue over the student strike movement, which he fears will undermine authority and disrupt education. "A tiny minority has no right to interfere with the majority who want to study." I told him that minorities are the media of change in any field. I asked whether, if the French students had not rioted in May, there ever would have been the Faure reforms of last week. "No matter!" He would change his mind. I can always win a argument with him on politics citing his own case and the history of modern Israel. On these two great contradictions of order, stability, and authority, much of his life is built; they make all of his defenses of authority and majorities vulnerable.

"What do you think of Onassis?" I asked, to change the subject. "Who?" Onassis, and Jackie Kennedy. "Oh! I tell you that I think it is a second assassination of Kennedy." Beautiful, I thought, either way. His idea is that of all the maudlin sentimentalists, Kennedy-dead worshippers, the sanctimonious, the suttee-ists. My way, it is revenge for a not too great love, followed by the maddening experience of suffering all of this cant and sick reverence. All of these mass-media addicts were hoping she would end up with a crew-cut college sophomore from Princeton. So she picks the ugly old Greek pirate, and I am personally pleased. The Hollywood and Madison Avenue brainwashed crowds have their fairy tale exploded once again. I know that people live off of these fairy tales; that is what makes valid history and rational politics impossible for them. Perhaps I should feel sorry for the great boobery, but I am diabolically pleased with Jackie's revenge upon them. And upon JFK too, with his harrowing political life and difficult character and mistresses. What is there to insult in his memory, I ask myself, and what business is it of old ladies and shop-girls to define her husband. "Onassis, I don't know the gentleman. Probably they like each other. I wish them happiness." *Basta*.

We returned to majorities and here is how he defined the Jewish majority in Palestine. "Over history, the dead of the Jews are a majority in that country. They live in that tradition wherever they are." Voting the dead to make a majority, like the Confederate southerners do, or the bosses of "rotten boroughs" in the northern cities. *Grüssgott!* What would V. say to *these* majorities and so many others that are *alive*, as well. But Israel is the *idée fixe;* facts are the dependent variable. Indeed, as I have known for as long as I have known him, the *idée fixe*, the highly conventional, traditional literal interpretation of and respect for the Biblical passages: from this conservative position spewed forth in all directions the most radical theories.

Deg's journal, October 25, 1968

Reflecting upon the failure of our infant foundation to launch an archaeological expedition at El Arish last summer, I think it may be well to set down my view, which contrasts somewhat with that of Velikovsky and Holbrook. V. was too willing to accept rumors about Prof. Philip Hammond and placed too strong a weight upon adverse facts. V. had no right, as I told him bluntly, to destroy Hamrnnond's possible role as leader of the expedition on the grounds that Hammond was pro-Arab and that he had a mistress who would accompany him. Holbrook, whom I regard highly and even warmly, with all his youthful arrogance, was too ready to accept V.'s evaluations and then afterwards the positions expressed by the Israeli authorities, to wit, that we could not afford to support the diggings and that the political situation was dangerous. I fell that we had gone so far in our adventure that we ought to have let Hammond himself battle with the Israeli. He might, I think, have outfaced them and dragged in his crew and equipment over their grumpy dispositions. I doubt that we would have uncovered anything of great significance in a few weeks, but we would have planted our flag. We would have moved on from there.

Deg's journal, November 2, 1968

Met with Velikovsky this afternoon. He is back from a triumphal tour of lectures in Texas. We argued over plans for the foundation. Juergens was present. I asked him pointblank to pull out any materials he might have that others had sent him and might be used as articles for the proposed journal. He did so. [There was almost nothing.] I asked him also to pull together all his address lists and to let us place a man in his house to build up a list of friends with whom we might communicate. He agreed. I was most pleased. I borrowed V.'s manuscript on *People of the Sea* to read again, and left with everyone in cordial spirits. What a difficult man but what an enormous grasp of everything, intellectually and physically!

I must set some probability theorist to work on some of V.'s proofs. They are strong as they stand in their conventional historiographical form. But an application of mathematics would do much more, e.g. the chances that the Greek letters on the backs of Ramses Ill's tiles might be some 'flowing' or shorthand hieroglyphics.

The Foundation spent the fall of the year, following the El Arish fiasco, in some small constructive matters and in self-destructive self-appraisals prompted by V.'s misgivings. Ralph Juergens addressed the Board of Trustees extensively on November 13, writing *inter alia:*

1. ...He [Velikovsky] is concerned that funds collected, as it were, in his name, as gifts intended to further his own researches, will be diverted to other purposes. Among such other purposes he includes such FOSMOS projects as the Institute in Connecticut, the journal Cosmology (...). To the doctor's way of thinking, only two projects thus far discussed would be legitimate applications of such donated funds: a) the El Arish dig, and b) the hiring of Princeton graduate students to carry out library and/or laboratory research under his direction. 2. Dr. Velikovsky is aware of our plans to launch a direct-mail campaign early in January and he is offended at not having been consulted in the preparation of mailing pieces. (...) He insists, at the very least, that literature sent out make absolutely clear to the reader that he is not a power behind the foundation and that he will not be a recipient, direct or indirect, of any funds collected by the foundation. (...)

It seems to me... that some rather fundamental misunderstandings remain to be cleared up, not only between Dr. Velikovsky and the Board of Directors, but perhaps also among members of the Board. In the first place, there is confusion as to the purposes of the foundation. It may be that Dr. Velikovsky has never seen a copy of our By-Laws, which seem to make the point that the foundation is to serve as a clearinghouse for a variety of information, not all of it necessarily related in any obvious way to Dr. Velikovsky's work. This would appear to leave us free to tread ways not yet probed by the Doctor. And of course we thus face the danger of becoming what Dr. Velikovsky would call a clearinghouse for cranks. But our statement of purpose at least broadens our horizons to the extent that we cannot think of our orgainzation as a 'Velikovsky' foundation.

Or can we? The confusion seems rooted in the fact that we members of the Board, almost to a man, have been brought together through our common desire to see his work get a fair hearing. Do we really intend to operate a "Velikovsky" foundation in spite of our more abstractly stated purpose? If so, we must accept certain consequences, e.g., foregoing a tax-exempt status and placing absolute veto-power — quite properly — in the hands of the Doctor. If not, I suggest that we make haste to disillusion ourselves and Dr. Velikovsky.

On November 22, Deg writes a harsh letter to V.:

November 22, 1968

Dr. Immanuel Velikovsky
78 Hartley Avenue
Princeton, New Jersey 08540

Dear Immanuel,

As you have no doubt expected, your succession of favorable and unfavorable comments concerning the progress of the Foundation has created a crisis of morale among the Trustees. For years you longed for just such an organization to dedicate itself to the testing and propagation of your theories, and now that we have constructed it you are undermining it.

You trust nobody, delegate nothing, and have, partly therefore, no capacity for administration. You also do not wish anyone to speak in your name but wish help to drift down like manna to dispose of as you desire. Actually, we shall be trying to do both things — administration and help — in spite of you, if you do not disrupt the process.

The Board of Trustees has unanimously pledged itself to an independent course. Whatever the Board of Trustees believes to be useful to the advancement of science, it will seek to foster. It cannot bargain with anybody. If it chooses to do one thing rather than another, it does so, not out of friendship to you but out of respect for the work that you and others like you have done.

In order to make demands of others, both inside and outside of the Foundation, I have to make demands of you. You should cease making accusations against the Board, even if only among the inner circle. You should cease bargaining over your Archive and the materials that you do not intend to personally use, and let the Foundation work with a copy of them as soon as it can arrange to do so. You should accept what we can offer you (or reject it) in good spirits, knowing that we are doing our best in a complicated setting over which we do not have complete control and that sometimes we must obtain indirectly what we cannot gain directly,

The men on the Board are your friends. If you have better ones, let them step forward and we shall welcome them. The men on the Board are not the best scientists in the world and, if you know better ones, we shall welcome them too. The Board has to finance the Foundation's activities in whatever ways it deems appropriate. If you have the names of persons who, you believe, might contribute to its work, we shall be happy to receive them. If you wish to reserve the names of certain individuals or groups for your personal solicitations, please let us have their names and we shall not approach them, whether in your name or in the name of the Foundation. If you disagree with the policies of the Foundation, we would value your opinions. But you cannot have a veto over anything that the Foundation does.

If you do not wish to relate to the Foundation in all of these ways and want to dissociate yourself from the Foundation, I believe that you should do so, either by a personal advertisement in a journal or by letter to all those of your acquaintances who matter. I shall then put a resolution to the Board to the effect that the Foundation will go ahead with its philosophy and plans. If the vote is positive, we shall go ahead; if not, we shall dissolve the Foundation, an action which will disappoint me and give me immense relief at the same time.

Of course, if you do not desire to take any such measures, I would assume that you are basically pleased with our work and will work in tandem with us.

With warm personal regards, as always,

Sincerely,
Alfred de Grazia

V., Deg learned from Elisheva and Ruth, was upset. Then he proceeded to put some of the blame upon Juergens, where it most certainly did not belong.

Dear Ralph:
Yesterday morning, as you know, I received a rude letter from de Grazia with unfounded accusations and it shocked me. Suspecting some provocation, I called you. You disclosed to me that already on November 13 you have sent a memo to him and to the members of the Board of Fosmos. Next I was surprised to read the memo and its content being your interpretation of a discussion we had at one of our meetings. I wonder why you have not checked with me on the correct presentation of my views or at least mailed me a copy of the memo. Giving it yesterday to me, you gave me also a covering letter. Your intent was good — you must have suffered

observing that I am under wrong impression based on oral declarations made to me, whereas the Board assumes a different policy; and it is good that you brought the situation into the open.

Your memo, however, is full of inexactitudes; knowing you for pedantically accurate, I wonder at your rendition of our conversation. The only explanation I would know, is psychological: your opposition to the idea of the Foundation — or only to the dichotomy (you use the term 'duplicity'), and that can be a subconscious urge during your writing. (...)

The sentence in your memo that obviously outraged de Grazia who repeats it is "veto power." Nothing of the kind was spoken between us or between anybody else. There is a wide gulf between a "veto power" and being kept in the darkness, as several instances in this letter testify. (...)

If time permits, I shall also put in writing what I exactly expect from the Foundation. As to yourself, you know how I value you; you are also at this time the closest. To you I always opened all my files. I wish you would be the one to organize my archive. I never promised Alfred anything concerning the disposition of it, though we discussed its lodging at Princeton University. Most offensive to me is his reference to my "bargaining": I never responded to his many approaches...

Juergens then writes to Deg and passes along a never-sent but typed letter to Deg from V. with the hand-written notation *"This transcript of a letter drafted was not mailed nor typed — it dates from probably 1967. I.V. November 26, 1968."*

Dear Alfred:

Yesterday evening when I was already preparing for sleep I had your telephone call. Elisheva listened too. You told us of your plan to incorporate a Foundation for studies in modern science. At your last visit about a week ago you first mentioned of some steps taken by a partner of yours to charter a search along the lines pioneered in my books, thus to exploit possibilities now neglected because of the inertia or even opposition of scientific groups or the entire scientific establishment to new approaches and especially those embodied in my work. You told me yesterday of the founding committee that you intend to convoke in a few days — two names out of the business world, unknown to me, but also Livio and Ralph, and a few more. You indicated that I should at some point assume honorary presidency of the new venture. A new publication should be one of the projected activities. Organizing of my archive, another project.

I was through with my sleep at 3 a.m. when Elisheva that did not yet fall asleep came to discuss the project. Her thoughts and mine (crystallized by the sleep) were very similar.

The positive in your plan needs not to be recapitulated by me for you. But here are the adverse conditions.

For over a quarter century, since 1939, when I came to this country and dedicated my time to research in ancient history, I carried the material load of existence and study and writing with their concurrent expenses entirely by myself. This, at the end, gave me great satisfaction since alone and a stranger in the land and facing since 1950 the concerted opposition of faculties, scientific societies, and scientific publications, I now find myself in a changing climate, even though animosity in some circles, or among some individuals is even more vitriolic than before, but this can be recognized as a defense mechanism.

Should your Foundation and money drives be instituted, the following will occur:

1. My adversaries who tried to present me as a charlatan but could not point to any unproper action on my part, would be supplied with ammunition — a money collection [sentence unfinished]

2. Scientific organizations like American Philosophical Society or scientific publications, like Science of AAAS show recently some changes of heart; this mimosa-like attitude would be very sensitive to any activities [sentence unfinished]

3. Also many of my friends and followers would experience some shock if they should feel that a monetary pursuit under whatever guise accompanies my work and I would feel embarrassed.

4. I am most averse, even afraid of being made affiliated with other, so numerous, unorthodoxies. Through these years I am under an incessant barrage of such proposals to study the works of others, and in some instances what is known as lunatic fringe. The Yale Scientific issue caused a flow of letters to the editors from various individuals with appeals to have their theories given similar handling to that given to mine. I found often in letters claims that the writer is in the possession [of ways] to prove me right (as if I failed in this) or to improve my work by modifying it.

There are, no question, other worthy unorthodoxies. But I wish to continue my progress not burdened with the defense of others, like say, the organon theory of the late W. Reich. A foundation for studies in new [word missing] cannot close doors to new ideas; I, however, cannot and wish not to become a pope for all malcontent.

5. Organizations, like foundations, from the start or after a while, institute salaries, incur liabilities, oblige itself [sic] for grants etc, and should the organization be intimately connected with my name, it may disband under conditions of insolvency, after a promising start, causing an irreparable damage to my cause.

6. The small organization of Cosmos and Chronos groups is given to my close supervision and I feel quite comfortable in separating my scholarly pursuits from the work assigned to Cosmos and Chronos extending it to [sentence unfinished].

I know that S. Freud and to even greater extent C. Jung made use of donations, usually by their ex-patients, to establish schools of their respective modes of psychoanalysis or for publishing magazines. But their activities were not in the form of solicitation of funds.

In the morning after your call I drafted this letter to let you know how I feel.

Deg's journal, November 30, 1968

Yesterday was one of those fine mornings when most things seem to go wrong, but I didn't much mind. The mail brought a batch of documents from Ralph Juergens — the gist of which was that Velikovsky was deeply perturbed by my ascerbic letter to him of ten days ago. V. had promptly asked to see Ralph's memo describing V.'s thoughts. Then V. wrote a letter indirectly answering mine, and implying that Ralph has misstated his position, etc. etc. V. added a newly typed version of a letter that he said he had once written me but never mailed, full of forebodings concerning my establishment of the foundation, together with a letter from Arens of Gimbel's of Philadelphia, also full of doubts about the wisdom of proceeding with a foundation. All of this was to justify V. in the face of my attack. I know V.'s pattern of responses so well now that I could tell there was nothing new about the whole business. He writes everything down to have it on paper for some future stratagem. He warns against everything to be ready to be proven a prophet should things go badly. He cannot let go of any power over things or people, but plays upon every means of entrapping and embroiling them, sucking them in and pushing them off as he feels the one way or the other in his succession of mobilizing-for-action and trust-nobody moods.

I phoned him and visited him in the afternoon. I brought him the copy of *Etruscan Tombs at Sesto Fiorentino* which Prof. Nicola Rilli had inscribed to him, and he surlily carped at every point of Rilli's development that I brought out. 'Very risky,' 'I don't think much of him from what you tell me.' 'He does not seem to be a scholar.' 'He has very little evidence for what he is saying.' We finally got to the sensitive subjects of the flurry of documents. He claims his position has never changed. I said, 'Very well, you need not have anything to do with the Foundation, but if you wish to write articles for it or refer people to it, or receive support from it, you are welcome.' He agreed. (He will, of course, not keep his agreement, but will intervene at every opportunity.) I offered also to turn the Foundation over to him completely and let him designate someone to carry it on, but he refused that. I said, 'Please name those men and foundations whom you do not wish us to approach for support.' He would not do that. I promised that his name would not be used in support of the Foundation, which satisfied him. I know what he would like to see happen: the Foundation helping him in every possible way, but he criticizing it constantly for its faults. And provided it does not demoralize others, I do not mind. I have from my first meetings with him concluded that I should do what I thought he basically would want and weather as best as possible the gloom, the negativism, the wounded shouts, the suspicions, and the ingratitude.

We drank a glass of dry white wine (the Israeli wines are becoming excellent), and he showed me a few late letters, as he usually does. With some emotion he declared that, for all I have done for him, he was going to give me sooner or later the whole history of the case — the reception of his ideas by science and the public. I didn't feel as grateful as I should, for I need nothing so little as another pile of documents and a book to write, though it be the richest such case archive in history, and I thanked him. I prepared to leave, bidding Elisheva goodbye, and he stepped into the next room to get something. When he came out, I stepped close to him and said 'You know, there is nothing that you can do that will drive me away.' He said ' I will read you a line of poetry that you wrote' and quoted "the most opposed will most believing be." 'Not a bad line,' I said, smiling, and bid them goodbye again.

Deg's journal, December 1, 1968

The Foundation Trustees met today and perused the volume of recent correspondence relating Dr. V. to FOSMOS. They agreed that his conduct was sick. Still Juergens and Stephanos are under his thumb. I pointed this out and questioned whether the Foundation should not slow down its program for a year until everyone clarified their position, especially Dr. V. But we decided to move ahead anyhow, and suffer V.'s conduct as well as possible.

The more I think of his behavior, the more indignant I become. Every kind of evidence comes out in his letters, actions, and the experiences of others. Today he told Juergens that the Foundation should get another box number, because he wishes to go ahead with his absurd, presumptuous, and self-glorifying Cosmos and Chronos 'Clubs' (of which, in truth, none exist). Day before yesterday, he tried to buy my loyalty by the gift of his papers and documents on how science received his works, 'only for you, not for the Foundation.' A great collection, but I wish it for others to use, not myself. He is incredibly obtuse on some matters, I try to love him for his faults, but they are too numerous and large to embrace.

On Dec. 1, the Board of Trustees met in Princeton at Deg's home, without the important presence of Mainwaring and Holbrook. Nor were Kallen and Hess, who played no part in these proceedings anyhow, present. Juergens carried a new letter from V. to the Board, divorcing himself from the Foundation, which, as he asserts, he had never been married to in the first place but with which he is hoping for good relations nevertheless.

I repeat the following from the Minutes of the Meeting:

> An extensive discussion developed around the subject of the Foundation's relations with Dr. Velikovsky. Juergens reported that Dr. Velikovsky was of the opinion that FOSMOS' aims and activity were to deal only with such work as concerned him directly and as he might approve, and that FOSMOS was changing its direction since its inception.
> The President moved that, after examining the record, the Board resolve that the Foundation had not deviated from its original aims, which remain unchanged and are reflected in the following description offered by Stecchini, plus the subjects of 'communications of science' and 'science of science':
> The Foundation is concerned with conducting and aiding in the investigation of theories
> A. That the geophysical and astronomical history of the planet Earth has been characterized by sudden changes;
> B. That these changes have taken place in historical times and, as such, are documented by historical records, archaeological findings, mythological traditions, religious practices, and scriptures; and
> D. That these changes have affected the human psyche and affect contemporary social behavior."

Afterwards, Deg addresses V. once more, to tell him that the Foundation agreed with him and had always pursued the course that he now was advocating.

And then Deg receives a rather surprising letter from Stephanos who now becomes the instrument of V. in a new way; he lists his benefactions from V. as if he were under hypnosis, and declares:

> ...I must state that I find your letter to him [Velikovsky] misdirected (it should, perhaps, have been addressed to another), and in its tone, totally unjust and unwarranted. I believe it could be damaging to the interest we all claim to share, the acceptance of Dr. Velikovsky's work, and capable of great personal harm to him and to his good name.
> Since I was privileged to receive a copy of that letter (...) I want and do here deny its content as my experiences allow, and respectfully request, as a member of the Board, that you write a retraction to Dr. Velikovsky as soon as possible...

Deg replies to him:

> Dear Bob:
> I am afraid that your letter to me of December 5 and the circumstances of its preparation tend to confirm the contents of my letter of November 22 to Dr. Velikovsky.

It also indicates that Dr. Velikovsky should probably not have circulated a personal letter.
But thank you for your concern, I am sure that all will end well.

    Sincerely yours,
    Alfred de Grazia

  It did end well enough, except for poor Stephanos. The Foundation moved along cautiously, doing only small projects such as disseminating materials on the Velikovsky Affair, supporting Eddie Schorr's work on the Greek Dark Ages, and soliciting memberships. It was disturbed by a new attitude that V. had taken toward Stephanos, hitherto his most faithful and welcome disciple. He seemed to believe that Stephanos had encouraged persons from the lunatic fringe to become followers of V. and was giving them inside information of V.'s activities and archives. V. wished to dissociate himself from Stephanos and expected the Foundation to do do, too. Sizemore stuck up for Stephanos in private conversation with Deg, who sensed no great loss should Stephanos resign. Then he saw Sizemore's point — Stephanos should not be sacrificed to V. — and did nothing. Stephanos resigned anyhow. By the following Spring, Deg was withdrawing, too, as this Journal entry of April 19 seems to indicate.

  On occasion, Dr. V. and I have discussed a biography in dialogue form. But the three occasions on which we went to work with a tape recorder were disappointing to me. He becomes stiff, even more aware of his role and audience, and, though I try to break through my informal comment, he remains fixed like a peasant before a camera.

  I have not seen him in several weeks. My own problems with women and children are many and my book *Kalos* cries for completion. Immanuel's magnificent self-centering is not consoling or even rational, under the circumstances. I have ceased completely to work on FOSMOS, in part because of the foregoing, but also because the members of the Board were not up to editing a Bulletin, or raising funds. Bill Dix [Director of the Princeton University Libraries) told me, too, that the Velikovskys, during V.'s illness of December, had sought to give (with tax deductions well in mind) V.'s archive to Princeton University. Yet FOSMOS was to have been the beneficiary.

  Holbrook took over active management of the Foundation, working out of his new office in Washington. He did not succeed in developing it well, and, by general agreement, it was dissolved several years later.
  V. was doing well enough as his own majordomo as we discover when we read Deg's Journal of October 7, 1972 in Princeton;

  I borrowed Jill's bicycle and rode it to the Velikovskys. Francie, whose memory of me hardly dims with my long absences, loped alongside. Velikovsky was issuing directions to a University representative on how to set up the stage for a forthcoming lecture to the Graduate School Residence Hall Club. He spared the man no detail, prescribing publicity releases, and his desire to have his full first name spelled out rather than I. Velikovsky (is there a wish here to conceal the I, egoist, or the normal desire to spread out one's own name, as he said?). He requested that all his books and even a copy of *Pensee* dedicated to his work be on sale at the University Store beforehand; asked that two parking spaces be kept for his car and that of his daughter; wondered, since the British Broadcasting Company would be video-taping the show, whether the President of Princeton might not come if invited; denied a suggestion that a local radio station broadcast the speech but insisted that provisions for a televised relay into an adjoining hall be provided for people who could not crowd into the banquet hall. He stipulated that some announcements reach New York and Philadelphia so that disciples might come from those places to hear him. The young bald impresario left the Presence dizzy with details. V. is many things but he is also a master impresario. He has had to be; his overwhelming need to be recognized for what he is can only be satisfied by mobs of admirers under instructions which, given his detachment from the Establishment machinery, only he can provide, or by some wonderful stroke of recognition, a great prize like the Nobel Prize, the Fermi Prize, or an invitation from a head of state to deliver a series of lectures. I believe that he would then retire from his promotional labors and give himself over to finishing several important books.

I thought so yesterday as I watched him masterfully, but yet exhaustingly, promoting himself and his work, and later privately conveyed this thought to Sheva, when he had gone up to nap. For when the door closed on the graduate club representatives, he sat back, listened to me for a few minutes, ate an apple, and began to doze. I enjoyed the chance to talk to Sheva; she can tell me less flamboyantly all that has happened on their trips and where all the characters of the drama of recognition are at the moment — Mullen and Schorr and Bucaloe and so on. I borrowed a book and biked home to Mom. After dinner, Immanuel called to apologize for falling away from our conversation and I assured him that I was delighted that he could sleep well and hoped that he would always behave in exactly the same way. I had mentioned to him that I contemplated a little book of forays into myth, science and our adventures over the past decade of our friendship; he wondered how I could write it without his archives. I can imagine how I might, but if he would dig into them a little, my work would be greatly improved; I did not, however, suggest that he give me materials. I shall show him the table of contents when it is sufficiently elaborated. Then, if he wishes, he may find some material that would help me.

Deg is living in New York City, and only visits Princeton on occasion now

Deg's Journal, October 23, 1972

I telephoned Velikovsky at 10PM to see how he was. He was well. We talked of the book I intended to write. When I said that I was investigating *Hermes* he warned me against starting to repeat his work of 20 years. I guess he'd like me to ask for his files and then trap me into an endless affair. I said, don't worry; I have only in mind making several penetrations in depth, at widespread points, to show the method that should be followed to mine the ore. He said that he couldn't "approve" my book unless he read it. Of course. And no doubt there are some bouts ahead. In general, he likes the idea that I will write the book.

Then I gave him some firm advice. I said "you must finish *Peoples of the Sea* and the *Ramses II* volume promptly and publish them. You must not lecture and run around. Ten people can go around lecturing about you but only you can finish these books. Furthermore, you must not work on the Einstein book, or *Stargazers and Gravediggers,* or *Ash.* These can be finished by someone else. You must write something, if only 30 pages, on your theories of what happened in the skies before Venus in 1500 B.C." He agreed, "You are right!" He added, however, that he must write his autobiography because nobody knows him really or how he did his work. He only let out a few facts here and there. Alright, I responded, add that to your required list, following the ante-Venusian article. But that's all. "You're right!" he said again, with unusual accord. And so we left the matter, saying good-night.

P.S. V. told me that Harlow Shapley had just died at a nursing home in Colorado. After reading the extensive obituary in the New York Times, V. concludes that Shapley, always a great self-promoter, had seen to it that the Times possessed his own account of his life. Thus Shapley hurls his last insult to V. from the grave.

Again on November 9, Deg exhorts him:

Had long telephone conversation with Velikovsky. He was in a grim mood. I tried to cheer him up. I also read him the list of chapter titles for my projected book. He said a few approving things but generally he was critical, full of admonitions, careful of his own sources of information, making no generous or even modest offer of assistance, wondering how I could have any new ideas (though he did not say this explicitly) when he had had them all, and in some manner had *published* them all.

I don't know how he expects ever to encourage serious efforts to follow or parallel him. He beseeches this from the world but then denies in advance that they can either be original or important.

I tell him to move rapidly on his theory of the pre-1500 catastrophes — to publish at least a synopsis of it, lest he accuse even his supporters of plagiarizing him. All I know of this work are a few remarks of John

Holbrook relating essentially the truth of the Greek theogony — Uranus, Chronos (Saturn), Jupiter.

I am telling V. that if he doesn't do something soon here instead of parading around the country *he* will become a successor instead of a predecessor of someone else. Further, *his* predecessor will probably do a poor job because V. has withheld his information and assistance.

And he is concerned whether V. will be elected to greatness:

Deg's journal, November 72

I.V. is running for election. The office he wishes to achieve is Premier of 20th Century Science. I believe that he has as good a chance as anyone up to this time of winning the election.

However, I am not a campaign manager. And though an election in science is unfortunately like a political election — in that a campaign biography should be written that will show the candidate in gorgeous lights — I feel I must pass up the chance to win glory as a publicist. My interest in biography is as Conant [President of Harvard University and chemist] once put it, to find the full meaning of science through its means of creation.

Immanuel V. as I see and know him is here and you must understand to begin with the fact that no person can fully know another one.

Problems of health depressed V.:

Deg's journal, December 22, 1972

Called V. He is gloomy. The doctors told him that he must go away to rest. His days are full of calls, visits, correspondence — too much to handle; his writing lags .I invited him and Elisheva to New York for a day of rest and walking around the museums. Maybe. I also suggested he might go to Yucatan and see the ruins there. He doesn't "want to be carried around by the tour buses." "Let the buses go without you. Stay at hotels. Then provide and make your own daytime itinerary." He wondered when I would be in Princeton. I didn't know. I told him I would think of what he should do and would call him back.

The "Apollo" program suffers severe cutbacks:

Deg's journal, December 23, 1972

Called Stecchini. He is feeling better after a gradual six months' recovery from an old back injury. He said V. may be depressed by the closing down of the Apollo Moon project which, whatever its premises and procedures, had brought forward some support of his views. The signs of volcanic activity are still being reported, though their time of occurrence is naturally placed conveniently far away — 100,000 years, 500,000 years, their freshness suggesting "recency," but recency being defined arbitrarily on the lengthy geographical scale. If 100,000, why not 3000? No answer. No question, in fact, by anybody, save the Velikovskians. Cape Canaveral (Kennedy) is already being dismantled. The scientific community did not rise to the occasion, said S. "I didn't rise, either," I said. "It was a great waste of world resources." He half agreed.

Deg worries both about V.'s health and his attitude towards a friend:

Deg's journal, December 26, 1972

Called V. again yesterday. He is more cheerful, but says his diabetes is moderate, not light. He is grumpy over the stricter diet he must follow. He asked me about all my children and I recited their whereabouts and conditions of life. He asked whether he could help me. I should have said, "Yes, let me read your pre-Venus notes and correspondence." I didn't. He wouldn't; not now. He would ask me to show him all of my ideas. I would do so, but he might well not reciprocate and, even though his materials must be better than mine on the whole, he might very well absorb them and simply lock the gate on me by putting me onto this or that matter stretching on endlessly. He cannot help himself. He is authoritarian. And he finds it difficult to think that anyone in the world but himself can supply anything but a few details nor indeed should until he has breathed his last word. This kind of game seems bizarre between friends, but the reason I am perhaps his closest friend and oldest one is that I recognize his character and am not vulnerable to shock by its exposition. As certainly as the sun shines (sic!) he would reject my work repeatedly, absorb all that he had not known, and accuse me in the end of plagiarism.

V. begins to exhibit alarming symptoms:

Deg's journal, February 10, 1973

Velikovsky Visit — V. not well at all. Extremely nervous, thin, paranoid cryptic references, taciturn, jerky movements from time to time. Is diabetic. Asked him whether 10 years of good work might reconstruct 10,000 — 600 B.C. He didn't have an opinion. He said he doesn't know whether deluge was 4000 or 9000 BCE.

Deg's journal, February 19, 1973

Called Velikovsky at 5 PM. Says he is feeling better, but is having troubles with "people." Has matter of importance (ominous tone) to talk over with me. If I want to hear it, I must come to Princeton tonight. I tell him it is difficult. Won't tomorrow night do. Maybe. "Who is it?" I ask. "Can't I help." "You come." etc. All remote, intimations of disaster, confusion of personal and the world and of all of past with the present. I try to talk of an article about Mars. 'The author believes in all miracles except yours.' He's not sure he read it. But uninterested really. He is involved in his personal huge caravan of suspicions, lawsuits on his house in Israel (so Ruth tells me to make clear his references), forebodings of catastrophes, possible suicidal impulses (my enemies wanted their martyr; now they have it). Nina hands me a note as she overhears me: "Do not try to get abstract conversation. He is trying to talk about himself." But he is uncommunicative. Finally, I leave it that I may come tonight or in the next couple of days. He is reluctant to close but finally I end the call.

Called Ruth Sharon. Father not feeling well. Diabetes out of control. She tells me not to go to Princeton. He will be better and there is nothing I can do. I tell her I fear he will regress irretrievably. She cannot answer to that. She says he may even resent me later if I see him in weakness. I tell her I am more concerned with whether he will be helped now if his situation is serious. Maybe she and her mother cannot suffice to pull him out. I ask her to call her mother and if they want me to come to call me.

8 pm. Ruth calls me back. She has talked to her mother but her father hung onto another phone throughout the conversation. She says, however, that he was feeling a little better and was thinking of driving out to purchase several articles. So I should call and give my regrets at not coming.

8:15 I called V. Sheva came on the extension phone. I said I had not finished my proofs that had to go to India and asked him to excuse me if I did not come this night. He assented. I said further that I did not wish to see him before I could show him an outline of my work on pre-history. He replied that he would have no time to read it, for he was so behind in his reading. Sheva interrupted gracefully to say that it was a short piece and I hastily agreed, saying that it was only a page or so. He said nothing then; I uttered a few additional inanities and hung up with the promise to see him soon. He sounded a bit stronger of voice.

V. then recovers:

Degs journal, April 4, 1973

I phoned V. this morning and found him much improved since my last call before leaving the country. Three weeks in the hospital had somehow restored him. I said, "Life without a telephone to bother you was good for you." "No, I had telephone. I took my calls."

Anyway, he is better and will drive perhaps to Youngstown, Ohio, for a speech next week. He is working on Ramses II again. He is pleased that Carl Sagan is writing an article for *Pensee* on Venus. He agrees that I shouldn't bother with book reviews for *Pensee* but should present a significant paper. Maybe I shall get down to preparing one.

He is hopeful. He speaks of particular tasks. He has even begun rearranging some files. It is a great relief.

Bill Mullen is getting ready to move from Princeton University to a new appointment at Boston University. He is glad to be away from V.'s moods. He writes to Deg:

August 12, 1974

...The summer has been curiously unproductive and jammed as far as Velikovsky is concerned. He has spent virtually all his hours talking about what he is not accomplishing and bewailing the magnitude of the battle against his enemies on all sides. I've contributed only bits of help here and there, otherwise being forced to concentrate on preparation of this fall's courses. Eddie [Shorr] has been of tremendous help, spending day after day in the library going through The Peoples of the Sea with a fine-tooth comb. But here too the result has not been of the kind to cheer Velikovsky up, since Eddie has found many minor errors which need correction. Nothing that shakes the reconstruction, just a lot more nitpicking work that really has to be done if the book is to be spared the dismissals by Egyptologists on the grounds of inaccuracy which are feared. In short, be thankful for the serenity of Naxos, Al, since little would have been gained by being close to Princeton this particular summer.(...)

* * *

But V. reorganizes his forces and this time calls upon Irving Wolfe, who graciously responds by addressing Mullen, C.J. Ransom, Juergens, Rose, Steve Talbott and Milton:

Dear Alfred,

I visited Velikovsky last week, along with Lynn Rose and Earl Milton. We discussed several matters with him, among which were
— the number of books he's working on at once
— his archives and related issues
— he wants people to submit and keep submitting articles on or arising from

his work to scientific journals, whether they will be accepted or not
— setting up a Newsletter, about which several steps are being taken
— public recognition for advance claims and theories

You will be familiar with most of these matters already, but I've drawn your attention to them because I think we need to get a number of people thinking about them and coming up with solutions, because Velikovsky can use help in all these areas.

With regard to the last item above, here is an example — the recent discovery of substantial quantities of argon and neon on Mars seems to puzzle scientists, as an article in *Science,* June 21, 1975, indicates. Yet Velikovsky predicted argon and neon on Mars as far back as 1946. Key scientists must be given the facts — dates of original advance claims, letters, confirmations, etc. — and urged to write the major scientific journals. Velikovsky feels he's too busy to do this himself each time, and so I've offered to handle it for him, telling him that, wherever a case like this arises, he's to send the relevant documents to me and I'll compose a covering letter and send it all out to the right people.

This is where I need your help — I want to make up a master list of key people, perhaps divided into two or three categories, to whom such things can be sent as each occasion arises....

Deg could imagine the huddle at 78 Hartley Avenue, planning the counterpropaganda campaign, the "truth squads" as the Republicans and Democrats had come to call their counterpropaganda teams. Next year, Wolfe was calling for an "alarm system" which he had worked out with Milton in Canada. It was to be a network, highly sophisticated, with members divided into generalists and specialists, with squad leaders who would call upon their assignees to respond to the alarm. Wolfe had been called by V. to activate the system, as he had promised the year before, and V. nominated as a test alarm the publication by Doubleday of *Immanuel Velikovsky Reconsidered,* which should exercise the network to produce reviews, letters, and public discussion.

This meant helping the Talbotts who were otherwise blacklisted by V. and several of his circle. "Regardless of what any of us feel about the Talbotts," wrote Wolfe, "I agreed because Velikovsky asked." (Actually, I doubt that Wolfe ever felt antagonistic towards the Talbotts himself; the plea was for others.) "He (V.) may feel that he wants to aid the success of that book because it will affect his own case." So the Talbotts and the inner circle were momentarily in bed together again, an event that had not occurred since the Talbotts' *Pensee* had collapsed. The results were not remarkable, and after a time they got out of bed.

There came a lull in attempts at general organization; V. continued to turn his attention and the minds of his several collaborating followers to the AAAS affair, a story to be told later. It is noteworthy how much time was taken up with all the maneuvering, research, writing, and wrangling connecting with a single sitting of an AAAS panel in San Francisco, much of five years of V.'s time and of the time of several others, the time too of Elisheva, but who counted that? — more hundreds of hours flushed down the drain; there the tragedy is marked, for she was a sculptress and musician of consequence.

She never complained, so I am reporting Deg's complaints on her behalf, unsolicited. Moses would have been pleased with her self-sacrifice; Deg was no Mosaist. When she lay dying after a long illness, and he had not seen her for months, he thought to write a poem for her.

Then came the infatuation of V. with Christoph Marx, and following upon Marx' return to Switzerland, V. addressed Lynn Rose, who was perhaps feeling both grumpy about the affair and pleased that suddenly V.'s attention was turned elsewhere. However, V. was writing in a euphoric mood, and one could see the alarm bells ringing around the world.

The letter to Lynn Rose is dated May 11, 1977, and I summarize it. Marx was to be "a central figure" on the European continent; Isenberg sends a paper he gave to a conference of science editors and V. urges him to send it to the major hostile magazines — *Nature, Science, New Scientist* and the *Bulletin of the Atomic Scientist,* "as coming from the convention." ...A letter from Langenbach, a supporting attorney working in the Harvard scene... A call to William Safire of the *New York Times,* a self-designated "great fan" to get advice... An announcement that Juergens has resigned his engineering job and would probably now work for him, V.... A hope to teach a course in Egyptology at Princeton University... A report of Deg's taking issue with Lustig of the *Encyclopedia Britanica Yearbook...* Last minute changes to the English edition of Ramses II... A carpenter-mason is building a room for guests and Elisheva's music... A letter from the widow of maligned Harvard supporter, Professor Pfeiffer... Mainwaring will be sending a complete file of all C14 communications with the British Museum and the University of Pennsylvania Museum... A conversation with Holbrook, once more in Washington... A gift of Czech rights to Jan Sammer who helped so well with Ramses II... Some minor foreign rights also to his early copy-editor Marion Kuhn, now ailing... Reporting plans to sponsor publication of Alice Miller's Index to his works... Detailing the distribution of 1000 free copies of *Kronos* to college libraries, financed by Jerry Rosenthal... Denouncing Steve Talbott for recommending in a pamphlet that all subscribe to *The Zetetic Scholar* which has recently defamed V... Urges that the five former associate editors of the now defunct *Pensee* "should make a common statement and try to teach the subscribers of *Network* (Talbott's serial pamphlet), deluded into believing that the *Network* is an organ to defend and protect my work.... Dr. Gowans of the University of Victoria "comes back to the fold" after consorting with the likes of Dietrich Muller of Lethbridge... An exchange of letters with Jacques Barzun... Reports that *Peoples of the Sea* just released has already outsold *Earth in Upheaval* (11 printings since 1955) and

Oedipus and Akhnaton (12 printings since 1960)... He resists Doubleday's efforts at putting *Peoples* into a book club as an alternate selection... Ramses II is to be delayed once more, this time by the publishers... He is happy that his British publishers, Sidgwick and Jackson, have given full prominence to his *Peoples* while somewhere in the nether pages "Patrick Moore is modestly displayed for his '1978 Yearbook of Astronomy,' and has to take this pecking order, he being the author of 'Do you speak Venusian?' presenting me as a King of Fools"... More letter exchanges... He doesn't want Rose to be distracted from their plan to write together "The Grand Ballroom" dealing with the AAAS affair which was already the subject of several books and many articles... "...The hammer of the builder sounds like a song... do you know that my real vocation is in architecture, and the years that I visited the Library on 42nd Street, I regularly visited also the room with architectural journals, watching for a chance to compete for a plan and construct a public building?"... "Keep well, act strong, Lynn."

V. was obviously in fine fettle. The Mastermind was back. He had a great deal going for him on two continents now, it seemed.

The euphoria subsided. The resistance to all of his ideas continued unabated. It seems that he could say nothing that would be right in the eyes of his opponents. His growing disenchantment with Christoph Marx was not compensated by new faces. (New ideas were out of the question: proofs were wanted, and defense.) He had now close to himself principally Greenberg and Sizemore; for them *Kronos* was not fun and games anymore. On June 3, 1979, Sizemore writes Deg, "This issue is going through hell — trying to get V.'s approval on Lew's article about the latest probes."

* * *

By now I believe that you and I know enough of the principal characters here to venture a more fundamental answer to the question which I dealt with unsatisfactorily at the beginning of the chapter: why did Deg stick with V.? It appears that the two men were close to each other even when separated and out of touch. I conclude that there was a familial relationship being reenacted between V. and Deg. It was not father to son, but older to younger brother. In significant ways V. was of the character of Deg's older brother Sebastian, and Deg was relating to him as he had to his brother throughout life but especially from two years to twenty years of age.

It was as Lasswell somehow discovered, a sibling rivalry between Deg and Sebastian, more intensely activating for the younger than the elder. No matter what Sebastian did, he couldn't put down his younger brother; and his younger brother, while trying to outdo him, was absolutely fond of him and set him up as a model for others, to be surpassed only by himself, and he was determined all the while that noone was going to put down

Sebastian so that there was a strong protective impulse going incongruously upwards — material and demanding — rather than downwards as one might expect.

V. had two older brothers, neither of whom he saw after 1921 and with whom communication was rare, if only because the "Iron Curtain" barred East from West and he said once to Deg, speaking of his scientist brother, Alexander, I would not want to jeopardize his position over there by reintroducing myself into his life.

And Sebastian and V. were of the same rawboned, tall and handsome physique, unlike Deg's more compacted form and features, and both were umbrageous, too. Both felt that Deg could do anything he set his hand to, but that he was always off on some wild goose chase when you needed him.

There were of course differences. However the song goes; "I want a girl —just like the girl — who married dear old Dad," no girl is ever quite like Mother; and so with siblings, no two sibling relationships are quite alike. The major differences were two: like Deg, V. was fantasmogene: he day-dreamed much and often and dueled with the universe of nature and men in his mind. Sebastian was not a dreamer. And, further, V. was there, in place, at home; for seventeen years Deg knew where to find him at Hartley Street whose number he could never remember, and that he would be welcomed like a brother, which, no offense intended, he could not always count on from Sebastian.

I think that the crux of the relationship, that which proved its psychogenesis, was the fact that Deg, unlike so many of the cosmic heretics, could be constantly critical of V. without risk to his affection for V. Then, too, while V. would never let Deg take away his toys, nor admit that he was equal, he would not stop him, short of outright usurpation of his position and place, which Deg in any event would never wish to do. Indeed, one of Deg's main virtues and weaknesses in human affairs, if it can be called that, was that he would often win a contest but could never administer the *coup de grâce*. Neither V. nor Sebastian lacked this capacity except in the case of their younger brother.

Sebastian never became friendly with V. but supported him quietly, just as he never committed himself to Deg's efforts on behalf of V. nor to Deg's quantavolutionary ideas. He engaged himself mildly one time in their futile effort to obtain an honorary doctorate for V. at Rutgers University. Another time, when Deg was abroad, Sebastian, perhaps prompted by his wife Lucia, thought of getting V. and Elisheva together with the Director of the Institute for Advanced Study, Carl Kaysen, Ambassador George Kennan, and their wives. Perhaps V. should be invited to join the Institute (which would in fact have been an ideal place for him and ideally in keeping, too, with the institute's professed aims). Elisheva and Immanuel were irritatingly preoccupied with the menu for dinner, however, and settled finally for a visit during the cocktail hour, which went off nicely.

***

Deg's communication lines generally thinned out in the years 1976 to 1983. Even his lateral communications in quantavolution dwindled as he pressed to break through with the several large studies underway. Here he is writing from Naxos to Professor Ernst Wreschner in Haifa on December 21, 1976:

> "I am returning from three weeks in Mexico as a guest of the government. I attended the inauguration of Jose Portillo as President, gave a paper at a special conference on the 400th aniversary of Jean Bodin's Six Books of the Republic (author of my least favorite doctrine — absolute sovereignty), and visited a number of Olmec, Maya and Aztec ruins and sites. It has been a good trip and I found a considerable interest in translating my political works and even some surprised involvement in my questions about mythology and catastrophes. I did not find the lost tribes of Israel but perhaps learned something of pre-"Atlantean" survivals. I also had a car wreck (I was not driving), had my wallet stolen by a large fat Indian lady with an overpowering smell that put me to sleep on the bus alongside her, and then later on my little camera as well (before I could turn around, the pickpocket had dived into the marketplace mass). *C'est la vie.*
>
> With luck, by late spring I shall have a general manuscript ready on the holocene destructions and human development and will send you a copy. I hope that my present letter finds Ella and yourself very well and in good spirits. I have resigned all teaching at NYU and am now free to give my time to research and writing and perhaps sometime to a visit to Israel, unless you meanwhile visit here.(...)

Deg showed his materials on Homo Schizo to Harold Lasswell who approved their significance. Deg wished he might get the famed polymath involved in seeking the origins of the human mind, even in contemplating quantavolution, for Lasswell was as much a fantasmogene as Deg. But not long afterwards, Harold Lasswell climbed into the bathtub of his apartment overlooking Lincoln Center, suffered a stroke, and spent two helpless days in the tub before his apartment was entered. His friends rallied around and attended the cheerful but now addlepated great man until he died. Deg hoped he had not been unkindly critical when they had last been sitting, drinking whisky and looking down upon Manhattan, for he had been suddenly seized with impatience when Harold spoke of a great new understanding overcoming the medical profession owing (by inference) partly to the introduction of techniques for better human relations in complex technical situations (in which he was playing a part, as always) inasmuch as Deg felt like raging — not only against the system of medical care, but also against the world at large for its frightful bungling.

When I went back in time for Lasswellian material related to quantavolution and the heretics, the latest was from November 4, 1972, when Deg's Journal reads:

> I met Harold Lasswell at the University Club at 7 and after two Scotches and 'what have you been up to' and 'what are families and friends doing,' we taxied to Washington Square, where Nina prepared dinner. She pulled out all the stops of her culinary organ and enthralled Harold with poached whitefish and freshly made mayonnaise, stewed hare, spinach and egg salad, Port-Salut, stewed pears in brandy, and a variety of wines and cognac. We talked until after midnight.
>
> He is looking as he has for thirty years. Still grey and pink, still ranging all over the world and talking upon every subject; the chasms of unintelligibility when he swings into Lasswellian sentences from time to time still enchant me. It was Nina's first exposure to them and she couldn't decode them.
>
> He described his unexpected walk many years ago up a set of 18-inch spikes hammered into the walls of Santa Sophia in Istanbul. He had a hangover from a night of drinking sweet Turkish liquor and could barely save himself from nausea, vertigo and panic. How I know the feeling. He talked too of a ride in a military plane from Paris to Vienna after World War II, where he sat on a metal bucket seat with two other men and watched a cargo of coffins creep through their bonds toward the freedom amidships.
>
> We talked of economists and he expressed his pleasure that the social sciences were being recognized for Nobel Prizes, particularly Ken Arrow and Samuelson, but his subtle manner of speaking, which one must watch carefully, indicated he was a little hurt that he who had achieved so much for the social sciences had not been recognized with such a prize. I agreed with him, without mentioning the matter; what a corrupting influence the Nobel Prizes are; they pretend to omniscience, in whose name, on what grounds; what presumptuousness.
>
> He is now working on a Policy Sciences Center, promotes a world university, heads a Rand Corporation Board, etc. etc. He was delighted with my stories of the University in Switzerland and would have gone the whole evening on the subject.
>
> His mentioning Arrow and Samuelson came when I reflected upon the betrayal of human economics by the economists. I explained my struggle with Scott-Foresman over publishing a chapter on economic policy and especially on a guaranteed income. Harold says that A. & S. and others just published a statement indicating their adherence to such in principle. I should use it to back up my attack on the subject.
>
> I mentioned my advice to Velikovsky to publish *now* instead of awaiting the 'no-mistake' nirvana; H.L., who feels a certain competition, insisted that I was right, that V. wanted to be God, that it was unscientific, that no man could expect his work to stand free of error indefinitely, that the courage to err was the glory of a true scientist.

Lasswell spoke of a book called *Chariots of the Gods* by a Swiss, who apparently believed in the depositing of inventions upon Earth by superterrestrial beings. I thought this was a modern version of the gods of the Greeks descending at will upon earth bringing discoveries as well as evil. I added that I am pursuing a theory that the flowering of certain early metal ages came in consequence of the showering of metals upon earth from comets and meteorites.

Probably I should add a chapter to my book on the descent of the Metals. If the metals are heavy, they should have sunk to the core of the Earth's molten mass, never to surface again. Why should in theory the earth's crust contain them? For noone says that the turbulence of the crust descends to a greater depth.

Before our last cognacs had been finished, we spoke of the family system, Nina presenting the nostalgic view of the extended family, Harold asserting that the blood family has little to offer any longer, while admitting her argument. He described his early family — he an only child, but with numerous relatives, now scattered from the Midwest to California and Florida, those graveyards of American families. I had been urging him earlier to write his Autobiography; he is silent about his past to an abnormal degree. He is noncommittal. Perhaps he prefers to remain a Great Man of Mysterious Origins. Very well, but a good autobiography is worth more than a large question mark.

Washington, 1979

In Memoriam

HAROLD D. LASSWELL
(1902-1978)

Harold! Greetings!
 Snifting bubbles, are you, this season,
  in the land of the tall drinks?
 Are they pouring you doubles?

Come back to Chicago, Vienna, Nanking.

 Sounding like we know it all,
  in tones serene as your very own,
 We slump in low divans
  and hunch over brown tables
Spilling smoothly the news about how

THE THIRD WORLD OF SCIENCE

you walked upon the Earth once.

Welcome back to Washington, New York and New Haven;
 your train is set to run on time.

You said straight what you saw
Without hee-haws oinks or meows
No winks, curtsies, or knotted fists
No cow-eyes, or stony gaze.
Viel Blitzen, kein Donnern,
No "Ho-ho-ho."

Pleasant, agreeable Hero of our times,
 "if-then" propositions cornucopiously emitted.
Two pounds of value-sharing for all men alive.
Mix one pound of deference, a dash of income,
 well-being and safety added to taste,
Be generous with enlightenment.

 Now that you're not in it,
 More Seasoning is needed.
 Some of the gusto is gone.
 In-put, out-go.

Hearing the world's secrets and ours nevermore,
You heard them all, and those to come
 that we must explicate ourselves.
Thanks for configurating the futuristics.
Please to stay warm at the North Pole
 under your gray hair, behind your glasses,
 in your midnight coat. Your gloves are too thin.

Come home again, if you get the chance—
 The New Year is here.

So long, Saturn!

* * *

Deg's journal, November 18, 1980

It's cold outside. I received a letter from Gilbert Davidowitz' sister telling me that my letter to him arrived but that he had died 'of a heart attack' last July. Poor lonely mad scholar. He was only fortyish. He must have committed suicide. Never an academic appointment. Nothing published. Brilliant worker in the origins of languages. I immediately wrote Charles Lee [Director of the State Archives of South Carolina, onetime President of the American Society of Archivists] who will be startled to hear from me after 38 years, explaining my memorandum on the archives of the dying and their total loss to our culture. I feel extra sad about Gilbert, because he was so alone and so incapacitated for everything except the history of languages. But what a fine capacity. If he might only have known when dying how I liked and admired him. He must have known. But he needed just then to be told so.

CHAPTER THIRTEEN

THE EMPIRE STRIKES BACK

The asininity of the attacks by the science media and conventional scientists upon Velikovsky was consistent with book reviewing and editorial practices generally. Sympathizers of V. had an ample data bank from 1963 onwards from which to demonstrate that V.'s critics were brash, dogmatic, imitative, narrow, selective, unprepared, precipitous, vulnerable, incomplete, pretentious, possessed, unversed, unserious, unselfcritical, prejudiced, unsystematic, inexact, unphilosophical, ideologically scatomatized, vague and irrelevant — to say the least. Yet withal Velikovsky was said to have been "buried" not once but repeatedly, and all of his supporters with him.

In a field so broad, hundreds of major statements and thousands of details offered in over a thousand published pages somehow emerged unscathed. Several scores of statements were indicted for ambiguity or rendered more doubtful. What everyone knew ahead of time could be reasserted: the prevailing theory of celestial mechanics would only make nonsense out of the data presented. In addition, planet Venus probably lacks massive clouds of hydrocarbon; if so, either such clouds were never there or they burned off over time, the latter being V.'s second line of defense.

All in all, this was so small a bag that V., when it came time to write his address to the San Francisco AAAS meeting, ended it with the words, "None of my critics can erase the magnetosphere, nobody can stop the noises of Jupiter, nobody can cool off Venus and nobody can change a single sentence in my books." He knew the last expression was bravado, but he felt like sticking it in, so unsuccessful did he consider his opposition to have been. He asked Deg's opinion: should it stay? Deg was happy for the swashbuckling septuagenarian. Besides there was enough truth in it to let it go as the last firecracker of a speech that crackled throughout; Why not? Fling it in their teeth. And so it stands. Since effectively it says nothing and says all, who can object to it?

I have given much thought to what kind of review might be tendered V.'s books, such that his supporters could not assail on substantial or moral grounds but would not please them. I consulted Professor Joseph Grace, a historian of science, and he kindly wrote a review for our pages, holding to a 700 word limit, such as is common.

> "Velikovsky is a highly skilled and erudite scholar, who works comfortably in several major fields of science and the humanities. He has a style, an attack, that is primarily humanistic. By this I mean to exclude social science, which today has a format often resembling natural science, complete with jargon. He writes more like Ignatius Donnelly, a predecessor of a century ago, whose style is even more pleasurable. There can be only mild objections to such a style, considering the undefined and exotic, even occult nature of some of the areas he must venture into and the non-existence of a scientific language covering so broad an area. Of course, we would lose much in clarity and orderly communication if our students were to adopt it in all manner of writing.
>
> Velikovsky sees prehistory and protohistory as frequented by stupendous natural catastrophes that call into question the stability of the solar system over long time periods, and therefore the gradualism of darwinism in biology. His evidence is limited and fragmentary, much of it anomalies that puzzle historians both human and natural. Most of his evidence must, and does also, serve conventional approaches, our received knowledge, although he insists upon viewing it as catastrophic.
>
> His most radical hypotheses, which he expresses far too confidently, propose drastic erratic movements and changes of planets, particularly the Earth, Mars and Venus, not to mention the lunar satellite and the giant planets Jupiter and Saturn. The mechanics, even the electro-mechanics of such allegedly historical events are, if conceivable, quite unknown and undeveloped.
>
> Here and there in his works one finds nuggets of valuable ore, some in history, some in legend, some in natural history. One finds these days a plenitude of studies of meteorites and comets, a few of which he cites. One finds, too, many good works on historical and stratigraphic chronology, and it takes more than innuendo to shake the solid foundations of radiochronometry. One must be impressed, on the other hand, by Velikovky's ability to discover anomalies and contradictions, especially in Ancient History. He may well be on the right track in discovering continuities between Pharaoh Akhnaton and Oedipus, and concordances between the Biblical Amalekites and the Hyksos conquerors of Egypt, and even is stressing a baffling absence of archeological material to fill in centuries of assigned time in Egypt, Greece, and elsewhere.
>
> The reader will find many entertaining and suggestive pages as well. As

for his general ideas, practically none of them can be fitted into contemporary scientific theory. The more heretical a theory, the more hard evidence must be found to support it, and Velikovsky's ideas of an electrically run universe, which he never develops, and his claims of planetary aberrations in early times, to which he gives a great deal of attention, are, to put it mildly, bizarre; there exists, that is, no astrophysical theory to support them.

I would not recommend his books to anyone. Their pretensions will enrage the learned and confound the ordinary reader. Every age has books like them. I can mention Donnelly and Mesmer in the nineteenth century and George M. Price and C. Beaumont in this century, but there were many more, which are best forgotten. The genre is well known to science and historians of the most ancient times, and one can judge the future of the books by what has happened to their predecessors.

The fact that a great many people read such works tells us little about their value as science or literature. No doubt, in time, such scientists as can be spared from other tasks or are involved with his specific hypotheses will build up what would amount to a total assessment. It is certainly too early to assert, as Prof. A. de Grazia did after only a dozen years, that he is one of the great cosmogonists of the century.

What can be said for this review is that it gives a general impression of what is talked about in the books and how, and it does not challenge their right to be published, nor dismiss them as anti-scientific, nor berate the author.

When researching on the Velikovsky Affair, Deg stimulated V.'s interest in the techniques of suppression, putting into a framework the host of items which protruded from V.'s archives. Deg told V. of a favorite old book, Henry Thouless' *Straight and Crooked Thinking* and explained how it might be applied to V.'s experience. V. was excited by the idea and prepared a handwritten list of "70 ways of suppressing a theory," which the two men discussed. The list that follows is largely in V.'s words and idiom. It was not included in the published work. Each item is based upon one or more concrete instances that can be documented and dated. Later on V. wished to engage Lynn Rose in fleshing out and publishing the list.

Actions of Established Scientists
and Cohorts Aimed at I. Velikovsky and his Book
Worlds in Collision (1950)

1. Refusal to read or examine the manuscript.
2. Charging it was not presented to specialists before publication.
3. Refusal to help with inexpensive tests through established facilities.
4. Accusation that work was not offered for testing.

5. Assertion that work has been disproved by tests.
6. Efforts to discourage printing.
7. Demands for censorship.
8. Engaging in censorship.
9. Boycott of the book.
10. Boycott of all textbooks of the work's publisher.
11. Threats of reprisal against publisher by not offering manuscripts or withdrawing books.
12. Threat against associated publishers without text books.
13. Appeals to the scientific community.
14. Efforts to influence reviewers in advance.
15. Appeals to mobilize hostile reviewers.
16. Efforts to suppress favorable reviewers.
17. Efforts to supplant regular reviewers with volunteer authoritative writers as reviewers.
18. Checking the allegiance of scientists and officials of scientific organizations.
19. Firing of unaligned scientists and officials.
20. Punishment of book editors and firing.
21. Demand that there be a public recantation by publishers.
22. Refusal to print author's papers about his books in scientific magazines.
23. Return of supplementary papers unceremoniously without reading.
24. Refusal to reprint answers to distortion of facts in reviews.
25. Misquotation from the book, and quotations out of context.
26. Copying of wrong figures into a quotation used in the book.
27. No correction of erroneous statements in reviews by anybody in the scientific community.
28. Use of knowingly false argument.
29. Dogmatic statements and accusations.
30. Setting up and knocking down "strawmen."
31. Dishonest rejoinders.
32. Defamation and discrediting abuse.
33. Promotion of antagonistic critics.
34. Appeal to religious feelings.
35. Guilt by association.
36. Treating work by association with other ridiculed or denounced books.
37. Use of fallacious statistical method to decide whether a genius or crank wrote book.
38. Writing reviews and criticisms without reading the book.
39. Copying from other reviews (even of those who had not read it themselves).

THE EMPIRE STRIKES BACK

40. Innuendos that unneeded counterarguments abound.
41. Refusal by scientific periodicals to advertise the work.
42. Warnings against readers' inability to judge work.
43. Assuring the reading (and book-buying) public the book is dull and worthless.
44. Accusing author of using methods not actually used.
45. Denials of acts of suppression, compounding perjury.
46. Omission of credit or of footnoting the work when offering "new" theories elsewhere that are contained in the book.
47. Refusal to give credit for discoveries confirmed ultimately in tests.
48. Refusal of information to author.
49. Refusal to engage in communication with author or allies.
50. Suppression of news of disputes or debates won by author.
51. Deprecating value of crucial tests favoring author's theories.
52. Concocting stories that "1000 wrong predictions" were in book.
53. Defamation in letters and intimidation of potential support.
54. Use of great names (e.g. Nobel Prize winners) for defamation.
55. Whispering campaign; private letters.
56. Intimidation of students, both undergraduates and graduates.
57. Elimination of the name of the heretic from books of reference.
58. Removal of the book from libraries.
59. Demands to place the book on the Register of Forbidden Books.
60. Pressure on scientific supporters by bribing with better jobs to abstain.
61. Grants given to disprove the book (no grants ever given to "prove").
62. Efforts, include fabrication, to show misuse of sources by author.
63. Damaging statements put in the mouth of deceased persons of influence.
64. Heaping of accusations without substantiation in quantities making any response impossible in the same media.
65. Insinuations of profiteering and other ignoble motives for writing the work.
66. Attempts at organizing character assassination and special meetings to dispose of the challenge.
67. Dissemination of selected damaging reviews.
68. Offering the readers arguments from specialized fields that they are unable to verify.
69. Generalization and complete disapproval on grounds of a single alleged error.
70. Accusation of lack of sources by misrepresenting the term "collective amnesia."

A service to the history and science of science would occur in the expansion and testing of the list. Deg wished that he might complete the list concerning V., then move to other cases in science, and then to all occupations to display the universal prevalence of misdemeanor, not so much to scandalize, nor to stop it all (an impossibility), as to expose to light the epidemic predicament.

When asked to place them into categories (for Deg was distressed by their stringing out aimlessly) V. divided them into: suppression of publication; punishment and rewards; examination of the theories refused; ostracism of a nonconformist; rewriting of history and scientific finds; control of criticism; unfair criticism; and unfair criticism continued by unfair rejoinders. Deg in his turn divided them into logical errors, moral offenses (cheating and dishonesty); factual errors; illegitimate demands; hyperbole; personal abuse; material sanctions; etc. V. was especially pleased with what Deg called "the absent footnote technique," which with disastrous effectiveness eliminates an undesired line of ancestors, such as V.

Stecchini in the 1970's pointed out that Schiaparelli was a leading astronomer but could not get aceptance of his idea that Venus was scarcely rotating in relation to the Sun, showing an "Earth-Lock" as it comes closest to the Earth. The "Earth-Lock" was proven a century later, but although it supported V.'s view of a young and closely Earth-related history, other reasons were sought for it and V.'s position was not even mentioned, when, for example, the *Encyclopaedia Britannica* (XIX, 78) connected the phenomenon with "unsolved but very significant celestial mechanical problems connected with the origins and early histories of the planets." Here is a case of partial incorporation of quantavolution with the help of the "absent footnote technique."

The tricks used against V. were all commonplace in the scientific world. Since his work was so widely publicized and since he collected evidence so carefully, the tricks were simply more completely displayed. The more basic causes of resistance and opposition, which spawn tricks, have been discussed by Bernard Barber, with a wealth of examples. V. was not a sociologist. Allegations of meanness and non-rational thought exhausted his repertoire of analysis, except for his handy notion of collective amnesia of ancient catastrophe, which, he began to think, was the essential cause of the opposition to his theories; people, including scientists, could not bear to admit to open discussion their own suppressed terror of the original events.

But, of course, resistance to new ideas occurs whether the new ideas are catastrophist or uniformitarian, and with ideas that are false as well as with very true ideas, which Barber has shown in the cases of Helmholtz, Planck, and Lister, among others. As Deg has argued, the great fear of the poly-ego in the normal schizoid human determines memory at the same time as it demands forgetting (or resisting memory), and ancient catastrophes were materially grafted onto this human mechanism; but the resistance to V.'s theories can be only slightly assigned to the peculiarities of his catastrophism.

Deg prepared another list in 1978. He was making up this one out of disgust with politics: he was gloomy over the practical impossibility of finding persons in the world who were capable of organizing, agitating, and contributing to beneficial and benevolent movements. But he saw that the list applied also to getting support for scientific ideas and movements.

"Why Doesn't Somebody Do Something?"

    Noone wants to follow
    Helplessness
    Hopelessness
    Incompetence
    Hardheadedness
    General Disbelief
    Indifference
    Too busy, no time
    Can't afford to, financially
    Hurts somebody
    Meets opposition
    Arrogant to tell someone what to do
    Timidity
    Fear
    Fickleness
    Inattention and distractedness
    Leave it to the experts
    The crazies you have to deal with
    Hard work
    Resentment against being ordered about
    Ignorance of particulars
    Disbelief in use of force or any form of manipulation
    Hatred of those to be helped
    Lack of foresight
    Interested only in the moment
    Can't believe a few voices might prevail
    Things will work themselves out (laissez-faire)
    Fear of being corrupted
    Distaste for manners of other activists
    Have to work with inferiors
    Suspicious of potential collaborators
    Fear of physical harm
    Fear of failure
    Fear of being responsible for effects
No wonder nothing ever gets done!

* * *

In 1978, Dr. Henry Bauer, later Dean at Virginia Polytechnic Institute, offered the first full-dress anti-Velikovsky manuscript and the Director of the University of Kentucky Press asked Deg to read it with reference to its possible publication. Cutbacks in funds and programming forced the Press into giving up the manuscript or finding a $5000 subsidy of its production. The University of Illinois Press was finally to have brought the work out in late 1984. Meanwhile one can have a review of it by way of Deg's Readers

Report of January 10, 1979:

To: University of Kentucky Press, Attn. Mr. Crouch
From: Professor Alfred de Grazia
Subject: Reader's report of Henry H. Bauer, *Beyond Velikovsky*

In my opinion, Dean Bauer's manuscript should be published. It is the first generally adverse criticism of the work of Immanuel Velikovsky by a single author. The author has researched practically all available public sources. He is aware of and also adversely critical of the failings of many of the critics of Velikovsky. The book, strangely, is a likable book, which probably reflects the author's character more than the contents, which must prove annoying to a hundred people.

The book will be controversial. There is no avoiding this. Feelings run high on the scientific and sociological aspects of Velikovsky's work. The most incisive criticism is bound to come from the supporters of Velikovsky, for they are much better informed on all aspects of the controversy than the opponents of Velikovsky. These latter are usually cut down quickly. Dean Bauer realizes, though, that it is not easy to address the issues, and has the advantage of four hundred pages to explain himself and balance his analysis.

Because of the scope of the book, not only Velikovsky but also a number of his supporters will be motivated to respond. And one cannot doubt that they will have good grounds to enter the fray. Let me take myself as an example of what may very well happen with others. On p. 236 the author mentions my "utter conviction that Velikovsky is right." Right about what? I am favorable to his general theories, his genius, and his defense against the almost invariably misplaced attacks upon him. Bauer might well stress his distinction between the 'True Believers" and the scholarly supporters. Among the latter, there are many differences, the atmosphere is highly critical and, if they seem overprotective of Velikovsky, it is because the enemy outside is so massive and aggressive. It will add greatly to the clarity of the analysis if the author distinguishes the scholarly supporters and the lay supporters. (The word "public" is better but unfortunately has several meanings.) The scientific opponents of Velikovsky have also their scholarly and lay supporters. As for disputes among the scholarly supporters and Velikovsky, contrary to Bauer's statements, there are dozens, beginning with Juergens, Hess, and Stecchini and ending with the young writers in the current (Nov. 1978) issue of the *Society for Interdisciplinary Studies Review*.

At the bottom of p. 237, Bauer shoots from the hip at both Juergens as an absurdity and myself as a political scientist, while favoring physicist Kruskal's scornful attack upon Juergens. This does not accord with Bauer's many comments upon dogmatic remarks and against extolling specialized authority. Apart from whether he understands Juergen's theory, which he does not bother to demonstrate, and whether I understand Juergen's theory as well or better than Kruskal, he takes up a vulnerable position: what qualification, one might ask, does Bauer have for writing a book of sociology, history, ethnology, and political analysis, not to mention meteorology, geology, astronomy, etc.? Does he regard himself as a greater polymath than any of us?

Then again, he contradicts my analysis of Margolis and a group of Yale reviewers, claiming that his own count in the first instance is at odds with my own. Perhaps he should reproduce, in a couple of pages, the Margolis article with my comments, adding his own. Such would be the better way to damage my conclusions. The readers might then judge.

And so on. To say only of the distinguished group of scholars who passed on the *ABS* special issue on the Velikovsky Affair that none was a scientist gives a completely misleading idea to the reader. Lasswell was one of the founders of quantitative method in behavioral science; Cantril was a distinguished psychologist and opinion analyst; etc. Nor does he stress that Harry Hess, who is sometimes regarded as having been the leading geologist of the past generation, was a thoroughly sympathetic friend of Velikovsky. Hess and I talked on two or three occasions of Velikovsky, and Hess was as eager as I to see Velikovsky's scientific ability respected. Hess recommended that his students at Princeton read *Earth in Upheaval*, for example. These are but a few of the hundreds of points of contention in the manuscript and yet I feel it should be published with only modest changes, because it might otherwise take years to redo it and I am not at all sure that the public functions of the book would be greatly assisted. Perhaps I am saying that the book as it stands invites a full rocket display and, in the process, the public, science, and students will become better educated. I doubt that any amount of revision will make it a definitive and conclusive answer to the rapidly developing body of work sympathetically or willy-willy aligned to Velikovsky's books. I have four books in process myself that are more controversial and upsetting to the established doctrines of contemporary science than those of Dr. Velikovsky. But I have the impression that I shall not encounter the same type of opposition as Velikovsky if only because the intellectual atmosphere has changed so much and in part because of the Velikovsky Affair.

Readers perhaps will little note the criticism directed at myself and some others in the book, but they will be alert to a number of points respecting Velikovsky, and I would suggest that Dean Bauer reconsider them. He is attacking Velikovsky in 1979 partly on the basis of a pamphlet that Velikovsky published in 1946 ("Cosmos and Gravitation") and which Bauer even appreciates is not pushed by Velikovsky himself or scarcely anyone else. True, Velikovsky hates to recant, but the pamphlet is not a necessary prologomena to the later books. Indeed, Bauer's often insightful views about Velikovsky's character and motives should make him wonder whether the pamphlet was not merely a brash preliminary exercise, which vanity demanded be published as advance claims. Further it has become fashionable now to predict the doom of the concept of gravitation, and Velikovsky's musings were in a way the fashions worn in 1946 for anti-gravitational thought. This might be said also regarding the model of the atom as resembling the solar system. Only lately has that idea become discredited. Are we to dump all scholars who early in their careers exhibited what was currently believed? Then everyone will have to walk the plank.

Bauer sometimes abuses Velikovsky, contrary to his professed aim, generally observed, of avoiding inflammatory and *ad hominem* statements. It should be easy to revise such expressions as "astonishing ignorance" (. 159), "supreme ignorance" (p. 154), p. 161 etc. I think that he would reap rewards if he, or an editor, were to erase fifty to a hundred non-functional adjectives or phrases.

And, in respect to Velikovsky as a knowledgeable scientist, aside from "who is a scientist besides the self-elect," Bauer underestimates Velikovsky totally. Let him ask Burgstahler (chemist), Motz (astrophysicist), someone like myself who knew Hess (geology), Hadas (linguistics), Lasswell (psychiatric psychologist), Cyrus Gordon (Near East studies), Einstein (physics), Juergens (electricity), *et al.* Every last one will or would say that Velikovsky is not only a good scientist, but an imaginative one, and at home in a number of fields. I wonder why Bauer did not take the step to include himself in this group by interviewing the subject of his book. Velikovsky may be in error, but he is a scientist.

Also, I would recommend dropping the discussion of whether Velikovsky is a crank. Bauer admits that he himself is a crank, about the Loch Ness monsters. It's unworthy of this book to waste itself on this unscientific concept. I would, as Dean Bauer appears to believe, devote only several necessary paragraphs to exposing the term "crank" and kicking it out of bounds.

On p. 248, I note a striking contrast betewen a group of pro-Velikovsky publicists and a group of anti-Velikovsky scholars of distinction. This is a "foul blow." Either let both be publicists or both be scholars.

So, I should conclude that the off-hand abusive terms ought to be excised since they take away from a book some of its good air of casual and pleasant inquiry. Cut back the section on cranks. Perhaps dispense with the sections on "Cosmos and Gravitation" save for a simple statement of its inappropriateness and its inelegant foreboding of things to come. The admirably clear piece on gases should win Bauer an excellent contract for an elementary textbook in general science, but may not belong here. Perhaps other paragraphs can be removed here and there at the instigation of a generally well-educated lay reader.

The style is clear at the college level. Many, many things are said that need to be said about both sides: about how scholars are just (simply) people; about how the general public reacts to controversies in science as to political struggles, baseball games, etc.; and about the foibles of Velikovsky (though perhaps not enough, regrettably, about how these foibles have had something to do with driving him on relentlessly and with good effect). And I think that Dean Bauer might even, in the end, bite the bullet and state that on the whole it were well that Velikovsky's books were published, then bad that they were mishandled by the press, scientists, and disciples, yet good that a million people began to read into history and science. Finally take the word of the author himself (p. 366) that an astronomer's statement that "Velikovsky's scenario was impossible on grounds of celestial mechanics was just not so." That is worth something and will win the author a medal for courage, after all is said and done.

To avoid rumor-mongering or a delayed denunciation Deg told V.'s retainers of the existence of the work and of his recommendation. "Why?" he was asked, meaning why didn't he stomp it. It's not bad, he answered, you'll see, and it will keep the dialogue going, even improving it.

Meanwhile, those who were termed by the anti-heretics "devotees," "followers," "disciples," "supporters," "sympathizers," and were consigned to the limbo of science as "benighted," "anti-scientific," "occultists," "astrologers," "fanatics," and so on, unendingly — from these who were seriously considering his work as well as doing work of their own, came the discovery and reporting of his errors, qualification of his statements, essays at quantification, adduction of contrary materials, tempering, amending, and explaining. We need not go into the question, "Whose mass of supporters is better — yours or ours?" We are saying precisely that the effective scientific criticism of Velikovsky came from those who were sympathetic to his work.

It was the heretic scholars who designed alternative scenarios, in geology and astronomy, who upset V.'s chronology beyond the Eighteenth Dynasty of Egypt, who pointed out correctly evidence of pro-Biblical bias, who disputed his identification of the astronomical bodies implicated in certain legends, who pinned down the sources of numerous uncertainties, who reduced vagueness, who found and accommodated predecessors in the esoteric and difficult literature of catastrophism, far beyond the sporadic dark hints that "nothing new" was being proposed.

To be blunt, if you want to know what's wrong with Velikovsky, ask his friends, as much as his enemies; ask his admirers, as well as his detractors. You must know the literature of quantavolution and catastrophe. It is contained by now in many books and hundreds of correctly postured articles, many old, many new, many forthcoming. One can think no longer, if ever, that by "not believing in Velikovsky" science will proceed on its customary paths; a growing parade of many different kinds of quantavolutionaries is finding its own paths. The parade cannot be dismissed by uttering an imprecation against Velikovsky.

268　　　　　　　　　COSMIC HERETICS

* * *

The *Bulletin of Atomic Scientists* had been established in the triumphant days of nuclear physics following the blast at Hiroshima and was dedicated to voicing the responsibilities felt by scientists. But after a time it began to lose its halo and was seeking a larger audience. Like the playboy college student who excused his poor grades on grounds that his college was anti-semitic and who persuaded his father that his nose, his curly hair, and his name ought to be changed, whereupon, his grades remaining poor, he had to confess that "us Gentiles ain't very smart," the *Bulletin* did change its name for awhile and had the same old problem so it changed it back again, but at this time, around 1964, was trying to boost its popularity by exposing what Editor Rabinowitch regarded as scientific impostors, and his chosen weapon, a science publicist named Margolis, settled upon Velikovsky, whence was published a cavalier article entitled "Velikovsky Rides Again."

Deg's larger and more detailed refutation of the offensive article is reproduced in *The Burning of Troy*. So here I may introduce a letter in the same vein from Eric Larrabee, a publicist and early supporter of V., later head of the New York State Arts Council.

April 21, 1964

To the Editor:

The "Report from Washington" by Howard Margolis in your April number is a mixture of intemperate accusations and misstatements of fact. Margolis dismisses as "hokum" the work of Immanuel Velikovsky, which he has demonstrably read without care and judges without experience. He claims there is "no scientific way to examine" books which abound in references to physical fact. Their author had furnished specific scientific tests of his theory and on all of them to date, according to Professor H H. Hess of Princeton, he has been vindicated. Margolis brushes off Velikovsky's successful predictions as "science fiction" and offers instead the results of his "few hours" reading in philology and history.

He can apparently read neither French nor Hebrew. If he could read French he would not speak of the "actual" inscription at el-Arish in words from the outdated English translation of 1890 instead of the modern French translation of 1936, which is plainly cited in Velikovsky's footnote. The French translation gives the name Pi-Khirote. Margolis is flatly wrong in stating that Velikovsky "alters" the text, either here or in the case of the biblical Pi-ha-hiroth (so spelled by Velikovsky in *Ages in Chaos,* p. 44). If Margolis had read even the English translation attentively he would have found "King Tum" (the French gives "*le roi Toum"*). This is the text: *"Voici que Geb vit sa mère qui l'aimait beaucoup. Son coeur (*de Geb) *était négligent après elle. La terre _____pour elle en grand affliction."* It goes on to describe "upheaval in the residence" and "such a tempest that neither the men nor the gods could see the faces of their next." The inscription is shown to be historical by the fact that the King's name is written with the royal cartouche.

Velikovsky's reasons for suggesting that *bkhor* (firstborn) in the Hebrew text might be a misreading for *bchor* (chosen) are given at length *(Ages in Chaos,* p. 32-34) and are not essential to his argument that Exodus and the Egyptian sources refer to the same natural catastrophe. He uses the word "obvious" in proposing that the phrase "to smite the houses" refers to an earthquake in view of the fact that Eusebius, St. Jerome, and the Midrashim all confirm this interpretation. Margolis' sarcastic repetition of the word "obvious" is wholly without justification.

Margolis accuses Velikovsky of saying that St. Augustine puts the birth of Minerva at the time of Moses whereas Augustine "says the opposite." This would be a serious charge if true but it is doubly untrue, both as to Augustine and Velikovsky. The relevant passage in *The City of God* (Book XVIII, Chapter 8) reads that Minerva was born in the time of Ogyges and Velikovsky quotes it *(Worlds in Collision,* p. 171) in those precise words. In support of the damaging assertion that Velikovsky alters evidence, Margolis alters the evidence from both sources.

Margolis cannot even read Velikovsky correctly. He says that Velikovsky "can cite no description" of Venus growing larger in the sky despite the fact that on pages 82-83 and 164-65 of *Worlds in Collision* it is so described from Western ("an immense globe"), Middle Eastern ("a stupendous prodigy in the sky") and Chinese ("rivaled the sun in brightness") sources.

The sociological interest of the Velikovsky case lies in the willingness of scientists to dismiss the work of a serious scholar as "hokum" on the basis of slipshod, inaccurate, and abusive criticism. Margolis has proved once again that the interest is justified.

Eric Larrabee

Deg was in an ornery mood and had threatened the *Bulletin* with a suit for slander. V. was all for the idea and consulted his friend, the libel expert, Philip Wittenberg. Deg also consulted Herbert Simon and adopted Simon's view, as expressed in the letter below:

Dear Al,

I have read the materials you sent me about the Velikovsky matter. (Incidentally, I lunched with Velikovsky last week, and we are going to have him back to the campus next autumn for a lecture.) I have a few comments to offer on the matter of strategy.

As I am sure you know, there is a doctrine in the law of libel known as "invitation to comment." Anyone who performs publicly — and that includes publishing a book — invites critical comment, and has no recourse if he gets it unless he can show actual malice. The critic does not, in general, have to sustain the burden of proving truth. (I may have forgotten details, but your lawyer will tell you that that is the general idea.) Two consequences follow from this: (1) one should not publish books — or issues of the American Behavioral Scientist devoted to the Velikovsky Affair — unless one has a thick skin; (2) when one is flayed by a critic, one should almost never threaten legal action, however righteous one's feelings.

The opponents of Velikovsky are not malicious, they are indignant. Nothing about the Margolis article seems to me libelous, however much I disagree with it. We certainly do not want to imply that *we* wish to suppress his right to hold, or even publish, these opinions, however much anguish they cause us. Hence, if I were editor of the Bulletin of the Atomic Scientists, I would politely but firmly reject your request that I "withdraw my support" from the article. He might even point out that to an anti-Velikovskyite, some of the language in the September American Behavioral Scientist might seem quite as offensive as Margolis' language did to you. *C'est la vie.*

When you receive the refusal from the editor — as I am sure you will — I would advise that you then request an opportunity to have three pages in BAS to reply to Margolis (perhaps offering the same number of pages in ABS for a rebuttal to the September articles). There is nothing to be lost by a public discussion of the issues, especially the issue of freedom to publish, and nothing to be gained by defending that freedom through threats to suppress it.

With best regards,

> Cordially yours,
> *Herbert A. Simon*
> Professor of Administration
> and Psychology

After much deliberation and testing of the winds, Rabinowitch wrote Deg:

> 25 June 1964

Dear Mr. de Grazia:

In answer to your letter of May 12, I do not see why, and in what form, the *Bulletin* should "withdraw its support from the article of Mr. Margolis." I do not understand what you mean by "your contributors and advisors urging you to take action to remedy the wrong done us." The responsibility for the contents of the articles published in the *Bulletin* rest (sic) with authors of the articles. It must be obvious, of course, that the magazine cannot disclaim legal responsibility for any defamatory statements, but I do not see in the article by Mr. Margolis any statements of such nature with respect to yourself or to the contributors of your journal. If all polemics over matters of scientific competence would end in court, this would be bad indeed for the climate of free discussion in this country. In our society, the enemies of evolution can call scientists, espousing this theory, ignoramuses, or heretics; the enemies of fluoridation can call the medical authorities supporting it whatever like names they might choose — short of character assassination — and the proponents of fluoridation can do the same to their critics. This is as political processes should be in a democratic society.

In his article Mr. Margolis, after dealing briefly with the astrophysical difficulties of Velikovsky's theory, expanded on the interpretation of ancient texts. From the point of view of the *Bulletin* the physical and astronomical evidence is crucial, and the considerations of what Velikovsky calls "experience of humanity," can only be subsidiary. Physical evidence is simpler and more unambiguous; while interpretations of old texts and hieroglyphic inscriptions is an inevitably tentative and often controversial matter.

Since Mr. Margolis brought up the paleographic evidence in his article, we must in all justice, permit Dr. Velikovsky (or a spokesman for him) to point out the errors, if any, in his argument. This should be done by someone with first-hand experience in the field — either Dr. Velikovsky himself, or even better, some independent recognized authority in Biblical history and ancient languages. We are willing to publish such a letter in one of the forthcoming issues (giving Mr. Margolis the opportunity of answering it, if he desires); but, we will then terminate the discussion, since Egyptology or Old Testament studies do not represent a field of the *Bulletin's* major interest.

As far as physical possibility of the events suggested by Velikovsky is concerned, I mention the names of Menzel and Shapley because I remembered that they did analyze Velikovsky's theories at the time of their publication. I would be glad to have any other recognized astrophysicist or geophysicist (including the Princeton and Columbia astronomers who have pointed out in *Science* the correctness of some of Dr. Velikovsky's specific predictions), to present in the *Bulletin* briefly what they think of Velikovsky's theory as a whole.

I believe it is a mistake to accuse modern science of intolerance to the theories which destroy its accustomed frame of reference and force it to revise its foundations. Einstein proposed a revision of Newton's conceptions of time and space; for a few years, there was some resistance of the type suggested by you, but it was silenced by Einstein's explanation of the precession of the perigee of Mercury, and his prediction of the bending of stellar light in the neighborhood of the sun. If the correct predictions by Velikovsky, pointed out by Hess and others, do not change the general rejection of Velikovsky's theories by scientists, it is because changes in the laws of celestial mechanics and revisions of well-established facts of earth history, required by Velikovsky, are quite different from the subtle, but logically significant and convincing changes in the scientific world picture suggested by Einstein (as well as by Mac [sic] Planck, when he postulated the atomic structure of energy, or more recently by Lee and Yang when they postulated a physical difference between a right and left screw, object and mirror image). Modern science has learned to be open-minded to revolutionary suggestions, if they are brought up with strong scientific or logical evidence. Reluctance to go along with Velikovsky's *Worlds in Collision* is, in my eyes, evidence not of stubborn dogmatism of "official" science but of the physical and logical implausibility of his theories.

Your letter and its request misinterprets the position of the *Bulletin*. To conclude, since Mr. Margolis brought up paleographic evidence, fairness requires the *Bulletin* to give space to a letter disputing this evidence (provided this letter is not more abusive than Mr. Margolis' criticisms). If Dr. Velikovsky can suggest a recognized authority in astrophysics or geophysics willing to discuss his theory as a whole in the light of recent verification of some of his predictions, I would consider giving space in the Bulletin for a brief discussion of this kind.

It is in this spirit of scientific argumentation that the whole problem should be resolved.

Sincerely yours,
Eugene Rabinowitch
Editor

During the next few weeks Deg drafted a brutal reply to Margolis's article and prepared a letter to accompany the critique. However and meanwhile, V., ever hopeful of access to and acceptance by the authorities of physics, prevailed upon Harry Hess to submit on his behalf to Rabinowitch an article he had prepared on his Venus theory in the light of new findings. It would serve as a counterweight to the Margolis article, without reference to the libertarian and legal issues involving the *Bulletin*.

In September Rabinowitch wrote to Hess, returning V.'s manuscript without having read it and saying, "the *Bulletin* is not a magazine for *scientific* controversies — except on rare occasions (e.g. in the field of genetic radiation damage) when they are directly related to political or other public issues... Neither is it the function of the *Bulletin* to provide an outlet for scientific theories not recognized by professional authorities in the field." He explained the Margolis article as an attempt to undo the work of "behavioral scientists" in aid of V. whom, he said, they "championed in the most violent way."

In October, the *ABS* published Deg's critique of Margolis, and Deg sent it to Rabinowitch along with the letter that he had drafted three months earlier.

November 12, 1964

Dear Mr. Rabinowitch:

Please permit me to answer frankly your letter of June 25, which asks *why* and *in what form* you should "withdraw your support from Mr. Margolis's article about us.

The *why* should be apparent in the attached analysis of Mr. Margolis' writing, entitled "Notes on 'Scientific' Reporting." This explains in detail the errors, the malice, and the legal offenses of Mr. Margolis. Unless you can by the use of evidence and reason erase those 54 notes, you are bound scientifically, morally, and legally to "withdraw your support."

In what form should you "withdraw your support"? You should "withdraw your support" by expressing in seven columns of space in your magazine (1) your acknowledgement of the excessively large number of factual errors contained in Mr. Margolis' article, and (2) your regret for the incorrect unjustified slurs upon the character and motives of Dr. Velikovsky and the contributors and editors of THE AMERICAN BEHAVIORAL SCIENTIST, together with your hope that your readers would join you in repairing in the course of time such damages as was caused by this article.

My present letter could now end, as might have your own at the same point. However, you go on to make further comments that require answer.

You say that it would be "bad indeed for the climate of free discussion in this country" if "all polemics over matters of scientific competence would end in court." I answer that "all polemics" are not at issue, but only *one* polemical action. (You are, of course, at liberty to universalize its meaning.) Moreover, "the climate of free discussion" that you mention has been clouded and cannot be logically cited as a reason for staying out of court. It is precisely to get people out from under this cloud that the law and courts are built. The courts enable an objective determination to be made of a matter in certain cases where free discussion is impossible. They permit and require the calling and interrogating of witnesses under just conditions. They prevent and remedy the abuses that you have presumably endorsed. The law of evidence and the rule of law, Mr. Rabinowitch, are the grandparents of the scientific method. They are not its antithesis.

You say that in our society, disbelievers in evolution can call scientists espousing evolution ignoramuses or heretics. You say enemies of fluoridation can call medical authorities supporting it like names and *vice versa*. You are defending your magazine evidently for assuming the privilege of such name-calling as opponents of fluoridation and evolution employ. Very well. Your readers must judge you for that.

"Character assassination", you say, is *not* permissible, however. The issue here is of course just that. I call to your attention the numerous instances, well-noted in the aforesaid memorandum on "54 ways", in which your magazine is guilty of character assassination, slander, and libel.

Your next paragraph is logically queer, for you say that the *Bulletin* is largely concerned with the astrophysics of Velikovsky and not with the humanistic evidence. (I will not tarry with your incredible distinction between physical and humanistic evidence.) But then you go on to admit that the *Bulletin* reversed itself and abandoned its chosen field in this case. (Apparently, any and every policy can be reversed to get at Velikovsky. How true we were!) And you say you want to see the historical evidence argued. Argued — but not too much you state, for you have to get back to your major interests! Like fluoridation? Like UN affairs? Like scientific freedom? You may go back to your affairs, Mr. Rabinowitch, but not before we are done with the matter.

Now you would graciously permit Dr. Velikovsky or an "independent authority" of the classics to answer Mr. Margolis by a letter, to be followed by a reply from Mr. Margolis, and then *stop!* Two-to-one is bad enough. But how does Mr. Margolis deserve this reply? By his own expertness as a biblical scholar, specialist in ancient languages, and classical historian? I submit that this exchange might be equal and appropriate if I might delegate my daughter who is majoring in archaeology at Bryn Mawr to take up your invitation to reply.

A general appraisal of Dr. Velikovsky's theories in your paper would be a good idea, as you suggest, and I think you should find a set of scientists to make such an appraisal. I would not go to Drs. Menzel or Shapley, whose participation in the Velikovsky case, as documented in *Harper's* and *The American Behavioral Scientist,* has been most unbecoming. Your hazy remembrance of their posture is scarcely a firm basis for risking the reputation of your magazine and colleagues. Besides the balance of evidence has continued to shift between 1950 and 1964. Do read that document; you must take the time: you and your writers cannot decently continue to ignore all the factual record of the case.

Still, all of this is not the central point, which is the behavior of scientists, and you do well to return to it in your last two paragraphs. There you first say that modern science is not intolerant of unorthodox theories. This is not so; even the case you site, Einstein, was in your own words victim of "some resistance" of the type the ABS described. But even if it were so *generally,* why would you unscientifically and dogmatically refuse to recognize an "unusual" case of resistance when it loomed before you?

How can you say that the actions taken concerning Velikovsky and his theories were tolerant? Please state one procedure, whose value you would defend, for the reception and consideration of new scientific material, which was followed by the leadership of science in the Velikovsky case. Show us that he was given one key to the kingdom. I believe, as you seek to do so, you will gradually eliminate from consideration all the decent and rational procedures that are supposed to govern the behavior of scientists. In the end you will either be indignant or a cynic. You will not be the Rabinowitch whose letter I am replying to.

I must end in laughter, which I hope you will forgive. For you conclude by permitting Dr. Velikovsky to answer by letter "provided this letter is not more abusive than Mr. Margolis' criticisms!" I am not clear whether you are here defining the outer limits of abuse, or whether you suggest pursuing scientific truth by balancing two sets of slander.

Go back to my beginning, sir; you will find our two requests to be generous offers made in the veritable "spirit of scientific argumentation" that you appeal to.
        Sincerely yours,
        Alfred de Grazia

Dear Mr. de Grazia:
    Thank you for your letter of November 12th. I can only add my appreciation that you published the full Margolis article in *The American Behavioral Scientist.* Your readers may judge.

        Sincerely,
        Eugene Rabinowitch
        Editor

December 3, 1964

Dear Mr. Rabinowitch
    We acknowledge your appreciation of our fairness. Does your appreciation mean that you, too, will be fair to us and present our rebuttal before your readers?

        Sincerely yours,
        Alfred de Grazia

The rebuttal was *not* carried by the *Bulletin.* A great many scientists had their prejudices reinforced at the expense of V., Deg, and the *ABS.* In the final analysis and many years later, Deg's indignation seems overdone, and it is doubtful that he ever had the intention of suing, but he was up to his typical game of driving home contradictions and pounding away at the basic homology between legal and scientific procedure. Furthermore, while discounting his rhetoric, I should also call attention to specific instances of the damage caused by irresponsible behavior in scientific circles tied directly to the *Bulletin* article: one on the matter of fluoridation, one an exchange between Urey and Deg, and two to be treated in chapter 15 on "The Knowledge of Industry" involving the Sloan Foundation, Moses Hadas, and a project of Deg in economics.

* * *

July, 17, 1966

Dear Professor de Grazia:

Since writing you earlier in connection with my review of "A Struggle With Titans," I have been reading the various documents cited in "The Velikovsky Affair."

One that particularly "struck" me was the article by Howard Margolis in the April 1964 issue of the *Bulletin of the Atomic Scientists* that you so ably dissected in the October 1964 issue of the *American Behavioral Scientist*.

What came as an even greater surprise, however, was the article written by Margolis about fluoridation in the June 1964 issue of the *Bulletin of the Atomic Scientists*. By failing to take note of published reports of toxic effects from fluoridated drinking water, he constructs a very favorable case for fluoridation and makes his opponents appear to have no scientific grounds on which to oppose it! Since you were able to show that Margolis is not a good philologist, I thought it might be worth pointing out that he also has not read the fluoridation literature very thoroughly. The major documents he cited to support his view are guilty of omission just as he is. The one that was prepared in 1955-1956 is hardly relevant to "current" findings, while the "Select" bibliography is no more than a compilation of proponent research, with virtually no mention of contrary results reported by others, especially in relation to clinical findings.

I realize your interests lie primarily in the area of the "sociological" aspects of a subject like fluoridation, but the strong scientific evidence against fluoridation has been kept so heavily suppressed that there is a close parallel to "The Velikovsky Affair." Our own local public library, I might add, has refused to accept or acquire a copy of "A Struggle With Titans" on the grounds that the standard reviewing media have ignored it —just as they are ignoring "The Velikovsky Affair"!

>Sincerely yours,
>*Albert W. Burgstahler*
>Professor of Chemistry
>The University of Kansas
>Lawrence, Kansas

June 2, 1964

Dr. Alfred de Grazia
The American Behavioral Scientist
80 East 11th Street
New York 3, New York

Dear Dr. de Grazia:

I am sorry to see that you have gotten mixed up in the Velikovsky case. Velikovsky was a charlatan. There is just no doubt about it at all. It is not true that outstanding astronomers would not welcome a truly original man with constructive ideas. We would put him on the staff of the University of California San Diego. I do think that you should try to withdraw from this controversy as gracefully as possible and not continue it. I assure you that every physical scientist of my acquaintance will rise to defend the *Bulletin* against anything you do.

I am terribly concerned at present about the lack of control in scientific publication. Science had always been aristocratic. Not everyone could get his ideas published in effective journals. Articles to the scientific magazines have been carefully edited, and unless they conformed to reasonable scientific standards they were refused. Today anyone can publish anything. In the first place, very second-rate scientists can get jobs somewhere — with industrial companies, government agencies, the space program, etc. They all have their private printing press in the back room, namely a reproduction device. As a result, papers of all sorts are sent out. Also there are new journals springing up with no decent editorial control whatever. The result is an enormous amount of confusion. In fact, as I have stated and I now repeat, there is often so much noise that one cannot hear the signals.

With best regards,
Very sincerely,
Harold C. Urey

Deg's journal, June 29, 1964

...Velikovsky had palpitations last week. For several days his pulse was irregular. He has gone into a three day period of rest and is taking a little tranquillization by drugs. He has been travelling too much and spending too much time trying to direct strategy in his scientific defense. A letter I received from Harold Urey depressed him greatly. Identifying as he does with authority, V. is hurt when a Nobel Prize winner for chemistry refers to him as a charlatan. What can he be expecting? I have not been able to educate him to the sociology and political science of science. He believes in rationalism and that other experts only by odd mistake "because they haven't read his works," treat him so contemptuously and with hostility. V. wrote what he thought should be my reply. (Sometimes his presumption becomes arrogant.) It was a strange letter, full of pathos and humble remonstrance. I could not and would not use it. It is an interesting document about V. himself. It would do him no good even if I were to use it. Yet he was deeply perturbed when I informed him I was sending my own letter of reply. He claimed that his was a perfect letter, which he was proud of and felt must be sent. It was then I learned of his palpitations. The thought occurred: the strangeness of this letter goes with a nervous disturbance. He desperately wanted me to send *his* letter; he mailed it by special delivery to New York where I was and phoned to press me about it. In a week or two, when his illness is passed, he may be secretly pleased that I went my own way.

I spoke later to his wife. She seemed displeased with me too. She too will come around. She confirmed how "hurt" he was by the Urey letter.
Urey is a-------------! What better could come from him. His letter to me is a disgrace and I mean to call it that. "

July 8, 1964

Dr. Harold C. Urey
School of Science and Engineering
University of California, San Diego
P.O. Box 109
La Jolla, California 92038

Dear Dr. Urey:

Thank you for your letter of June 2. I appreciate your concern that I may "have gotten mixed up in the Velikovsky case." Since everyone whose attention is called to the case has gotten mixed up in it, in one way or another, I guess that I am in good company.

Your second sentence is that "Velikovsky was a charlatan." He neither "was" nor *is* a charlatan. Resort to your nearest dictionary will satisfy you on that score. If you insist that you have not made a linguistic error, then you must give me *one*, just *one*, bit of evidence to support your allegation. Indeed, your next sentence is "There is just no doubt about it at all." Since you are a scientist and know the nature of proof, you must have a great many pieces of evidence, adding up to certainty. If you cannot cite such evidence, then you must apologize to Velikovsky, or you become yourself a charlatan and slanderer.

You may refuse this challenge. Very well. We do not usually carry substantive discussions of factual theory in the *American Behavioral Scientist,* but if you will honor us with one significant error of fact or logical contradiction in Velikovsky's works we will print it and let it go at that, for we are not concerned to solve the problems of physics and astronomy, or politics and economics in our pages. I know that you will have no trouble with this small matter; I could probably manage it myself; that Mr. Margolis could not succeed, nor some others who tried, does not prove that the works are flawless.

Then you say, "It is not true that outstanding astronomers would not welcome a truly original man with constructive ideas." I am afraid, Dr.

Urey, that you will be hard put, in the light of the history of science, to maintain this statement also, unless you would again resort to evasive semantics, defining the words "truly original" and "constructive" to suit your ends. Your saying that "we would put him on the staff of the University of California, San Diego" could be regarded as an idle threat if it were not for the well-known anxiety of certain Californian colleges to discover warm bodies wherever they may be.

You thereupon urge me to withdraw from the controversy. Actually, I had done so; but the stupid brazenness of the *Bulletin of Atomic Scientists'* article brought to me a sharp realization *that many of your kind simply will not learn.* "Ontogeny recapitulates phylogeny:" every error of the scientific mind and spirit in the history of the Velikovsky case was by almost preternatural skill recomposed into a few columns of the *Bulletin.* This you ask me to swallow!

The controversy will continue. You say that "every physical scientist of my acquaintance will rise to defend the *Bulletin* against anything you do." Perhaps you will not have as many acquaintances as you claim and they will not be willing to act as your troop if they, or at least several of them, were to read the pages of the *American Behavioral Scientist* and compare them with the article of the science correspondent of the *Bulletin.* (Isn't it interesting that the scientists' *Bulletin* should have to hire a non-scientist to write about science for them?)

You have, it is clear, a rather horrifying vision of science. You gently threaten me, you promise to bring in your gang, and then you begin to reveal the Utopia that occupies your mind. "I am terribly concerned at present about the lack of control in scientific publication," you write; "Science has always been aristocratic. Not everyone could get his ideas published in effective journals. Articles in the scientific magazines have been carefully edited, and unless they conformed to reasonable scientific standards they were refused. Today anyone can publish anything."

I, too, Dr. Urey, am concerned about scientific publication. I am not, however, concerned about the lack of control by the scientific oligarchy, as you are, but by the lack of communications, the haphazard and chaotic situation that is caused as much as anything by a defective leadership in the sciences. Your kind of scientific aristocracy is precisely the reason why your subsequent claims are laughable: if there is any villainous theme in the history of science, it is the continuing attempt to deny a voice in the organs of science to iconoclasts, outsiders, and just plain *kleine Menschen.*

You will be responsible for retarding the progress of science if you succeed in reestablishing the old system of information controls. You should turn your attention to organizing scientific information rather than to suppressing it.

Similarly you should be pleased that more of our working population today are scientists, rather than coalminers or ditchdiggers. Indeed you seem to be angry with them for pretending to perform the same operations as are practiced by you happy few. "...Very second-rate scientists can get somewhere — with industrial companies, government agencies, the space program, etc. They all have their private printing press in the back room..." Einstein with his patent-office job, Da Vinci doing his civil engineering, Freud setting up his own printing press, Darwin idling on his patrimony — there certainly are a great number of these second-raters, without university chairs, not content to eat common fodder and let their intellectual ambitions expire peacefully!

I am beginning to see your point. You would wish only first-rate scientists such as Howard Margolis, formerly a science writer for *The Washington Star* and now correspondent for the *Bulletin of Atomic Scientists,* to have freedom of scientific expression. Your idea would be to have a kind of Empire such as Alice discovered in Wonderland where the knighthood of science is conferred by your power elite and the Sir Margolises can be sent out to harry any peasants who may have the temerity to poach upon the truth.

Your conditions for peace are not acceptable, Dr. Urey. Our condition is that science be open and public, and remain so. If you wish to alter your conditions substantially we would be pleased to hear from you again. Meanwhile, with regards to your work on tektites, I remain

        Respectfully yours,
        Alfred de Grazia

* * *

The special magazines given over to reporting and supporting V.'s do ings have been *Pensee, Kronos,* and the *Review* of the Society for Inter disciplinary Studies. Each of these has carried extensive materials on the preliminaries, proceedings and aftermath of the American Association for the Advancement of Science convention panel dealing with Velikovksy's ideas at San Francisco in February 1974. According to astronomy Professor Ivan King of the University of California at Berkeley, it was Carl Sagan who suggested the confrontation. It was intended that the panel be divided into supporters and opponents of V., but over a period of months, the pro-V. nominees were weeded out. This was suspicious, and I am inclined to cast suspicion on both sides.

In the first place, both the establishment (for it can be called such also on these occasions when it puts on a face) and the heretics chose a deceptive yet revealing title: "Velikovsky's Challenge to Science." V. would never allow himself to be called a non-scientist; yet, to have his name in the limelight, he allowed himself to be juxtaposed to science. Simultaneously, the establishment (that is, the government *ad rem* in charge of the state of science), in order to isolate the heretic, allowed the personalization of the panel, in itself an abuse of the scientific method which addresses itself to ideas, not men. Might not a better title have been "The Validity and Prospects of Neo-catastrophism"? Then with eight papers, four on each side, the topics of the mechanics, the electromagnetics, the historical record, and the reception of neo-catastrophism in science could be taken up.

Did V. want to appear without support on the stage, keeping the spotlight, whether for the hero or the martyr, upon himself, and therefore did he not fight hard enough to ensure himself that support? He ended up with two neutral parties, the opposition of a biased chairman, and three convinced antagonists eager for the fray. Surely there must have been some masochistic force at work in him, coupled with an extremely clever Machiavellism: a pro-Velikovsky paper would do nothing for V.'s image as a great scientific loneer and martyr.

If the one man who knew the Venus historical record best, Lynn Rose, had been present, he could have devastated, on the spot and forever after, the presentation made by Huber. It would have been ineradicable from the book that followed, entitled *Scientists Confront Velikovksy*. If Juergens had been forced into the panel by V., then Mulholland would have been finished off. If Deg had been invited, he would probably not have gone, but if he had, he might have effectively harried Sagan and Storer, considering what these two ended up by saying. Then V. would have been off and running.

Instead, it was a gruesome exercise at V.'s cost, then and thereafter. He behaved magnificently, like Samson dragging down the temple of the Philistines upon himself. He won the crowd. The press, ignoring the crowd, and incapable of reading the papers, pronounced him dead. V. did not really go to San Francisco to have the crowd be with him. He went there to gain scientific recognition. Or did he get mixed up and rely upon the crowd, and hope for a victory against impossible odds while cultivating the fantasy of martyrdom?

The establishment — and Professors King and Goldsmith, the official sponsors, found themselves irresistibly playing the roles of the establishment — was quite pleased to let the panel develop into an over-kill of V. It could not even conceal its hope when explaining the public presentation of the symposium. King, who was the Chairman of the panel, explained privately that he was so anxious over the responsibility of presenting V. at a scientific forum that he had to persist in saying that the purpose of the symposium was to refute a set of ideas that science had proven absurd. Actually he said so publicly beforehand:

> What disturbs the scientists is the persistence of these views, in spite of all the efforts that scientists have spent on educating the public. It is in this context that the AAAS undertakes the Velikovsky symposium. Although the symposium necessarily includes a presentation of opposing views, we do not consider this to be the primary purpose of the symposium. None of us in the scientific establishment believes that a debate about Velikovsky's

views of the Star System would be remotely justified at a serious scientific meeting.

Now I would like to quote the economist Shane Mage's booklet, *Velikovsky and His Critics,* because of its elegant conciseness. Besides, he was present at the occasion, and neither Deg nor I was there.

What took place in San Francisco was... the beginning of a real debate, even it often seemed to those of us in attendance like a donnybrook. Of the six invited panelists, one, Norman Storer (Prof. of Sociology, Baruch College of CUNY) disavowed competence in any aspect of the subject but nevertheless managed to conclude that the mistreatment of Velikovsky, though abstractly deplorable, was also an "understandable" response of the "scientific community" to a perceived "attack by right-wing forces in American society." Velikovsky himself presented a short paper outlining the basis of, and some of the evidence for, his *Challenge to Conventional Views in Science,* and often took the floor vehemently to rebut specific criticisms. His views on the importance of electrical forces in celestial mechanics also received strong support from Professor Irving Michelson (Mechanics, Illinois Institute of Technology), who described his paper *Mechanics Bears Witness* as "an act of objective scholarship," intended to be neither pro nor anti-Velikovsky.

The polemic against Velikovsky was conducted by two Professors of Astronomy (Carl Sagan, Cornell University, and J. Derral Mulholland, University of Texas) and one Professor of Mathematical Statistics (Peter Huber, Swiss Federal Institute of Technology). Almost all the media coverage of the panel consisted of favorable citations of these three contributions, especially Sagan's very long essay entitled *An Analysis of Worlds in Collision.* In the absence of Sagan, who left before all papers had been read in order to attend a taping of "the Johnny Carson Show," a vigorous discussion, involving audience as well as the remaining panelists, continued for almost two hours after conclusion of the formal presentations. Both sides claimed victory.

The logical next step was publication of the symposium proceedings, but of the panelists only Velikovsky was willing to permit publication of an integral transcript of the speeches and the floor discussion. Lengthy negotiations failed to arrive at a mutually agreeable format, and ultimately the two parties decided to publish separately.

The anti-Velikovsky case was presented by Cornell University Press under the title *Scientists Confront Velikovsky* (hereafter referred to as S c.V). In addition to revised versions of the AAAS papers by Sagan, Mulholland, Huber, and Storer, this volume also includes a paper by Prof. David Morrison (Astronomy, University of Hawaii), prepared, in its original form, for a 1974 conference sponsored by the editors of *Pensee.* There is also an introduction by Dr. Donald Goldsmith, editor of S c.V and organizer of the AAAS panel, and a foreword by the novelist and authority on heresiology Isaac Asimov. From the proclaimed standpoint of "scientific orthodoxy," Asimov begins by raising the question "What does one do with a heretic?", with specific reference to Velikovsky; goes on, with unimpeachable orthodoxy, to write that Velikovsky's proposed physical explanation for catastrophic events recorded in the Bible is a "far less satisfactory hypothesis" than is "the hypothesis that divine intervention caused the miracles"; and concludes that "Velikovskians" are totally impervious to any amount of "mere logic." (S c.V, pp. 8-15) He does not, however, recommend that they be turned over to the secular arm...

The AAAS volume is presented by its sponsors as "a full scale critique" (Goldsmith, S c.V, p. 27) which, according to the review commissioned by the AAAS Journal *Science,* accomplishes a definitive refutation of Velikovksy's "downright preposterous" heresy. The essays in this book "utterly lay waste his theories," Sagan's paper "is amusing, acrid, and totally devastating...his essay alone is sufficient to reduce the Velikovsky theory to anile fancy," and "Velikovsky is flatly and totally disproven... As far as Velikovskianism is concerned, it is dead and buried. The final nail has been driven." *(Science,* v.199, Jan. 20, 1978, pp. 288-9)

Was this appraisal accurate? Referring to the trial by press, yes, V. was further damaged in the eyes of scientists everywhere. Speaking of substance, whether of the symposium or of the papers, it was not true. The arguments of Sagan, Mulholland, and Morrison were mostly well-known and those of Huber (the surprise amateur of ancient Babylonian tablets) had been long ago considered by Stecchini and Rose. Additions and revisions allowed to the writers did little to bolster their defenses when it came time to publish the book *Scientists Confront Velikovksy.* An early analysis of the enemy dispositions appeared in *Pensee;* then, in two issues of *Kronos* (III2 and IV3), and in pieces appearing elsewhere, supporters of V., forced to waylay the establishment speakers in the alleyways, stripped them of their arguments. The Cornell University Press, a willing captive of circumstances, which might have published a fascinating, meaty volume on the issues, published one poor lopsided volume, and sold paperback rights to W. W. Norton Company. The heretics remained in the alleyways. Scarcely any reviews (except those of the heretics) put the opposing volumes side by side and compared them judiciously, or even savagely.

I shall not go into the several dozen points of contention here, and will take Deg's word for it that the substance of the full arguments did more good than harm for a considerable range of quantavolutionary hypotheses, including some precisely attributable to V.

Shane Mage, in appraising the speeches against V., uncovered in them several important concessions that had been apparently achieved over the years. First, the book *Scientists Confront Velikovsky* "disavows and repudiates the entire 'scientific polemic' of the 1950's and 60's, both implicitly and explicitly" Next, both the sponsor, Goldsmith, and Mulholland assert that V.'s ideas and arguments are not "un" nor "anti"- scientific, whatever the press and then the scientific community presumed to draw from the event. Furthermore, the legitimacy of cosmic catastrophic hypotheses in science was acknowledged both by Sagan and Mulholland, but the specific hypotheses of V. were attacked (and obviously the scientists are in confusion as to how they can work historically and empirically with the hypotheses that they admit.)

In line with my earlier suggestion, a different and more proper title would have brought these most important areas of agreement to the fore. If these would have been the subjects of the panel, and if Velikovsky had been only one out of eight panel members and authors, four of whom would have adopted positive positions and four adversary positions, then the world of science would have been much impressed and enlightened, and the heretics might have surrendered their weapons with honor. V. himself would have acquired many scientific allies and be better received from then on in discussions among scientists; hundreds of hours of anxious and resentful negotiations and dispute would have been avoided; and many fresh minds might have been inspired to enter the newly opened field of quantavolution. The AAAS affair was a great opportunity lost to quantavolution by V. and the establishment agents.

Deg disliked the word "heretic." I mentioned so earlier. Perhaps I should have renamed this book. To him the word was un-American. It was one more useless nuisance for indulging V.'s self-image. True, the dictionaries include it with its modern meaning, "one that dissents from an unaccepted belief or doctrine of any kind," but in a modern democracy, he said, the occasions for heresy are innumerable, while, without severe sanctions, the historical pitch of the word is absent.

Whereas V. called himself a heretic both in respect to religion and to science, he chose to stress science as the offending authority. In his day, in Western Europe and America the idea of heresy hardly held meaning for the larger society, although it could be effective in the ambiance of, say Catholicism or Presbyterianism; even here one had to lay claim to authority heretically within the group itself.

V. was determined to be a heretic from within science but to do so one had to be a scientist in the first place, and one of the childish games played between the scientists and V. had to do with whether he was indeed a scientist and therefore properly within science's jurisdiction to be adjudged heretical. Logically, we are back with Alice in Wonderland and not the least of the skits from never-never land was the massive attack upon V. launched in the name of science and culminating in the book, *Science Confronts Velikovsky*.

Here, from the beginning, the scientists promoting the event at the AAAS meeting in San Francisco were befuddled. Yes, they felt, they had to defrock V. but to do so they had to admit him in their canonical court.

But to admit him they had to claim jurisdiction over him; that is, they had to legitimize him by allowing him to debate his ideas with them. One can perceive this strain and stress clearly from beginning to end of the touted confrontation over a period of years. The promoters, King *et al*, would say, we are not meeting to discuss V. but only to make it clear that he is not speaking as a scientist. And then, of course, they proceed by the only modern way science knows, to refute him as a scientist in public argument.

When the time came to publish *Scientists Confront Velikovsky* the establishment, operating by queer contradiction, obtained the good services of Isaac Asimov, the most famous popularizer of science and science fiction author, to introduce the work, admitting *ipso facto* that its contents alone would not fulfill the contract put out on V.

Then what does Asimov do but to fall into the pit of scholasticism by spending his precious few pages as an instant expert on heresy. He accepts the fractured word and further mangles it. He concocts and improperly applies a distinction between two kinds of heretics, those who commit heresies from inside the system and those who do so from the outside. The first type can be sometimes correct, the second never. V. was the never-correct type. Says Asimov, "Public support or no, the exoheretic virtually never proves to be right. (How can he be right when he, quite literally, doesn't know what he is talking about?)"

Lest he be pilloried for such bold statements, Asimov has insured himself by the most vulgar kind of verbal trickery: he makes insiders out of outsiders if they have "reached the peak of professional excellence" whatever that is. So naturally — once again he says it — "the exoheretic... is virtually never right, and the history of science contains no great advance, to my knowledge, initiated by an exoheretic." There is no arguing with such foolishness. The foolishness, I must add, is compounded by self-contradiction, for is not Asimov's gun hired to introduce this book because he has a large public that buys books? So here is Asimov, the outsider, depending upon the public which, he says, is always wrong, to follow him in his denunciation of heresy.

But matters become worse for Isaac Asimov. He says that the scientific establishment (calling it the "scientific orthodoxy") is "completely helpless if the heretic is not a professional scientist — if he does not depend on grants or appointments, and if he places his views before the world through some medium other than the learned journal." That is, the establishment can withhold grants, appointments, and publication from its own heretical members, but cannot from "exoheretics" or outsiders. That leaves the public as the only outlet for the exoheretic's views, but Asimov says that the public is never right; "the appeal to the public is, of course, valueless from the scientific standpoint." He does not seem to realize that he is condemning himself and science, for he seems to approve this situation while granting that in rare instances an inside heretic is incorrectly punished. I cannot easily believe that the two publishers (Cornell University and W.W.

Norton) and the several authors, especially not the clever Carl Sagan — but how can one watch out for everyone's business? — did not read carefully the few passages that prefaced their great act.

* * *

In the years of which we speak, Deg had a part to play in the establishment and it was not a bad life. He turned up in Washington from time to time. He lunched with the Executive Director of the Political Science Association, where he was for a time a Council Member, or at the Senate or the Cosmos Club with friends; William Baroody was funding some of his writings from the American Enterprise Institute for Public Policy Research. Earl Voss and Tom Johnson there were pleasant companions; it was a smallish show, then, close to the Republican Presidents and Conservative Democrats in Congress, not the big and famous show that it became later, after his direct relations with it ceased. Deg knew a number of Senators. He had access to the U.S. Office of Education when Frank Keppel of the Harvard Graduate School of Education had gone to run it, for he had worked with Keppel at the Harvard Graduate School of Education and had been offered appointment there. He consulted with the Department of Defense when "winning the hearts and minds" of Vietnamese was top priority, and went to Vietnam on a panel requested by General Westmoreland, then Commander-in-Chief. He had acquaintances who were in the top echelons of half a dozen great companies, and half a dozen of the largest foundations, others who were millionaires, UN ambassadors and bureaucrats, New York politicians, and so on. He helped leaders like Nelson Rockefeller on occasion (without compensation). He went as a delegate to UNESCO. He helped the Publisher of *Life* magazine to help the American Jewish Committee to establish better relations with the Vatican, and was shoved by a wily Spanish priest for a moment into the ample arms of dear old wobbly-eared reformer, Pope John XXIII.

The New York University President, James Hester, also from Princeton, was as friendly as he could be to a faculty troublemaker. The departmental faculty itself was to Deg's ways of thinking too petty, unintellectual and anarchic to launch upon large schemes, and moreover his giant University was always in a state of imminent financial collapse. After his first year there, he had to bring in practically all of the funding for his projects from foundations and gifts, which is not so difficult when one is in the swim of things. His middle-level university income from his tenured appointment was supplemented by consulting fees, honoraria, and grants. He spent all the money that he could spare on his *American Behavioral Scientist,* which was felt to have a good influence on social science research, and gave him editorial influence, whether critical, or to help friends, or to assist students and up-and-coming scholars to get ahead.

Publishers were easy to come by. Advances were generous for

textbooks, subsidies for the others. Complimentary books flooded his library, he could stop at practically any university in the world and be invited to lecture, dine, discuss. He travelled abroad often, always with jobs to do, always funded at least in part by some agency (never The Agency) or foundation.

To hear him tell the story, he could have gone on and on this way with *la dolce vita,* spreading his wings of influence over more and more people, things, and activities. He could have dawdled more with attractive women, driven a new car, worn new suits, written books with ex-Presidents, etc. Why this was not actually his way, his route, his fate, could have been foretold in childhood. I doubt that he fully realized it. But perhaps enough of the reasons become evident in the pages of this book to preserve us from going back to the "Roaring Twenties" of Chicago, Ill., U.S.A.

There seems little reason to doubt Deg, however, when he cites his friend Ithiel de Sola Pool's analysis of networks. By a calculus of probability, given an unstructured society, the chances of any person knowing a person who knows another person who knows any other particular singled-out person in the society are very high. Theoretically, given the relatively sharply structured society everywhere, he could be introduced to anyone, even in the worldwide society. Deg, in his old notes on Pool's manuscript, figures that he practically needs know only his own widely differentiated acquaintances to know anybody in the top elite, and needs but jump one more acquaintanceship to meet just about anybody else. He even made a parlor game out of his theory, picking by chance a name, address and occupation from any directory, and proceeding to say who whom he knew would know this person. This occurs because a person who knows 2000 people is in a position to know the, say, 500 acquaintances of each of these, and this million, with its 500 acquaintances each, exceeds the population by far, but since the population is stratified, the number falls short of total success until the chain is extended.

There are applications of network theory to the workings of science. Conventional science, we know, is not a juggernaut, a palpable monster, a solid phalanx, a disciplined corps of bureaucrats, a theocracy, or even an organized political party. It is — it must be, in order to avoid its own contradiction — a subtle, diffuse, often impenetrable, often disguised, often unconsciously composed network of relationships.

Marxist scholars would readily comprehend this fact and would tie the whole network to the economic production mechanisms of the capitalist system. The Chicago School of political science would see in it promptly the manifestations of Mosca's "political formula and ruling class" and Deg's "ideological imperative."

Discriminated against indifferently in American Society, evangelical Christians such as many Baptists, represented in a growing movement of "Creation Science," but usually acting individually from their nooks and crannies in the system, would also be characteristically alert to the operation of the scientific reception system. So would the large number of individual American and British heretics who compose a disinherited, not formally qualified, keen, and occupationally and characterologically diverse "watch and ward" network, ready to suspect the worst of the establishment. Resembling these latter would be many a disenchanted student, not yet amalgamated into the conventional system.

All of these together, plus the simply curious, might readily muster the kind of crowd that assembled to witness the Velikovsky panel convoked by the program committee of the American Association for the Advancement of Science at San Francisco. The audience, well over one thousand persons, was by far the largest of the Convention.

Let me now explain how it happens that the scientific network, or establishment, might in this case, as it has often done in the history of science, be acting against its own presumed interests and hence to repress new correct theories. How does the ruling formula of science triumph over challenging ideas, making them heretical, and chastising their proponents?

Every field of knowledge is nowadays organized. It has therefore leaders. Some of these leaders are parochial. Others have connections with relevant social networks and organizations of other fields and other segments of society. These leaders acquire fame (which already represents the same circular system of the generation past, advancing for instance a Menzel, who inherits from a Harlow Shapley, or a de Grazia, who inherits from a Charles Merriam.)

The mass media, though it hardly reports science, seeks out or gives access to fame. Reporters, woefully unprepared, interview the leaders. Educational media, including widespread fund-seeking alumni magazines, turn to their exemplaries of the famous. The occasional television, radio, and magazine concerns about the knowledge industry result in reports that are favorable to the same group. Foundations appoint from the same leaders to their boards of trustees and consulting committees. So do scientific and political government agencies, although other interests can intrude more here. The leaders, and now we are speaking of some five thousand persons, give awards disproportionately to each other, as do generals and admirals. Government foundations, such as the National Science Foundation, are even more susceptible to network influence than private foundations.

In the area of book publishing, the ideas of the leaders largely determine what manuscripts shall be published as textbooks, and on what kinds of books the university presses should spend their small resources. Trade book publishers for the general public have almost no viable interest in serious scientific or humanistic work. Usually what they publish in these areas is meant to blossom quickly and die, to challenge no strong interest, and certainly not to offer alternatives to major scientific paradigms unless they would join the ranks of somewhat disreputable and financially insecure publishers. Thus, if Velikovsky had published with Lyle Stuart's firm instead of the Macmillan company originally, the opposition would never have gathered. They had to have as their target a press that would seek to avoid censure for "conduct unbecoming a gentleman."

The scientific and professional magazines that report new knowledge are governed by boards and editors, who are acceptable to the leaders and are watched rather carefully by them. Fading away from the specialized periodicals are magazines of popular science, few of which are financially secure and all of which are dependent upon the good will of the leaders. The *Scientific American,* for example, would never wittingly go beyond the activities of the core elements of a science. When a troublesome or controversial theory surfaces on its pages, evidencing a conflict between two leader-led theories, it seeks to appease both sides by a second article or letters of comment. Its need to seem "original" is fed by lavish illustrations, a feature it shares with the *National Geographic Magazine,* the *Smithsonian, Discovery* and other periodicals. By editorial tricks, all such magazines lend their materials a glamour and adventurism that they usually do not in reality possess.

The network of leaders extends down through the public secondary and elementary schools from the colleges by way of lesser sheikhs, supervising boards, and *hoi polloi* of the fields. Not even the threat of teaching "creation science" in some state will excite overly the nabobs. The legal and journalistic techniques for handling anti-Darwinism have long been known, and a legion of educators moves efficiently into battle on this front with little direct participation of the national leadership. Private secular schools — the Lawrenceville Academies and Grotons — would never wish their pupils to utter the wrong titles or theories in anticipation of entering the halls of learning hallowed by the leadership. The Catholic schools are deintellectualized; nor has the Catholic Church yet retracted its judgement against Galileo.

A word, finally, about the corporate world, where so much applied and some pure research is done, from which, too, funds must flow increasingly into the coffers of the universities. Their corporate images, hence their profits, depend upon the skills people come to believe *(via* advertising and public relations) that they command and engross. Like university presidents, leaders of science dip into corporate treasuries on occasion as consultants, board members, and officers. Just as retired generals are common in the aerospace and engineering industries, highly placed scientists, even without the need to retire, are frequently positioned in corporate structures.

Immersed in this and in all that has gone before, a leader of the establishment network has almost no incentive whatsoever to take up a new controversial theory, much less to originate one himself. He is himself subject to disciplinary actions, often quite subtle, should he stray from the fold.

The network can be most simply presented as a list of institutions through which the leaders of science operate or upon which they exert influence. The influence is continuous, is intensified on crucial issues and in my opinion, is generally beneficial and should be enhanced throughout the system. Meanwhile, however, the influence needs consciousness-raising and built-in mechanisms of reform.

LEADERS OF SCIENCE
extend their influence into:

1. Audio-Visual Media
   (fame; reportage)
   a. TV and radio Networks
   b. Public Broadcasting
   c. Documentary films
2. Popular Press
   a. Scientoid Magazines
   b. Science Fiction
   c. Publicity (columnists)
   d. Newspaper and newsmagazines
3. Book Publishing
   a. Trade
   b. Textbooks
   c. University Press
4. Scientific Journals
5. Universities
   a. Secular Schools
   b. Religious Schools
6. Scientific Associations
7. Foundations (private)
8. Governments
   a. Executive offices, commissions
   b. Legislatures
   c. Government Foundations, Prizes, etc.
9. Corporations
   a. Research and consultation
   b. Board of Directors

The leaders of science in the English-speaking world can be numbered from 50 to 10,000, depending upon where you wish to draw the line of influence. They are fairly concentrated geographically in the Northeast Megalopolis, Chicago, Washington, and the San Francisco Bay Area, with a small English contingent, fairly closely in touch.

An extraordinary fact is that this immense scattered network ultimately engaging the whole world is composed of what in business or government would be regarded as absurdly small units. They are like the oldtime Piggly-Wiggly small grocery store, owner-operated network, not fully centralized, bureaucratic establishment. Furthermore, it is largely subconscious or scarcely perceived. Nevertheless, in the end — and merely to picture the network — the librarian in Juneau, Alaska, the student at the University of Tampa (Florida), the editors of the *Times Literary Supplement,* CBS, PBS, NSF, the Ford Foundation, Harvard University Press, the Board of Education of the City of Chicago, the engineers of Western Electric, the science section of the *New York Times,* the editors of *Science Magazine* and its popular offshoot *Science 84,* the National Academy of Sciences, the curators of the Museum of Natural History in New York, and many thousands of other "nerve endings" of the science system of communications and influence respond to cues and jiggles of power from the elite group.

Surely, it is one of the most benign elites of the world. It probably rules easier and can rule less than almost all other elites. Its punishments are relatively light. It stupefies people but all forms of rule stupefy their clients or subjects; here, indeed, the science elite is more enlightening, in its double function of stupefying and enlightening, in its S/E ratio, than most elite or influence networks. But its exists, and it is effective. To evade or avoid or attack the Scientific Establishment, to invade its inner sanctum and transform its Holy of Holies, its ideological center, its paradigms, *Weltanschauung,* ruling formulas, or whatever one might wish to call its heart, is the work of decades and, at least before, of centuries, and, in the words of Lasswell, almost always involves the process of "partial incorporation," by which is meant that before the revolution is won, the elite changes its behavior to concede the victory and keep out the revolutionary personnel.

Thus the monarchical regimes of Europe incorporated in most cases the key ideas of the French Revolution before the republican revolutionaries conquered them, and the capitalist regimes went "welfare state" before the socialists could take power; so that, if the quantavolutionary movement were to seriously threaten the ruling elite of Newtonian stabilitarian and Darwinian gradualist uniformitarians, these would be reacting, as in fact they are acting now, to incorporate the quantavolutionary formulas and outlook.

Meanwhile the quantavolutionary movement would be formed out of mistakes of the existing regime, out of apostates, and disaffected scientists and engineers, occult publishers, little presses, small personal foundations, religious creationists, maverick legislators, fugitive publications sliding out of xerox machines, and a motley public crowd of dissenting readers and talkers. Sooner or later, according to Roberto Michel's "Iron Law of Oligarchy," the Scientific Establishment would be modified in attitude, beliefs, practices and personnel but would still be the oligarchy, or, let us say, "a better and more enlightened class of leaders."

CHAPTER FOURTEEN

THE FOIBLES OF HERETICS

For his first half-dozen years on Naxos, Deg stayed in a town apartment that Venetians had built in the 13th Century; then he moved out to his stone house on the isolated promontory of Stylida. In these places, much of the Quantavolution Series was written. Deg's permanent encampment at Stylida was of marbled stone and primitively equipped, not a cabin, neither a villa. Antiques jostled useful junk on the marble tables and shelves. He pounded nails into the walls and from them everything dangled. Empty plastic bags were stuffed behind shelves for further use, empty bottles were hoarded. String, cord and rope in odd lengths were saved and hung up. From this frugal perch on the hill, he contemplated the serene seascape before him and the battling cats of the world beyond, not excepting the heretics.

Saving rope reminded him of Frank Knight, exemplar of the *laissez-faire* Chicago School of Economics who, in his office at the University of Chicago used to store the string he too saved. According to an eyewitness, he was mounting a train for the East one day when he called out to his waving family, pointing, "There, get that piece of string!" His highly regarded economics, thought Deg, were nicely encompassable by Homo Schizo theory.

Knight's colleague, the very liberal U.S. Senator Paul Douglas, was dining in Manhattan another time with Robert Merriam, Assistant to President Eisenhower, and with Deg, and Douglas told of a Republican Senator who had ridiculed the incessant internecine fighting among the Democrats; "like a bunch of alley-cats" they were. Whereupon Paul had risen to add, "That may be true, but what in the end is the result — many more cats!" And while they were laughing, the waiter handed the distinguished-looking elderly gentleman the bill and they had to laugh more as the Scot, Quaker, economist, and statesman, and foe of loose spending, winced, grumbled, and paid.

The cosmic heretics, bereft of resources, collected pieces of string to build bold systems. Coming out of nowhere, and without structure or discipline, they fought like alley-cats. Rebuffed by the world of the press and science, they often became morose.

Deg's journal, January 25, 1970

> I spoke to Immanual on the telephone. He is feeling poorly and he intimates both a throat ailment and sinister external moves as the source. We are all suffering vague symptoms in the world. For months, I have felt this and that pain and scarcely know to what to attribute them? There are thirty physical and psychical causes all intermingled and the physical uneasiness is appropriately vague. So many millions in the world are, I think, similarly affected. It is as if the germs of diseases were directed by a mastermind, who says to them, "Now man has learned to be specific and special in his therapies, so you must now be as vague as possible, so that he will not know what he is suffering from."

Deg might as well have gone on to talk of the generalized "germ" of schizotypus, which suffuses human nature and finds a great many ways of emerging in disease, now specific, now general. It may be no coincidence that in this decade two reciprocal kinds of slogan clashed with each other in the mind of society, the one aimed at pandemic expressing of paranoia, the other at fighting off paranoia, so that everyone was "unavailable" and "by appointment only," and "fill out the form" while people were telling one another "reach out and touch someone." Highly special acts of terrorism increased around the world as highly general public opinion surveys showed the public regarding every group of leaders and every special group as untrustworthy, including their own national and world leaders.

"The most despicable of all ways of suppression is denying to me the originality and correctness of my predictions." So said Velikovsky at a philosophical panel at Notre-Dame on November 2, 1974. He was directing himself at the moment to Professor Michael Friedlander. Friedlander had announced: "One of the things I'm not going to do is to attempt to defend the many foolish, and intemperate, and venomous statements that have been made by scientists over the last 25 years." He proceeded then to incite Velikovsky's outburst (which one might also call "foolish, intemperate, and venomous") by addressing himself to V.'s astronomical scenario of the Venus encounter with Earth.

> To be useful a prediction must be derivable logically and unambiguously from the model. If the prediction bears only a tenuous relation to the model, then the validation of that prediction may in fact say nothing about the

model.

In rebuttal, V. pointed to the details of his own early claims: that Venus was incandescent in historical times; that the planet had to be very hot to carry the gaseous hydrocarbon clouds that he believed to be there; and that he had declared the first announced temperatures of 600 degrees to have been too low, and in fact they were.

What constitutes a prediction gives grounds for incessant quarreling and name-calling. Deg was convinced that scores of his own prognostications in sociology, economics, and politics could be culled from his own books and shown to have been realized. For instance, he had predicted at one time that the achievement of equal population districts ("one man — one vote"), so stoutly advocated by the cities of America, would result in heavier political weight for the cities' chief frustration, their own suburbs. He was not surprised nor did he put in a claim) when the prediction was fulfilled. He never got around to predicting when the world would end, but, should it end, he could in the thereafter cite some highly probable estimates.

I did not know when Velikovsky got onto the claims and predictions "kick." I am guessing that the famous letter by Bargmann and Motz got him going. It was the first nice thing ever said about him in a scientific journal. The letter was V.'s idea and he provided much of the contents. It asserted that V. had suggested radio noises were emanating from Jupiter and were discoverable; they were discovered serendipitously by Burke and Franklin over a year later. Further, in 1950, V. said that the surface of Venus must be very hot, and, sure enough, by 1961 the heat had been discovered by reliable instruments.

Practically nothing was said of the method employed to arrive at these advance claims. But so guilty are scientists in the matter of "claims" and "priorities" that V. profited greatly from his cryptic and general utterances. And, no doubt, had he been guiding NASA research, these items would have been systematically uncovered.

The practice of advancing priorities is childish and the idea of proving a general cosmogony by a race of claims is ludicrous. There can be no crucial test or event. Even if Venus were to slip its moorings and drift toward Earth tomorrow, the historical scenario would not be proven. If the cosmogony is accepted for working purposes, the prediction (or test) will have meaning; if the cosmogony is not accepted, the prediction cannot be stated. This is shown by the resilient way in which the great heat of Venus has been claimed as a greenhouse effect by Sagan and others.

A member of the audience at the Notre-Dame panel made the most fitting remarks:

> Each side has constructed its own version of what would count as a crucial test, and has constructed its own judgement as to how that test has been

passed or failed. This is a singularly sterile manner for resolving disputes... As far as rational dispute is concerned, we have to begin by saying we might be wrong... to say what would count against us in our own book.

It would certainly be appropriate, within every scientific work and in a discussion of it, to confess its weaknesses, to argue its null-hypotheses. We are bound to do a poor job of attacking ourselves. And, of course, disputation may overburden issues to the harm of clear presentation of the theses. Nevertheless, Deg, in writing *Chaos and Creation,* was anxious enough about excessive positive argumentation to give over a chapter to the Devil's Advocate.

In one sense, the cosmic heretics in the Velikovsky case were a conservative group, asking for law and order in science, demanding even that the letter of the law be followed, all the more because their substantive ideas
— erratic planets, forceful electricity in space, short geological time, etc.
— were deemed untrue. In fact, like the typical heretical group in politics or religion, they had logically to deny that the word "heretic" could apply to themselves; for theirs was the truth. To those who like myself believe that science enjoys only hypothetical and useful "truths," a scientific heresy is logically impossible. Heresy is an excrescence of authorities.

Heretics typically are intolerant of other heretics, if only to hold together their highly vulnerable and unruly group within a mishmash of ideas. We find a push-pull phenomenon occurring: the heretics are pushed out of conventional science and attract or pull in the religious, the occult, ESP, "Ancient Astronauts," UFO's and astrology, the eccentric, and the revolutionary types. All of this provides a hustle and bustle on the fringes of science. And scientists are normally neurotic about their fringes. Only the wisest (read "self-aware and self-knowing") and self-loving of them could understand and sympathize with what they saw going on.

Onetime, in the fall of 1976, far from the scene of action, Deg heard distant sounds of strife and the name called out of his old friend, Professor Paul Kurtz, a pragmatist philosopher and Editor of the *Humanist* magazine. Besides many pleasant hours working together, Deg remembered how Kurtz had let him introduce a scatological remark into an article of this well-mannered publication. He wrote Kurtz a tender of good offices, suggesting attention ought to be given to neo-catastrophism, and sending a privately printed essay on Homo Sapiens Schizotypicalis.

Kurtz replied (in confidence, for he was a careful keeper of the peace) explaining that the fracas had generated out of a single sentence against Velikovsky in an article by Sprague de Camp, a detested figure among Velikovsky's cult. Kurtz said that even if he had wished to do so, he could not censor de Camp. He was startled by the vehement and even menacing letters that he received arising first from publishing the De Camp article and then from a possibly garbled quotation of him in the *Washington Post.* At the same time, Kurtz acknowledges, "The followers of Velikovsky claim that he was unfairly treated by Shapley, etc — with which I fully agree, I remember full well your justifiable concern." He was, he said, open-minded, aware of general disbelief in V.'s theories, but not conversant with them, or with Deg's, for that matter, and he wanted to know Deg's theory of evolution: "Your thesis is most creatively provocative. My major question is what does it do to the theory of evolution?"

Deg told V. of Kurtz's letter, V. spoke to Greenberg, and Greenberg fired off a letter to Deg, wondering how he had come to be in touch with Kurtz, and retelling the story as he saw it: "Kurtz may be your friend, but we are certainly not enemies." Deg could only wonder once more at how Greenberg could turn any situation into a personal threat and from this into an aggression.

The Humanist did publish an article by V., defending himself strongly against the then current voices of his opponents. Possibly the pressure of anger unjustified impelled *The Humanist* to give V. his say; after all, isn't the lesson of democratic politics that a group needs anger, not justice, to make its point?

V. was lucky enough to have a few opponents who made a hobby of him. They kept an eye on the news about him and cast enough aspersions his way to keep his more diligent supporters in fine fettle. In keeping with the history of ostracized movements, nearly all of the heretics worked part-time at the job. Most were poor, although they did not reveal their poverty, like the old Parisian bohemians. They were, too, mostly unreliable, partly because of their busy-ness and hand-to-mouth existence, and because they were not under the lash of the dollar, but also because they were often afflicted with intense inner struggles. I would quote Nietzsche regarding them, "It takes a chaos within oneself to give birth to a shooting star." "That's it, they're crazy," one might say, which is a fraudulent pretense of those who are crazy-normals.

Astronomy professor George O. Abell of U.C.L.A. writing in the| *Skeptical Inquirer* says that the followers of V. "are actually following somebody who may be a bit crazy. For isn't there something psychotic about a person who claims that he alone in a field with which he is unfamiliar, can fathom the pure truth, while hundreds of thousands of specialists with lifetimes of experience behind them are muddling about in the darkness? And doesn't the popular acceptance of such a scientific-religious hero suggest a problem, or at least some kind of an unfilled need, on the part of the follower?"

Deg's Journal, Princeton, December 27, 1978

Warner Sizemore here yesterday, 10.45-1.30, discussing many affairs.

He reported that not only Greenberg and others were angry at the SIS magazine group in England but that Velikovsky was upset because of their caviling at points and their undermining his theories instead of developing them.

Further V. ordered Sizemore and Greenberg to drop Peter James as Senior Editor from the editorial board of *Kronos* in three months, or else he would give them no further material of his own to print. James is associate editor for the historical content of *SISR* and also on the *Kronos* board.

Then, says Sizemore, V. reconsidered and told them that he didn't mean what he said. Sizemore did not guess whether this was a conclusion of principle or of expedience. (There are several reasons for expedience: the scandal, the harm to Sizemore and Greenberg, as well as *Kronos,* etc.). In the later case V. would remain guilty of the very behavior of scientist upon which his own case of persecution is based in part. If his retraction of his order was in principle, then the action may be partially excused because it was withdrawn.

It is not the first time that V. has come perilously close to practicing the behavior of his enemies. He is by character domineering and suppression of the opposition would come easily to him under other circumstances.

V. had been called a charlatan but there was nothing to it. Deg asked himself, how could anyone use the word? And that they used it as others use curses and obscenities. At most, on occasion and like most men, he believed suspiciously hard in ideas that were not so firm, but noone, thought Deg, in this sense had ever written a thoroughly honest book and none ever could, by the very limits of language, for language is fundamentally a compendium of psychic tricks, played upon oneself and others, fraudulent in a sense.

But now, I think, reflecting upon the heretics, that fraud is a remote cousin of pretension. To lay claim to something is a human necessity. Yet whoever has any claims must be a fraud. To say "I am alive!" is a pretense and a fraud, a boastful claim to what after all is a delusion about nature, a question begged. We are all such frauds.

There is something else, too, another kind of subtle fraud, a fraud in the too delicate sense of being wronged, and this V. had. One who feels that he has been defrauded is a fraud, as, for instance, in criminology, many victims of fraud are engaged in attempted fraud to begin with, making money out of nothing, etc... And then persuading others that one has been defrauded, is also a fraud. At such persuasive tactics, V. was a master.

He could persuade by overpowering belief and documentation that he had been defrauded on a grand scale. He could persuade the most pathetically defrauded people that he had been defrauded more than they, and the defrauded turned their purses of energy and sympathy over to him. For he had converted his defrauding into the collective conscience and was collecting retribution and returns on his defrauding because his supporters neglected their own suits in order to pursue his suit but received no more than abstract justice.

It was as if all the gas company's customers thought they were cheated and put all their energies into the case of one of them, making the case landmark, but the favorable decision on behalf of the test case resulted only in the vindication and compensation of that person, while the rest could not afford to sue, and the gas company hardly changed its practices.

* * *

Now the time had come for Deg to print *Chaos and Creation*. It was 1980. An outsider, innocent of the sociology of heretical groups, would expect the publication of *Chaos and Creation* to be welcomed. The field would open up further. Fresh material would offer itself for discussion. The implications of the work of V. would be extended. New possibilities would be manifest. There might even be some personal congratulations in order, for no one had yet produced any considerable work in the format of a book that could be readily assimilated to most of what the readers of *Kronos* were versed in and attentive to.

Not at all. When the book was in page proofs, it induced the dormant strain in relations between the directors of *Kronos* and Deg to rupture into hostilities. The occasion for the hostilities came, as it often does in human relations, whether personal or international, out of a situation promising well. Executive Editor of *Kronos* Sizemore and Deg were meeting weekly out of friendship. They ate, drank, walked and talked together for hours on end. Sizemore was enthusiastic about Deg's manuscript of Moses, and had also been reading *Chaos and Creation* as the proofs arrived from India.

At the time, Deg and Ami had largely abandoned Manhattan and were living in a tiny apartment in Princeton, writing their books, and spending as little money as possible in order to pay for the production of *Chaos and Creation* in Bombay. When Warner came to visit, they would huddle their sizeable frames together amicably amidst piles of books and papers for a-while, until Ami would retreat to the second room to write upon the kitchen table between the sink and the small bed.

The Indian production was nightmarish. A thick file of correspondence attests to the pains engendered by cultural and physical distance. A perfect book was out of question. The work was being set in hot type, linotype, which, unlike the word processors of today, lets new errors creep in as rapidly as old mistakes are expunged. For weeks a strike of Indian paper mills stopped supplies to the printer. The quality of the paper, never good, worried Deg, too. The poor Indians were trying to conserve their old machines and paper and ink and Deg could not tell from the proofs whether fonts were broken or the paper was refusing the bad ink, and, worse, whether the final printing impression would be uniform on the pages. The book was loaded with proper names of extreme diversity, with illustrations, and with hundreds of citations, three most common sources of typographical, printing, and formatting mistakes. Deg had known the same printers from a decade before; they had printed *Kalos: What is to be done with our World* and *Kalotics;* he had been to their shop; he liked the several owners and workers. But it was a different world, of different standards, and to convert it acceptably to American tastes, while keeping costs down and work within hailing distance of the schedule, was continually frustrating.

Warner, believing Deg would be pleased (and no doubt he would have been pleased) to see some portion of the work printed, sent (without Deg's knowledge) a photocopy of the page proofs to Greenberg, then in Florida, and spoke to Greenberg about the progress of the work in the course of their frequent telephone conversations. Greenberg was enraged by errors still in the proofs, or so the issue was presented to Deg by Warner. Deg, already upset by the defects and by the difficulty of close editorial supervision of the work, was angered by the report and asked Greenberg on the phone to be specific about the work being "full of errors." When the letter came, the little that was added to the mistakes transmitted by telephone was rushed off to India for correction. There were mistakes so slight as a compositor's misspelling of Greenberg's name in a footnote crediting him with work, and a wrong middle initial for Earl R. Milton, who received 'Earl S.', a complimentary psychological mistake tying him to a dear old professor of Deg, Earl S. Johnson, the same to whom *the Divine Succession* is dedicated.

Writes Greenberg:

> After going through half of the text of *Chaos and Creation,* the Citation, and Bibliography, I have decided to enclose a sampling of pages that is symptomatic of the entire work. The kind of repair help that you need goes far beyond any *gratis* assistance that I could provide. I have already spent the better part of three days reading your book and no relief appears in sight. Typos abound, names are misspelled, publications are improperly cited and dated, many dates are questionable and just plain wrong, not to mention glaring omissions from the published literature. The catastrophic sequence proposed by Velikovsky has been rearranged (Mercuria *precedes* Jovia) and work by people such as Warlow has been uncritically accepted, etc., etc.

He goes on to list various, mostly brief, articles, and certain contributors to *Kronos* that were not in Deg's bibliography (the longest and most complete that had ever appeared on catastrophism and Quantavolution), concluding "What you have done is downright insulting and I find it hard to believe that it wasn't deliberate."

Deg replies on April 2 from Princeton:

> You agreed to telephone me collect, later on, and to recite your list of such findings into my tape-recorder. You knew that the need for any corrections was immediate. I kept the machine by my telephone for six days more and

now here is your letter. Several additional typographical errors are indicated, two of which I wish I might change, along with the aforesaid. Otherwise your letter pullulates with grotesque exaggeration, unsupported allegations, hostility, and vanity. Dealing with paranoia makes one paranoid: could it be that you first promised and then decided not to offer corrections of the proofs because you want to be free to slander the book?

Deg was surprised at the rapidity with which the situation deteriorated. Sizemore, father, organizer; producer, financier, executive editor and trouble-shooter for *Kronos* let Deg understand that a selection from the book would not be printed and that the book would not be reviewed. Deg scoffed at this: how could it not be reviewed? Whose magazine was it? It would be a mockery of the pretenses of *Kronos* magazine, both substantive and libertarian, to suppress its mention. Warner unhappily suggested that the book need not be reviewed in *Kronos*. Deg insisted that Warner do something about the matter, to no avail. Their warm friendship abruptly froze.

Many months later, the book arrives from India. A review copy was sent to Greenberg. Other copies were sold to respondents from an announcement by way of the mails. One day in April of 1982, Deg received a letter from Stephen Franklin, whom he did not know. [I find that they exchanged letters many years before.]

Dear Dr. DeGrazia:

I wish to obtain a copy of your book *Chaos and Creation*. Please let me know whether I may obtain this directly from you, & if so, how much, etc. If not, where? I am enclosing a copy of a letter I received from *Kronos* since I feel you may be interested in how they are handling requests for information about your book....

Franklin was referring to a letter from Leroy Ellenberger, who had been promoted from a free-lance gadfly on V.'s opponents to Executive Secretary of *Kronos*. The letter was written on *Kronos* letterhead with a Glassboro State College address, and did not oblige Franklin's request for Deg's address. The letter follows:

Dear Mr. Franklin:

With respect to the book *Chaos and Creation* which is the subject of your March 25th inquiry, be advised that *KRONOS* has chosen, after examining it, not to be associated with its promotion or distribution. For your information, the book was published privately in India. Its author is in charge of its commercialization.

As a reader of KRONOS, you are no doubt aware that we are not averse to presenting a critical approach to Velikovsky and that we will entertain responsible alternative, and even opposing, views. Given our interest in developing a Velikovsky-based catastrophist alternative to unifomitarianism, we would be more than anxious to inform our readers of new, fruitful sources of information. The book in question leaves too much to be desired to merit, in our opinion, serious attention.

If your curiosity gets the better of you, so be it. *CAVEAT EMPTOR.*

Deg called Franklin, received authorization to use his name when raising the issue, and with malice afterthought, sent a letter to the President of the College, reproaching him for letting the College be a party to damaging slander through people who were pretending to be connected with the School. Official action and an apology were asked. Expectedly, there came no reply, but Sizemore was aggrieved by the step, calling it ridiculous and a charade.

Meanwhile, Deg chose out of the "staff" of *Kronos* several individuals whom he knew personally. He wrote to ask them their attitude respecting not reviewing his work. All replied sympathetically; still not one found the issue serious enough to deliver an ultimatum to *Kronos,* not Frederick Juenemann, not Cardona, not Lynn Rose.

Rose aroused Deg's ire for postulating an enmity between Greenberg and Deg which did not exist, and evaded the issue of Ellenberger. (Deg liked ornery characters like Greenberg more than suave types like Rose.) He wished to hurl at Rose a statement in *Kronos* made by V. against Storer of the AAAS panel: "One who maintains 'neutrality' between a gross offender and the victim of the offense does not give an objective account of the realities; the account is *biased* in favor of the offender."

Even Earl Milton who was so close a friend and collaborator did not take up a strong position. Irving Wolfe at the University of Montreal replied that *Chaos and Creation* should be reviewed and said that he would tell Greenberg so. Greenberg held firm, something he was good at doing; some of the heat was turned against Ellenberger, as if his letter had been a willful rash act, and a decline in his fortunes began, partly accounting for his retirement to his original home base in St. Louis. But Deg regarded Ellenberger and even Sizemore as toys of Greenberg in this instance. Toys for what? For psychiatric play-therapy, he insisted.

Many months later, as three of the "Staff" and friends including Deg sprawled about a sunny dock and swam in the August waters of Lake Kashagawigamog near Halliburton, Ontario, they talked of the affair and all seemed to agree (no vote being taken) that Lew Greenberg was acting the dog in the manger, that he acted so habitually, that Ellenberger was irresponsible, that the book should be reviewed, that Deg should cool down his reactor, and that *Kronos* would collapse if Greenberg resigned, as he frequently threatened to do. And if *Kronos* collapsed, where would its 2000 readers go, and where would its score of writers go to publish their articles? Dwardu Cordona, a writer and editor of hard opinion but essentially sweet character, asserted he would bring up the matter with Greenberg again. Deg was noncommittal. Later on, he did receive a letter of Cardona from Vancouver mentioning, *inter alia,* that he talked to Greenberg, who was still without remorse, and even still angry.

The past could not be recaptured, despite the restoration of a distant relationship, and the major issue remained (the refusal to review *Chaos and Creation).* Sizemore sent a note of condolences when Deg's mother died and then another note apologizing for addressing the first note to "Albert" instead of "Alfred." Deg had not noticed the mistake or, more properly, had noticed it and thought nothing of it. Now he apprehended that the printers' errors, which misspelled Greenberg's name in one place, etc., and the personal slips that made Earl R. into Earl S., and so on, might be compared with changing the name of Alfred to Albert, this involving a close friend of many years. Poor Sizemore, thought Deg, caught up in an object lesson; I should have thrown the fit of rage he expected.

Sizemore was at this time enormously busy. He had four major occupations, beginning with his professorship in philosophy and theology for one. Secondly, he was, as I said before, a creative artist who had put aside his larger skills to create a singular commodity, friezes in wood, copying in detail great (or lesser) paintings. And these he carried around to sell at fairs on certain weekends, and while sitting by his works he read books and articles and newspapers by the bag-load. Then he entered upon the national Amway corporation, and began to build a network of clients and customers to purchase a wide range of consumer goods; this entailed meeting upon meeting; much of the vast energy that had gone into advancing and promoting Velikovsky was moving into a truly American promotional enterprise — part crass materialist, part ideological fervor, a hybrid of love-thy-neighbor and get-rich-quick. Deg would not join him; he regretted the diversion of the intelligent energies that had placed Sizemore among the top dozen of no more than a few score active promoters of quantavolution in the world.

Yet he understood the figure of the missionary-capitalist, for he was reminded of the time he studied the leading Caucasian families of Hawaii, who had emerged from their work at Christian conversion owning a good part of the land, commerce, and industry of the Islands. He believed, unlike others, that Sizemore and his wife, who had never before plunged into an enterprise with him, might well make a fortune. Max Weber, Richard Tawney, Edward Shils, Sebastian de Grazia, Benjamin Nelson and their brethren of economic sociology would instantly recognize the puritan-capitalist nexus in Amway and in Warner Sizemore.

Nor, meanwhile, excepting his break with Deg, did Sizemore neglect his primary responsibilities in quantavolution. He still was the mainstay of Greenberg (and I do believe that Sizemore, were he to strike it rich, would generously fund *Kronos* and set up seminars, publish books, and promote the general development of the field); he still visited and helped Elisheva; he kept up with the field. He aided friends in need, as he did Sigmund Kardas, first when Kardas moved his house, and then when Kardas was nearly killed crashing into a wrong-turning trailer truck one midnight on the highway near Bordentown.

In October, 1982, upon returning from Greece, Deg was still needling Sizemore:

Dear Warner:

I hope that all goes well with your enterprise.

I trust that you have known of *Kronos'* decision last winter to not review *Chaos and Creation,* After your long history of interest in the book and its writing, this must have come as a surprise to you. Have you spoken to the staff about it?

Before leaving for Greece last Spring I submitted a note to Jan Sammer as Associate Editor of *Kronos* to read and forward for publication. I commented upon Velikovsky's Baalbek article. Sammer has since reported to me that when he told Greenberg about it, Greenberg said that he would not read it or publish it. This appears to be one more step in the recapitulation of the unconscionable techniques which, we say, were employed in regard to Dr. V.

Also, out of the blue sky came the enclosed letter from Ellenberger. [Not carried here.] I cannot afford the hours of rebuttal and psychiatric analysis that it calls for. What should I do with it?

Are you, or are you not, Executive Editor, father confessor, and angel of this mad show?

Sincerely yours,
Al

P.S. As you may know, we have been denied the privilege of renting *Kronos'* mailing list to announce the publication of *Chaos and Creation.* On the other hand, I have received in the mail on more than one occasion postcards advertising Leroy Ellenberger's Velikovsky T-shirts, beer mugs, etc., using *Kronos* addresses. I fail to appreciate the philosophical principle at work here; should you not consult with Lynn Rose and advise me on it?

The letter aroused Sizemore to stiffer opposition. He railed at Deg for trying to separate *KRONOS* from its Glassboro State College letterhead, and advanced two propositions. The first was that "factual errors" in

Chaos and Creation (which apparently he had not discovered in his intensive and enthusiastic reading of the manuscript and page proofs over a period of months) made its mention in the pages of *KRONOS* impossible "it would be difficult with such errors as would reflect upon our integrity." Second he rejected any analogy between the treatment which the reviewing media had meted out to Velikovsky and that which was rendered Deg by *KRONOS,* adding that V. had "not once in forty years of correspondence with his opponents" resorted to "invective or scorn." This is close to the literal truth, just as the fact that General Eisenhower never killed an enemy soldier.

Such ruptures of relations among heretics are common. In this instance the main material effect was to suppress attention to Deg's book for three years among a key audience for works on quantavolution, represented by *Kronos* magazine. By the end of 1983 Greenberg was intimating an interest in advertising and reviewing Deg's books.

I have come near to demonstrating that grand principles of morals and science can equally well be extracted from the dross of existence or flare out of imperial trumpets. The phenomenon of "self-destruct" is ever threatening in new movements of all kinds. Yet another phenomenon here deserves mention before passing on to other matters. It has to do with energetics, or more simply laziness. And I am fortunate for having spoken so much of Sizemore for he exemplifies the non-lazy, the antithesis of the phenomenon of limited energetics or laziness. The phenomenon has also to do with the motives of the persons in fringe movements, with what they want to get out of their belonging and in fact do get.

The cosmic heretics were fond of reciting the litany, Velikovsky in the lead, that if his new ideas were to be admitted to scientific discussion, the textbooks of most disciplines would have to be revised. Astronomers would have to correct their own lamentable errors, and also they would have to study electricity, geologists astronomy, anthropologists geology, historians mythology, and so on. At the same time, a number of cosmic heretics were solely Velikovsky buffs: they were incompetent and unfamiliar with other quantavolutionists. Some had never had, nor now wished to have, an education broader than that afforded by *Worlds in Collision.* They derived their political, moral, and intellectual sustenance from a couple of books and a man. They were housed in this comfortable concrete defensive pill-box from which they would sporadically fire and venture forth on forays and to scavenge.

To this type the threat of *Chaos and Creation* was as real as a full-scale attack upon *Worlds in Collision.*

To read another thick book? And more to come? A hobby would have to become a chore. Horrid possibilities in religion, geochronology, and human development had to be confronted. Much reading was required. A "snap-course," with its slogans, became suddenly a curriculum.

The format and style of the new book was itself a threat; it read well, but was organized like a text-book. The several hundred readers of its first year found even a chapter in it devoted to negative criticism. The chapter, called "The Devil's Advocate," was written by Deg under his dropped middle name of Joseph and an English translation of "Grazia" into "Grace" for the cognomen. He felt that a full self-critique, carried as he went along, would have been useful but would have doubled the size of the book. So he did his best to demolish his work in a single chapter.

That he succeeded with some is evidenced by an editor of Athenaeum Press who, in rejecting the manuscript, claimed to be persuaded by Professor Grace, and by a review in the newsletter of the Canadian Society for Interdisciplinary Studies, whose author wrote that much of what he had to say was well put by Joseph Grace. Deg did not like subterfuge and had foreseen that a reader who liked or disagreed with the chapter would soon enough catch on to the dodge. Still, Elisheva read it and was amazed by its being there and asked Deg who the writer was. That caused a laugh. And Leroy Ellenberger himself, even after hearing the explanation, was so suspicious and perplexed that he wrote to Deg to confirm that the writer was not a professor at Glassboro State College. Deg noted with interest that Leroy, who would not let the readers of *Kronos* hear of the book, was reading it, presumably having wrapped it in a plain cover after receiving the gift from Deg.

* * *

On January 17, 1982, Brian Moore is telling Deg about the difficulties the British Society is having with its publications and asking him to come and share a platform with Dr. Don Robins who is to speak on isotopic anomalies in radiochronometry. The Society would also like a talk on the past ten years since Deg published *The Velikovsky Affair*.

> Incidentally, mention of the Velikovsky Affair above reminds me of my current fracas with Lewis Greenberg which you may like to include in your comprehensive survey of the history of Velikovskianism (when you eventually come to write it). I had received permission from Dr. Hewsen to print in *SISR* his talk to the last Symposium at Princeton in which he criticized Velikovsky's use of his sources. Lewis, of course, would not print it in *Kronos* as it was too critical for his taste, but as we advertise ourselves as a forum for the Velikovsky "debate," we felt it could be a useful contribution from an informed Velikovskian. The result was hugely ironical; Greenberg has threatened us with legal action if we publish it as the words were actually spoken at his Symposium. To me it seems the ultimate sin for a Velikovskian to attempt to suppress views which he finds unpalatable, but when I put this point to Greenberg he avoids the question and suggests we terminate the correspondence! There the matter rests for the moment. Rather sad.

Deg notes to himself on the margin of Moore's letter: "Shall I send letter to Lew on this with copies to *Kronos* board?"

He does not do so. Instead, he calls Professor Hewsen, and later replies to Moore:

> I spoke to Hewsen about your fracas with Greenberg, also Sizemore. Neither H nor S is strongly interested in the matter; H confirms the offer to you but thinks G is serious about a suit; S would advise against such an action, which, to my mind, would be only taken up by a lawyer as nutty as G. H. never gave away any rights to publish. And, of course, the attitude of G is disgusting. I find G's policies and behavior frequently irrational and arbitrary, and have not talked to him in some time. S is occupied with a new commercial venture now as well as teaching, so sees into little. Ellenberger and G do the whole bit. I think that G would do battle with all the 1500 Kronos subscribers and all authors and with Mrs. Velikovsky and Shulamith Velikovsky and anyone else who would come into sight, especially all females; he is the most handsome rhinoceros in these parts and generally exhausted from his struggles.

And Brian answers:

> SIS still seems to be persona (prope) non grata with Mrs. Velikovsky. She would not allow us to put slips in the British edn. of *M in A* drawing attention to the Society. We are also excluded from the book itself though *Kronos* is listed. Warlow's book of course lists both organisations (though this has not stopped *Kronos* from berating him in their latest issue. With colleagues like this, who needs the Sagasimov?). Which reminds me — I mentioned the Hewsen Affair in my last letter and this obviously prompted you to enquire a little into the matter. I'm afraid this has fuelled Leroy's paranoia even more. When I last wrote to him I said I was not going to pursue the matter but he now thinks that I "asked" you to "intervene" on our behalf and gave me a little homily on hypocrisy to boot! Still, don't lose any sleep over this — such misunderstandings are endemic in our relations with *Kronos*. Leroy and I continue to collaborate on other matters so there is still a positive side to the relationship.

Greenberg and Ellenberger manage next to enrage Peter James, who has a sweet disposition but a sharp tongue. He resigns from *Kronos'* editorial board with a vengeance.

> Dear Lewis and Leroy,
>
> In view of the present shitty relations between KRONOS and *SISR* I can't see much good reason to provide Kronos with any further copy...
>
> Permission on "Darwinian Man" is withdrawn (or at least suspended).

The same applies to my *BAR* and Stiebing correspondence, and to the promised section on Carchemish from my Glasgow Conference paper. Whether this material has been set in type or not, permission is *firmly withheld*. I had also better tender my resignation from the KRONOS staff as well...

Frankly I don't see why Hewsen's paper has put the wind up you lot so much. On the other hand maybe I do. All Hewsen was saying is that we must not treat Velikovsky as a tin god, and that we would be doing far more service to the man's genius by admitting the weak parts of his work and sorting the wheat from the chaff. The KRONOS staff suppress his paper (yes, *suppress)*, at the same time protesting that they are not Velikovsky cultists. Give me one GOOD REASON why Hewsen's comments should not have the publication that *he wanted* them to have, apart from the desire of the KRONOS staff to suppress a point of view that doesn't exactly square with their own.

I am, to say the least, *disgusted*. I thought the name of the game was free speech and fair discussion. The "Velikovsky movement" has been crowing for so long about the suppression of Velikovsky's ideas. It makes me sick to see people who pontificate against Velikovsky's enemies do the same to someone who is basically sympathetic to Velikovsky's ideas. Go to the back of the class and join the Shapleys and the Sagans. You should both hang your heads in shame.

There was nothing untoward or irregular about Brian's letter to Hewsen. It was not going behind Lewis' back, conniving or in any way deserving the hysterical reaction we got. *Hewsen* wrote the bloody paper, a fact that seems to have been forgotten in this silly squabble, not Lewis Greenberg or Leroy Ellenberger. Brian quite rightly wrote to Hewsen about it, and *asked* him to clear things with LMG. There was no intention of "stealing" anything without KRONOS permission. Hewsen was asked to request KRONOS permission. Get that straight. Nothing criminal, nothing strange. The reaction? Sheer hysteria, and the usual childish threats of legal action. And why? You tell me why. Ask yourselves, have a good think about your real reasons for trying to suppress someone's thoughts...

I also find KRONOS' attitude to Peter Warlow rather weird. Why have you got it in for him? Answer: JEALOUSY, plain and simple. If he lived in the States and was one of your immediate clique you would be breaking your backs to help him find some answers to Slabinski, instead of running him down all the time as you do. Along comes the guy who for the first time produces a model and a mechanism for a Velikovskian event and publishes it in a well established physics journal, and you lot just try and jump on him. Rose, in his comments about Senmut's ceiling, doesn't even seem to be aware of Lowery and Reade's extensive studies, or Reade's later work on the Ramesside star-tables. What are you going to put in place of Warlow's model, which satisfies the mythological and geological evidence so well? Spin reversal? Crustal slipping? Go on then. Provide us with a model that will make Slabinski happy. You know damn well that Slabinski's calculations can't and don't take into account electro-magnetic effects. These are, after all, part and parcel of the Velikovskian view of celestial mechanics. So why do you take such great delight in Slabinski's calculations when they ignore them? Answer: jealousy.

I have taken a lot of stick from KRONOS staff for the criticisms I made of *Ramses II and His Time* in my review. Letters from Greenberg, Rose, and others made an incredible fuss as if my criticisms had come out of the blue, and I was told repeatedly that I was knocking Velikovsky's view of this period without putting anything in its place. On the 19th February 1976 I wrote a 5 page letter to Velikovsky, summarising several years work, pointing out my major objections to his equation of the Hittites and the Chaldeans, and the 19th and 26th dynasties. In February 1977 Velikovsky wrote back pretty well ignoring the points made, except to postulate an *ad hoc* invention of a second Neriglissar to get around problems in the Neo-Babylonian succession. In 1978 *Ramses II* appeared, and the major areas of problem which I had pointed out were almost completely ignored. The reader was left totally in the dark about key material that shows Velikovsky's scheme for this period to be impossible. So I felt perfectly justified in raising this problem for the benefit of *SISR* readers. It would have been intellectually dishonest not to have done so, particularly since I had raised the main points with Velikovsky two years before...

KRONOS no longer strikes me as a " magazine of inter-disciplinary synthesis"; it is rapidly becoming a cross between a Velikovsky fan magazine and an anti- *SIS Review*...

I am very sorry that it has come to this. But when KRONOS is filled over and over again with one-sided *ad hominem* piffle about Gammon, MacKie and Warlow, three of the most valuable contributors to the Velikovsky debate, and when KRONOS still continues to treat Velikovsky's work *in toto* as the proverbial sacred cow, then things have gone too far. I am only interested in having honest assessments of Velikovsky's work, to find out what is right and what is wrong. I am not interested in a silly KRONOS vs. *SISR* struggle which seems to interest you far more than the academic issues involved...

Peter James

But this is only part of the letter which I suppose might be summed up in the words of St. Paul to the Phillipians (1:15): "Of course, some of them preach Christ because they are jealous and quarrelsome, but others preach him with all good will."

* * *

The tenor of discourse among the heretics, we have seen, is often as vituperative as the salvos of heretics against the outside world. It is also more personal and intensely felt. There were times when Deg felt that Greenberg's tiny clique of *Kronos* was trying to make a sort of Trotsky out of him for advocating world revolution rather than "revolution in Russia" as Stalin would have it. He was consoled to know that the invectives and diatribes were the lot of other heretics and conventional figures venturing into the line of fire. Nor was he without blame; so that he could not but remind himself of the saying, "He who lives by the sword dies by the sword." Or "he who lives by the pen is poisoned by the pen."

By contrast with the heretics, the conventional scientists were most gentle among themselves on the subject of the heretics. It was almost unprecedented when once Robert Jastrow mentioned in print a serious statistical misapprehension of Carl Sagan in an attack on Velikovsky; Sagan defended himself vociferously. I do not mean to say that the conventionals are more fair or decent; they are nicer and more polite, and must go to print under institutional barriers against vehement expression. The heretic cries havoc and unleashes the dogs of war, and is often too distraught to tell friend from foe.

If all of this seems trivial, that is because the word "trivial" for a dispute is defined by contrast with horrible and bloody conflict. Or, I think, it is all trivial, even when there is horror and bloodshed. Examine the horror and bloodshed of history. Is it not very often over the trivial — a sentence of Marx, an oath to the King, a remark "against the people," a failure to salute the flag, the greasing of bullets with pork fat, these and a myriad of like trivia — which manage to bathe mankind in bloodshed and keep him in terror much of the time.

One can never tell from a virulent heretical letter or a smooth conventional reasoned critique whether, were the author possessed of the power, he would not exercise violent sanctions. The men and women who run affairs — in all spheres of life — are very often like the infant whose rages, so ludicrous, would be regarded with the gravest concern and even panic if *abracadabra* suddenly the infant sprang up adult and armed.

But that is the point of keeping the peace at nearly any cost: if people are kept from destroying themselves and each other, sooner or later they will be happy that they failed in their wishes. They will recognize that their aims are foolish, trivial, misguided, and mistaken, or that they would have been themselves erased, or that their enemies had agreed in principle with them, or that they and their enemies, alone or together, might find a better resolution of their mutual problem.

What has been shown here is that the establishment has violated most rules of logic and fair play in literary and scientific intercourse, but, further, I have shown that the heretics, in dealing with the outer world and among themselves, have also violated most rules of logic and fair play in their literary and scientific intercourse.

What then can be concluded as a matter of principle? Call down a plague upon both their houses? Go in search of honest men like Diogenes forever carrying a lantern to illuminate any rare finds? Favor the weak against the strong, the heretic against the conventional establishment? Continue to expose such illogical and unjust conduct wherever and whenever it appears? Psychoanalyze, especially in the sense of self-analysis, everybody including ourselves? Reform the scientific reception system by institutional inventions to bring about a rule of law, emplaced as part and parcel of the rules of scientific method?

The questions answer themselves. Each implies a herculean task. Yet each implies a remedy of value. The answer to each and all of these questions is a resounding "Yes!" All must be done, no matter that each in itself is, if not impossible, exceedingly difficult. In *Homo Schizo I* and *II*, Deg put forward a persuasive, if apparently pessimistic, analysis of human nature. Homo Schizo is incurable by nature. He can only be modified, constrained, trained, and controlled within limits. But within these limits stand at the one extreme the most horrible conduct and at the other extreme the most charming, endearing, and harmless conduct. The main trouble in the latter case is human unreliability.

Meanwhile, work was beginning on *The Cosmic Heretics* and I wrote Carl Sagan in 1981 asking for a meeting in the line of reporting first-hand something of Sagan's ideas about Velikovsky and about himself. A reply came, dated 9 November, 1981:

9 *November, 1981*

Belated but very sincere thanks for your letter to Professor Sagan asking if he might meet with you at some point while he is in New York City to discuss Immanuel Velikovsky as part of the background for the book you plan to write about Velikovsky. Unfortunately, Dr. Sagan is now totally immersed in science, having just returned to Cornell after an absence of more than two years. To his regret, he will not be able to accept your invitation. If you have not yet read it, you might wish to have a look at the chapter on Velikovsky in Dr. Sagan's book, *Broca's Brain*, published in paperback by Ballantine in 1980.

With kind regards,

Cordially,
Shirley J. Arden
Executive Assistant to Carl Sagan

I had indeed known of the aforesaid chapter, which had already appeared in at least three different publications and which had been mauled and dissected to the point of uselessness, Brian Moore's *SISR* review being perhaps the most nicely done of the valid commentaries upon the book. Perhaps a rebirth would come with the baptism of being "totally immersed in science" that would impel him to drive his own *Cosmos* TV series off the airwaves. Or to withdraw his book, *The Dragons of Eden,* from circulation, of which N.J. Macintosh wrote in *Nature* (27 April 1978): "It is inaccurate, full of fanciful and unilluminating analogies, infuriatingly un-- systematic, and skims hither and yon over the surface of the subject, unerringly concentrating on the superficial and misleading... profoundly unscientific."

Sagan was the latter-day Harlow Shapley for many a heretic, though Deg could never quite tell why. Sagan had denounced Velikovsky's suppression, criticized his work publicly, and at worst was slipshod and sophomoric. On Deg's last visit among the English heretics in 1983, and amid some chortling, Deg was told of one Michael "Mike" Saunders, a true-believing Englishman, who was representing interests in the never-never lands of the Gulf States sheikhdoms, and was ringing people up with "great" schemes, one of which was to win over Sagan by setting up for him a professorial Chair for Interdisciplinary Studies at Cornell University, counting upon him to sing a new song of solar space. After Deg stopped laughing, he opined that such things had happened before (see, e.g. the Morton Prince case, that is described in the next chapter), but that star professors are much too clever and ornery nowadays. Like the time when a large donation to the Psychology Department for the purpose of pursuing telepathic research was accepted by Stanford University but diverted to other uses, perhaps to construct bigger and better mazes for running rats. Apropos, unlike rats, professors avoid any mazes built for them and devise their own crooked ways. And some are quite principled, need I say?

PART FIVE

CHAPTER FIFTEEN

THE KNOWLEDGE INDUSTRY

Deg detested the new Bobst Library building at New York University from the moment he entered it on 16 December 1972 at 16.00 hours for a reception to celebrate its opening. The old central library had been in the basements of the Main Building. It was rumored that one could draw a book from there, and he did so from time to time. But now they had obstructed the view of Washington Square from his apartment to put up a casbah-red structure that from the outside seemed transported from the Near East while inside there was a giant space towering to twelve tall stories up, a roofed atrium around which wound narrow bands of shelving areas, obviously inadequate save for a few years of collecting, and already requisitioned on its top floor for the administrative officers of the University. The sensation was vertiginous; the building floated with its books tucked around its waist; how could a scholar study with his ideas precarious on the edge of exposed space?

A dance band was playing and he promptly envisioned how the design would permit its use by a Las Vegas concessionaire to bail out the near bankrupt school: a pavilion for dancing on the marble main floor, baths and massage parlors below, a bar on the second floor, social rooms on the third, a bordello for men on the fourth, one for women on the fifth, one for homosexuals on the sixth, then levels of gambling and a sky restaurant. One of the most expensive pieces of land in Manhattan had been used to roof empty space, the spectacle was dizzying. He rarely used the library.

When he was there he would ask himself whether it was hyper-critical of him to have such feelings, part of his basic envy of a world that rushed along without his consent, getting things done nevertheless; or was he simply observant of facts and aesthetics that most people, those in power as well as their subjects, could not see or think of. This happened often, that he would no sooner denounce something, privately or aloud, than he would reprimand himself for thinking that he could see truth and value and contradictions thereof that groups of intelligent people working in financial, architectural, legislative, and other task forces could not see.

He did not wish to believe only in himself; he would rather enjoy the warmth of consensus, the applause of the crowd, but it would rarely work out so. Everything he did, everything he got, it seemed to him, even under the conditions when he was boss, gave him not a whole loaf, nor even half a loaf, but a thin slice. (I am not speaking of material goods, but of the quality of the product.) The situation regarding money alone was bad enough; the incompetency of the rich society to obtain value with its money was much worse to suffer.

Throughout his career, Deg found that it was harder to get money, the better the cause. A wage for oneself was not difficult, a salary slightly more so, commercial money for an imaginative project easier the quicker the turnover and the realization of profit. The trouble with your ideas, Rodman Rockefeller said to him once while they were conspiring about the world, is that they do not involve things that people regularly consume in large quantities, like canned food and cement houses. Not that Rodman was spectacularly successful with his company, IBEC, which went progressively from more romantic to less romantic, from third world to first world projects, in those years. Deg wondered at how year after year Rod could go on administering — ever so comfortably to be sure — a business without breaking out more often into some of the more imaginative enterprises and social adventures that he obviously enjoyed visualizing. Deg blamed affable father Nelson for the suppression.

To continue on money: then longer-term money became harder, then money for a vulgar or fashionable charity, then money for important research or an extraordinary book. Money came hardest for a cause that one believed to be purely for the public good — unless it was a commonly recognized public good like the Bobst Library or some other building for a respectable university to house respectable and vulgar objects, or unless it was a concealed fraction of a public good (the thin slice of the loaf again), like a significant sociological question slipped into an advertising survey for dog food, or unless it was illegally obtained, wherefore some political radicals have robbed banks and others their families, and still others lived under miserable and dangerous conditions.

Deg made a dozen attempts in search of a teaching and study platform for catastrophe and quantavolution. Recall this was a period when all kinds of new courses were being pressed upon universities and colleges; standards were in general decline. Professors were wringing their hands and burying their files for safekeeping. Yet they consistently rejected the advances (never mind seeking the help) of quantavolutionists who had more respect for the traditional research materials of the culture — in classics, linguistics, foreign languages, history of science, philosophy, etc. and whose attractiveness to students would have erected a massive barrier against the anti-intellectual and book-condemning feelings rampant in student bodies everywhere.

A score of teaching heretics had managed to insert V.'s materials into their courses under various pretexts and in several cases could even carry his name in the title or subtitle of a course. The Dartmouth Experimental College at Hanover, N.H., invited V. onetime for two days of meetings with a seminar; at least six faculty members of as many different disciplines met with the seminar before and after to discuss his books *Worlds in Collision* and *Earth in Upheaval*.

V. was generally unhappy about the educational system, although he was displeased, too, with the student rebellions when they occurred. A dramatic polemic against the system of higher education finally appeared posthumously in three pages of *Mankind in Amnesia* (182-5). At least this statement is available to save him from reproach for never having attacked on general grounds (as opposed to personalized grounds) the foundations of authority or their institutions.

Before converting his own social invention course to a course on quantavolution, a one-time unauthorized change to which no official objection was made, Deg tried a frontal appeal. Here, in 1973, he addresses an assistant dean for curriculum, after discussing the matter with Bayly Winder, Dean and friend. He is trying to make as few waves as possible, by placing the course in the summer session (where "imaginative offerings" are encouraged). The proposal went to the Committee of Deans:

October 29, 1973

Memo to: Dr. Sylvia Konigsberg

From: Professor Alfred de Grazia

Subject: A proposal for a summer
           *Institute on Primeval Catastrophe
           and the Development of Human Nature*

A large and increasing public is interested in the theory that ancient astrophysical and geophysical disasters caused profound changes in the human environment and human nature. Much of the interest centers around the work of Immanuel Velikovsky and his school of thought. Wherever Velikovsky appears to speak, his supporters and critics assemble by the hundreds and even thousands. His sole talk at NYU drew hundreds of students and professors several years ago.

I have worked for a decade on problems raised by Dr. Velikovsky since the publication of my book, *The Velikovsky Affair* in 1963, and am presently going to press with another book on the disasters of the Homeric Age. A heavy flow of written materials and archaeological reports has begun and promises to be practically endless. There is a need for an academic center for presenting and discussing the problems they present to all fields. Excellent scholars are available to participate. I suggest that such an Institute might be held from July 1-20, 1974, at New York University. It would occupy three hours of classtime on fifteen days, would allow students not-for-credit, undergraduate students for four credits, and graduate students for the same (4-credits). The required readings would amount to 1200 pages and graduate students would prepare a research paper. It is expected that from 80 to 200 students can register for the Institute. Personnel for the course would include:

1. Prof. Alfred de Grazia, Supervising Professor, Full-time;
2. Adjunct Prof. Annette Tobia, Ph.D., Einstein University in microbiology and presently lecturer at NYU, full-time;
3. Prof. William Mullen, Ph.D., Princeton University classicist (one-third-time);
4. Prof. Livio Stecchini, Ph.D., JD, Patterson State College, historian of science (one-third-time);
5. Mr. Ralph Juergens, Engineer and astro-physicist, Associate Editor of *Pensee* magazine, (one-third-time);
6. Visiting Lecturers and Discussants (one day each): Professors I. Velikovsky; (general theory); Lynn Rose, SUNY, (philosophy); Frank Dachille, Pennsylvania State Univ., (geology); Edward Schorr, Fellow, American School of Classical Studies (archaeology); and possibly an additional person or substitute;
7. Prof. Nina Mavridis, CUNY, Political Scientist, administrative coordinator, full-time.

There would be fifteen primary one-hour lectures and 30 one-hour discussion meetings which would break the lecture audience into small sections of 25 persons. Related lectures and discussions would meet on the same day.

The titles on the lectures follow:

Primeval Catastrophes and
the Development of Human Nature

I. Time, Nature and Human Beings
   1. The Theory of Catastrophes — De Grazia
   2. Origins of Human Nature — De Grazia
   3. The Geological Record — D'Achille or Burgstahler
   4. Historiography of the Solar System — Stecchini
   5. Correlations of Geology and Astrophysics — Juergens
   6. The Synchronization of Prehistory — Mullen

II. Case Studies in Disaster and Development

7.	Case I: Atlantis	Stecchini
8.	Case II: The Age of Pyramids	Stecchini
9.	Case III: Exodus	Velikovsky
10.	Case IV: The Homeric Age	De Grazia

III. Origins of Behavior and Institutions

11.	Theology and Government	De Grazia
12.	Literature and the Arts	De Grazia
13.	Sexuality and Aggression	Tobia
14.	Technology	Stecchini

IV. Final Problems

15.	Is Human Nature Governable?	De Grazia

Discussion leaders: Professors De Grazia, Tobia, Stecchini, Mullen, Juergens, DAchille, Burgstahler, Mavridis. With 100 students, nine daily section meetings are required. If the number of students exceeds 100, we should add to the faculty.

Readings; In addition to several paperback books that will be required, the staff will prepare a collection of readings difficult of access, and xerox them. The basic readings will be *Worlds in Collision* by I. Velikovsky, the study of Homeric catastrophe and literature by A. de Grazia, and the collection of readings that will represent, among others, the rest of the faculty. A valuable and unique supplementary bibliography will also be provided, and, finally, a set of maps, drawings, and a special lexicon.

Continuation of Project: We would like to begin work on the project as soon as it appears probable that we would have 80 students, and to continue research in connection with, and to prepare for, successive Institutes. Therefore, it is suggested that 50% of the gross receipts from student fees (less additional faculty costs) for students in excess of 100 in number be placed in a special project fund in the University for continuing study and development of materials in the subject-area.

27 November 1973

TO: Professor Alfred de Grazia
FROM: R.B. Winder

The Committee of Deans discussed on Thursday, 15 November the proposal for a summer institute on primeval catastrophes as outlined in your memorandum of 29 October addressed to Dean Konigsberg. The consensus was that although the proposal might very well produce a large enthusiastic audience of paying customers, it probably would not do so from degree candidates. The Committee felt SCE might be interested in sponsoring the program, and I suggest that you take it up with Dean Russell Smith forthwith.

I do appreciate the drive you are putting forth for funding of various sorts and am only sorry that we felt this one would not work in the context proposed.

Nothing could be worked out in the unprestigious "School for Continuing Education. "My academic readers can practice a dry run on this proposal, or another like it as carried in *The Burning of Troy:* their own committees might well respond similarly. Practically all universities in America capture their students with "credit courses" and find "course anomalies" as distasteful as anomalies in science.

The New School for Social Research was not so impeded, although it too became divided into "non-credit" and "credit" areas. V. gave a successful series of lectures there in 1964. Clark Whelton also taught there a non-credit course on "the Velikovsky Question" in the Fall of 1979 and significantly some students kept in touch with him afterwards, interested in keeping informed and hoping to form an association.

Milton to de Grazia, February 15, 1980:

> Our department is being reviewed, and me with it. Trainor is one of the referees, the other is hostile. Yesterday he said, Milton is not doing physics because *Kronos* is not included in *Physics Abstracts* nor *Science Citation Index.* That remark deserves immortality. Hang in there, Al, we're winning.

Milton was a popular professor at Lethbridge University and was teaching and reading quantavolution in his general physics and astronomy classes. He was an intellectual force on the vast Canadian Prairie, in touch with the press and radio systems. He knew the vast skies there like a Polynesian navigator. His lifelong asthma kept him in a lifelong course in advanced nutrition, organic chemistry, and atmospheric science. Then he read into myth and legend, and there was no stopping him. In every picture he discovered fresh signs. Aside from his personal qualities, he could connect with the more than ordinary number of students there who had heard everything good about God and the Bible at home, but nothing at all, if not bad, about these subjects in "education." Even only to hear the Bible being used as a learning tool was exciting to them. One should recall, too, how low the estate of physics had fallen.

We find our Dean of science reporters, Walter Sullivan of the *New York Times,* admonishing us.

> "Physics is the most basic of the sciences, apart perhaps from mathematics. All phenomena, when probed to full depth, are controlled by its laws... Yet

physics is in trouble... Student enrollments in that science have plummeted... There is a public distrust of physicists that borders on revulsion and the physicists themselves are pursuing lines of research more and more remote from the problems of everyday life..."

Sullivan's key lines were the juxtaposition of two anomalies — public paranoia and physicists' schizoid remoteness of character, traits that do not marry well. The American Physical Society was discussing the low state of physics, and Sullivan wrote that generally the leaders thought that more money should be spent by the government. The British physicist and astronomer, Fred Hoyle, wanted even greater accelerators. He also wanted scientists to participate in politics. "You see why the world of politics is such an indescribable mess. Think of the opening of the baseball season. Think of the ceremonial first pitch. Think of what the baseball season would be like if that sort of pitching went on right through the summer. Then you have it — the present state of affairs." Presumably under Hoyle's new-age baseball, physicists would pitch and baseball would become nothing but home-runs as the batters perfect themselves to bang away at the invariable straight-ball coming right down the center. Or perhaps Hoyle was saying that physicists should join the pluralist republic, as the ethnic strain of physics, helping where they could. Deg was not sure this was "according to Hoyle," but he liked the idea.

Milton tied together the Eastern and Western Canadians, and the Canadian belt triangulated to the Princeton-Trenton-Philadelphia area where Sizemore, Deg, and Greenberg kept shop. In the Kronos network, besides Greenberg, Sizemore and Ellenberger, might be found Rose, Vaughan, Wolfe, Cardona, and Jueneman. Some say that there should be added Milton, Sherrard, Westcott, Hewsen, Ransom, Talbott and Sammer. It was a unifocal net, with Greenberg as the focus. Deg connected with London, Holland, Paris, Basel. Greenberg, losing Peter James in London, found Bernard Newgrosh as correspondent. Marvin Luckerman, a worker and doctoral student at the University of California at Los Angeles, founded a biennial magazine, *Catastrophism and Ancient History;* relations with Greenberg were cool, and the British were not much impressed with his first issues but praised the good try. Still he rounded up a thousand readers and began to improve his journal. The creationist groups stemming out of Los Angeles, Ann Arbor, and Seattle were quantavolutionary perforce, having been given only a few thousand years by the Bible to produce everything. Here and there were quantavolutionaries of orthodox connections — Gould at Harvard in paleontology, Ager of geology in England, and so on for several countries. The password that could readily cut these out from others was their answer to the question: "Has a planet moved?"

A very small group it all was, absurdly so when compared with the network of thousands of periodicals, scores of associations, and the mass media that served orthodox science. It makes one wonder whether the heretics were worth considering: certainly by the usual American standards of great size and multiplex technology they were not.

* * *

Deg heard when young from his democratic teachers how smartly the vested interests turned to minister to public needs, and was continually surprised when old to see how reluctant they had become to give themselves away. As his friend Lasswell put it, when writing with Abe Kaplan *Power and Society,* no ruling class gives up its goods without being forced to do so. This goes *pari passu* for philanthropoids and publishers, two industries affected with a public interest. The philosopher, artist, composed author, administrative innovator, and physical inventor, if he is to be creative, typically is driven to become a sneakthief, or revolutionary, or go mad, or all three. So says Deg, who worried only about becoming a revolutionary, because then he would have to spend his time among sneakthiefs and maddies as well.

"Of course the heretics would not get support, they did not apply for it. One must play the game by the rules. Apply and apply and apply again." Deg knew more about this than his heretical acquaintances by the time they had encountered one another. He had enjoyed the fleshpots and studied what motivated the foundations, publishers and universities. He could warn the heretics that they need hardly try — and V. was of this opinion, too — or, worse, in order to succeed, they must prepare themselves to spend much of their energies in trying, and he was insistent upon a point that few could appreciate, that only a peculiar type of masochistic personality could apply incessantly to the point of success without losing the vigor, freshness, profundity of his ideas and the vital energy needed to pursue them for their own sakes.

On a few occasions, the heretics would solicit funds from individuals in small amounts to disseminate a publication about Veiikovsky, but efforts at larger funding failed. The Foundation for Studies of Modern Science initiated a series a approaches, of which I have already spoken; still, I shall add one more instance.

Murray Rossant, Director of the Twentieth Century Fund, was reported by someone to be attracted to V.'s work. Because Deg and his brother, Sebastian, were already known and had been working with the Fund in very different fields, FOSMOS sent two fresh and handsome faces to meet with Rossant and his colleague Schwartz, Bruce Mainwaring and Coleman Morton, both enlightened businessmen. A friendly encounter ensued, the upshot of which was that, although the Fund had never gone into this area, the two officers were interested personally in seeking out other sources of funding, and when all was said and done, nothing happened. Nothing, that is, except that the Fund itself gave money to Giorgio de Santillana and Hertha von Dechend for research that they were doing on ancient and primitive myth and legend which, it was believed beforehand, would show that mankind was clever and scientific long before it was credited with being so, but also that there was no need to invoke catastrophism to explain the nature of mankind's early preoccupations.

This was recounted to Deg and the others by Stecchini, who was well acquainted with Santillana and von Dechend. The product of the research, *Hamlet's Mill,* was welcomed by the heretics, nevertheless, for its intimations of ancient quantavolutions, but, if the reader wishes to understand the rampant confusion of the book, he may simply apply the hypothesis: here are two great scatomatized experts trying to avoid mention of catastrophism.

Though they be liberal or conservative, foundations are unlikely to be creative. They think they are able to judge creativity, of course, and, especially if large, "creativity" and "the independent sector" of society are often included in their slogans. Their size and their bureaucracy correlate well.

> "But in any event," writes Deg, who had urged the Ford Foundation to apply this his scheme, "they are unlikely to make lists of all the people who lay creative claim to their bounty, and dispense it equally among a random sample of them. No, they put the applicants and petitioners through the hurdles that they learned in their first course in Business and Public Administration should be set up to employ typists and junior managers. So it happens that if all the people who ever applied for a Guggenheim Fellowship had given the same quantity of intense energy to a story, a painting, a song, or a study as they gave to applying, American culture would be up a notch or two over all its length and breadth. The waste of creative energies going into the national foundations of the sciences, arts and humanities is truly enormous; they use up at least a tenth of the country's creativity, with their sick games between the insiders and the outsiders. I would close them down and give their hundreds of millions to the colleges of the country — whatever their defects — in proportion to their budgets."

The cosmic heretics might discern that they were outlaws without going to the trouble of applying for their identity cards. But they could not help themselves: after all, they were educated in a way, bathed regularly, were fluent in the language, and found their interests carried in the index of foundation provenances. So they were tempted from time to time to try for a grant or subsidy. To my knowledge, they invariably failed. (I am not speaking of the occasional hand-outs tendered by friends and other heretics but of the system of lending a hand as institutionalized by the private or government foundations.)

Deg had enjoyed many experiences with foundations, small and large. The large were too "responsible" and proper to be bold. The small were generally pets and hobby horses of their founders.

Exceptions occurred that were interested in large social issues. A small foundation, the Relm-Earhart group, was a pleasure to deal with. It had a tough board, and was administered by James Kennedy and Richard Ware, both of whom bet on the man, not the institution, and did not try to make useless work for themselves and others. (The Cornuelle brothers, Herb and Dick, were this way too when they were in the foundation business. So was Bill Baroody.) Deg did a variety of economic and political studies with their help over the years. They were not occupied with ancient history or natural history. Since they lent you aid, they must be "good," I say to Deg sarcastically. Very well, he says, shall I give you some bad ones that have helped me? Never mind, I said, I'm in enough trouble with you already.

Yet the very deprivations and constraints that held Deg in his quantavolutionary trap made him more determined and passionate. Again Deg is writing in his notebook, perhaps to warn himself, like a politician warns himself to refuse favors or an infantryman warns himself to keep his feet clean:

> There is this in common among a gold miner, a terrorist, and a purveyor of new ideas; they often come to exist in a new moral dimension, called immorality and outrage. Lunacy, lying, cheating, contempt and inconsiderateness for others; misappropriation: the pandora's box of the creator spills these out.

Deg never committed such follies — almost never — and blamed his frustrations correctly or incorrectly upon his own character: he inspired himself but could rarely inspire enough of the all-important others. Society is run by networks and gangs, and you have to join a gang, stick with it, use it and let it use you, and if ultimately you fail or perish with the gang, well, that's the end of the trail, it's a life-term establishment. Most gangs and networks fail. Therefore skill and luck in getting into and out of the right gangs is often essential to success.

"We're working on an ABS issue about what needs to be done with the science of economics," said Deg to his colleague, Professor Arnold Zurcher, who was also Director of the Alfred P. Sloan Foundation. The Foundation operated in this area and Deg wondered whether they would provide support for the project in the neighborhood of $10,000. His colleague represented an approach to political science that Deg regarded as outmoded and was intent upon replacing. He was a jolly fellow and they were friends, and he knew that Deg was carrying the weak finances of the *American Behavioral Scientist* on his back. Do up the proposal, he said, I think that you have a good chance and I'll support it.

Not long afterwards, Deg received an official letter from the Foundation rejecting the proposal. He was surprised — the request was logical: it was for small money and enjoyed support. His colleague was apologetic. Al, he reported, the proposal passed from one vice-president to another, with Margolis' article from the *Bulletin of Atomic Scientists* about the Velikovsky Affair attached, and a big "No" scribbled on the face of your proposal. (Later on Bill Baroody of the American Enterprise Institute came up with some money to support the issue, and the economists were assembled and the issue published.)

April 22, 1964

Mr. Ralph E. Juergens
416 South Main Street
Hightstown, New Jersey

Dear Mr. Juergens:
    I continue to be amazed that sensible persons continue to give attention to the Velikovsky affair. I wonder if you have read the statement by Howard Margolis in the April 1964 edition of the *Bulletin of the Atomic Scientist.*

    Very sincerely yours,
    Warren Weaver
    Vice President
    Alfred P. Sloan Foundation

    Warren Weaver was a career philanthropist, wrote a good general survey on probability and, like many another, was a nice man. New York University named its Computer Center after him. (For a photo of it, in context, see Deg's *Politics for Better or for Worse.)*

May 4, 1964

Professor Moses Hadas
Columbia University
New York 27, New York

Dear Professor Hadas:
    As long-time subscriber to the *Reporter* magazine — actually since it started — I was very much interested in your excellent review in a recent issue of "Hebrew Myths: The Book of Genesis", by Robert Graves and Raphael Patai. I did draw a long, deep breath, however, when I read in the first paragraph that "in our own time Immanuel Velikovsky, who was maligned for making myth the basis for a cosmic hypothesis, appears to be approaching vindication."
    As a scientist, until 1960 a professor of chemistry at Columbia and an

admiring colleague of yours in Columbia College, I have always regretted the action of a few misguided souls who reacted 13 years ago to "Worlds in Collision" by attacking Velikovsky's publisher — I think it was Macmillan. The book, in my opinion, should have been classified as science fiction but, nevertheless, it was unrealistic, and humorless as well, to expect a publisher interested in profits, as they all have to be, to overlook an opportunity to make a few extra bucks. The reaction to "Worlds in Collision" and a subsequent book, the title of which I do not recall, was fairly violent but, as I remember, reviews by Harrison Brown of Caltech and a woman astronomer with a hyphenated name from Harvard pretty well disposed, so far as I was concerned, of Mr. Velikovsky and his theories of cosmology. But now along comes Mr. Howard Margolis to tell us in a recent issue of the *Bulletin of the Atomic Scientist* that "Velikovsky rides again."

Perhaps you have already seen Margolis' article, but if you have not, I think you may find the attached copy of interest and perhaps amusing.

With kind regards.
    Sincerely yours,
    L.H. Farinholt
    Vice President
    Sloan Foundation

To all medical psychologists: what is the vagus nerve syndrome that make a man "draw a long, deep breath"? *Re* Harrison Brown and the "woman astronomer with a hyphenated name from Harvard, see *The Velikovsky Affair,* Alfred de Grazia, Editor.

6 May 1964

Mr. L. H. Farinholt
Alfred P. Sloan Foundation
630 Fifth Avenue
Rockefeller Center
New York, NY 10020

Dear Mr. Farinholt,

Thank you for your kind letter and its enclosure. I can have no opinion about the validity of Velikovsky's work; his ideas may be wholly misguided, but I know that he is not dishonest. What bothered me was the violence of the attack upon him: if his theories were absurd, would they not have been exposed as such in time without a campaign of vilification? One after another of the reviews misquoted him and then attacked the misquotation. So in the Margolis piece you send me I read "Pi-ha Hiroth which Velikovsky has altered into Pi-ha Khiroth, further enhancing his evidence." But the two are equally acceptable transliterations of the Hebrew, and the latter is the more scientific. For the Egyptian name, Margolis, following old books, writes, Pekharti, but the Egyptian has no vowels, so that the correct form is P-kh-r-t, and of this Ph-khirot is a very plausible expansion. The *ha* in the Hebrew is merely the definite article. It is his critic, not Velikovsky, who is uninformed and rash — and so elsewhere also. The issue is one of ordinary fair play.

Yours sincerely,
Moses Hadas

May 31, 1966

Dr. Warren Weaver
Alfred P. Sloan Foundation
630 Fifth Avenue
New York, NY 10020

Dear Mr. Weaver:

I have harbored for many months your critical note concerning the studies of the *American Behavioral Scientist* on the reactions of scientists to Immanuel Velikovsky, thinking all the while of an appropriate constructive response.

We have recently published an enlarged version of the same studies in book form and I have asked the publishers to send you a copy with my compliments.

There are, of course, two issues in the Velikovsky affair — one, the conduct of scientists and the press; two, validity and utility of his theories. The issues are separable but an involvement in one naturally inclines one into a stance on the other. I think that you can help many people, including myself, find their way through these issues, granted that you may have neither the time nor the inclination to take on major responsibilities for the problems raised.

What I should like to suggest is that we get together for a day's conversation on the two issues in the company of several other men, with the sole end of educating each other. I have in mind persons such as Professor Donald Fleming of the Department of History and Science at Harvard University, Thomas Kuhn, Professor of History and Science at Princeton University, and Professor Harold D. Lasswell at the School of Law at Yale University. I believe that five would be the right number.

I have mentioned a reunion to none of the men named, and have an idea only of Lasswell's thinking about the subject at hand.

We might spend the morning on the question of validity (not "solving" it but working to understand it) and the afternoon on the question of treatment of unorthodox ideas in science.

I am quite at your disposition on the matter. Hoping to receive your opinion, I remain

> Sincerely yours,
> *Alfred de Grazia*
> Editor

There was no reply.

4 March 1974

Dr. Eleanor Sheldon, President
Social Science Research Council
230 Park Avenue
New York City

Dear Dr. Sheldon:

I have become increasingly interested over the past few years in the origins of human nature, prompted largely by a growing familiarity with some new ideas that Dr. I. Velikovsky has introduced in the treatment of pre-historic and ancient catastrophes befalling humanity. The field is not new, of course, and several disciplines in the social sciences and humanities currently share it. But a lively set of controversies with a considerable potential for new discoveries and new syntheses has begun to erupt here and there. Hence there may be occasion for the kind of interdisciplinary research-discussion efforts that are appropriate to the SSRC and ACLS or both.

Perhaps the eye of the cyclone moves around the question: Did homo sapiens become human and cultured in gradual steps, as received theory would have it, or was he compelled to think and behave humanly by the effects of natural forces so immense that factors such as sex, commerce, and "normal" invention must take a secondary role in explanation?

In preparing a monograph on the effects of disasters in homeric times, I have encountered and had to deal with problems that are central, not related incidentally, to the fields of linguistics, historical chronology, astronomy, physical and cultural anthropology, comparative literature, archaeology (worldwide), geology, fossil paleontology, soil chemistry, electromagnetics, astrophysics, sociology of sex, ecology, climatology, oceanography, theology, chemical and fossil dating, psychology of infancy and of stress, epistemology, the history of science, and political science for the origins of theocracy, bureaucratic systems and collective violence.

The problem of approaching the field is not as impossible as might appear from the listing. It can be stated as an excellent model for cross-disciplinary investigation and theory. The numerous sciences involved have been shocked and compressed, taken aback, you might say, and the time may be right for a reappraisal of where they all stand in reference to the question. I have felt continually the need for the kind of sounding board, stabilizer, consulting resources and motivator that I once experienced via the establishment of the first Political Behavior Research Committee of the SSRC, and its subsequent operations.

Should you be of the opinion that the subject might interest the SSRC and be within its jurisdiction, I should appreciate the chance to discuss it with you in some detail.

      Sincerely yours,
      Alfred de Grazia

April 5, 1974

Dear Professor de Grazia:

Thank you for your interesting letter of March 4, in which you suggest a possible role for the Council in exploring human socio-cultural evolution, particularly in the light of an hypothesis that posits discontinuous advances, following a massive challenge and response model, rather than incremental steps.

It is true that this kind of problem is inherently cross-disciplinary, is of potentially great interest, and needs strong guidance if it is to make progress. Also, I am aware that Velikovsky's ideas are receiving wide attention again — or, perhaps, at last. Nevertheless, the topic you outline, which demands a unified approach, is too enormous for the SSRC to handle, and even if the ACLS were to be involved (obviously, I cannot speak for the ACLS), it would still be unlikely that we could marshall the appropriate efforts. At the very least, the physical sciences, as you point out, would have to be closely involved.

As you know, the Council is now addressing itself to more than a full intellectual and administrative agenda, and I cannot foresee a way in which we could be helpful with this topic. It certainly deserves attention, however, and I wish you success in your capable efforts to bring that about.

      Sincerely yours,
      Eleanor Bernert Sheldon

In reflecting upon all that happened to V. and to Deg and the others, it would be unfortunate to keep one's eyes on the immediate characters alone. For they are all symbols, too, players in a drama, representing types of our civilization. If V. is subject of a hundred book reviews, these reviews are signs of the times that happened to gather electrostatically like fluff around his work.

J.B.S. Haldane, a noted biologist who also wrote on *Science and Ethics,* found V.'s *Worlds in Collision* a degradation of both science and religion, a peculiarly enraging combination, apparently, for a marxist and fellow-traveller, whom Deg, with a long nose for hidden political mazes, suspected might be showing the flag (red, that is) for his American colleague, Harlow Shapley; and when Deg, duty-bound to probe wherever necessary, intimated that political processes might characterize scientific processes by leftist politics, he was scolded by certain naive and intensely tender liberal consciences, as if external politics, or processes of political psychology, could never enter scientific processes. So he was amused when, in perusing an edition of Frederick Engels' *Dialectics of Nature,* a work which many Soviet scientists find it *de rigueur* to praise highly somewhere in their books and which contributes to biological science roughly in the same measure as Adolf Hitler's *Mein Kampf,* he had to note that the adulatory introduction to Engels' book was by none other than J.B.S. Haldane, who apparently could see contemporary marvels in the century-old work of a communist that he could not perceive in V.'s book.

Indeed Deg, in his typically optimistic manner (he would pick up a redhot stove), had conceived of the true interests of marxist theory as residing in catastrophism, not uniformitarianism. Why, he had asked himself, sometime around 1978, did Marx and Engels so strongly endorse Darwin, fashioning the pattern for marxists to follow ever since (the heresy of Lysenko in the 1950's being a significant incident thereto)? Perhaps, he thought, the model of catastrophism did not give them a broad natural inclined plane for the progression of history; it defeats man's greatest works in an instant. It plays hob with the development of the pure but reversed Hegelian dialectic of thesis-antithesis-synthesis in the historical process. It depresses man's will and capacity to build an ultimate Utopia. And Marx and Engels, despite their rejection of the Hegelian "will" and ideal, conceived of and nurtured the most fantastically strong human will, one that could overturn social orders and political regimes (of course, with the aid of history). So they needed natural change to back up social change — Engels waxing polemical on this need — but the change must not overturn catastrophically the works of revolutionary men.

Still, Deg thought also that the problem of arousing the masses was immediate and paramount with them, whereas the problem of nature and history (just mentioned) was less important. Now the masses must see themselves as the symbol or substance for a great tidal wave, storm, explosion, and destroyer. Therefore, the imagery of catastrophe would be more effective than the interminable gradual incremental changes of Darwin and bourgeois society. And indeed there are indications that Marx smelled an ideological rat in the theory of evolution. Furthermore, in reading Soviet studies pertinent to quantavolution, Deg could sense a slackness in their basic tie to Lyellism and Darwinism. In the back of Deg's mind there was an ulterior motive, to loosen the anchor of uniformitarianism (or "actualism" as the Europeans call it) in the marxist setting, thus to free up a flow of new quantavolutionary energy.

So Deg wanted to address himself to this problem, and he asked his daughter, Victoria, who was a professor by now and eminent on intellectual movements of the past century, and who said, yes, it did seem like a good idea, and she being much better atuned to the marxist mentality and avant-garde currents in the field than he, Deg promptly submitted a proposal to the political science and sociology section of the Natural Science Foundation. When the refusal came, he asked for and received the critiques of the review panel. He was a little dismayed to discover that he was illiterate and ignorant beyond his worst fears, even more so than most scholars must be on the measuring scale that the Foundation had provided conveniently to its panel.

But when he thought that he might judge the responses to his proposal better if he knew who were writing them, the request was refused, on grounds of "policy," and, of course, the policy was, as is usual, good for those who were in charge of the policy and working behind the defenses afforded by the policy. Momentarily Deg thought to investigate the law on the subject, and to have introduced a bill for laying open such matters, as an amendment to the federal law on freedom of information, or even to indulge in a lawsuit, seeking a mandamus to produce the records. He didn't do so, of course, because, as my readers by now amply appreciate, *ars lunga, vita breve.* Two years later, a postscript to the episode occurs in his journal:

January 20, 1980

A famous letter from Marx to Darwin is said to ask Darwin's permission to dedicate a volume of *Das Kapital* to him. Year before last, the National Science Foundation turned down my proposal to study the question why Marx and Engels, who perhaps should have been ideological quantavolutionists, not evolutionists — that is, catastrophists, not uniformitarians — would have so warmly accepted Darwin's group. (The anti-religious connection is, of course, obvious, but the Europeans were not so friendly to Darwin and were non-religious too). Then [1976] came the exposure that the famous letter had not been written by Marx at all and the mistake was traced back to its source in early communist revolutionary Russia. Marx could say once more "Je ne suis pas marxiste" (if he ever said it). I wonder whether he would also have said "Evolutionem non fingo." Probably he was content with two of the thrusts of Darwinism: materialism and historical progressivism.

* * *

But enough of foundations, lest I have no energy left for treating of publishers. The lesson that publishers learned from the Velikovsky Affair was the same as a first-term convict learns in jail, how not to get caught a second time. The unfortunate victim of the lesson was any author who was preparing a book in the field. Macmillan Company dumped Velikovsky's book, and Doubleday Publishers made a good deal of it over the years. All the nice people and the pundits and the heretics believed that Macmillan, Doubleday, and other publishers would have "learned their lesson" and now a new age in publishing would dawn. Controversial books would not be discriminated against, and so on. To Deg (I hope that I am not giving him too much credit for saying so), this was utopian thinking, and he ought to know, being a utopian, a "realistic utopian," he insisted, by which he meant precisely a person playing at a high risk game knowingly, because the game involved some worthy ideal. He said this to those who called his works on world order, "Kalos" and "Kalotics," utopian.

Publishers, on the contrary, did not venture into catastrophism, did not make any money out of the "pseudo-science" or "fringe science" of catastrophes. Ransom's *Age of Velikovsky* was privately published, and when later published commercially, sold only modestly. Patten's works were published privately and did well. Deg's *Velikovsky Affair* was handled by two small, high-risk publishers and sold under 5,000 copies, and later in England sold another 10,000 copies. David Talbott's *Saturn* did not repay Doubleday its large author's advance. Melvin Cook's book, *Prehistory and Earth Models,* published in England, sold very quietly and modestly; it was technically written, but an "acceptance" would have sold many copies in college courses, technological industry, and the *Scientific American 's* public. Hapgood's book on *The Path of the Pole* sold modestly. Milton's *Recollections of a Fallen Sky* failed to reach the American market from Canada.

Henry Bauer's book on the Velikovsky Affair took six years to be published and a University Press did the job (Illinois); since Bauer found little of substantive value in V.'s work, one need not wonder how a pro-V. work would have fared in the same circles. Dorothy Vitaliano's anti-catastrophic book on disasters in geology (Indiana University Press) enjoyed only a small sale. So it is not being pro- or anti-catastrophism that sells, but books on the subject are either unsellable, or the publishers will not bring them out, or promote them properly.

The most successful publisher attending to quantavolution was William Corliss' *Sourcebook Project,* a household concern, that culled the history of science and current reviews for worthy material, finding thousands, reprinting hundreds, all the while maintaining a nicely neutral position.

What was true for book-publishers held also for magazine publishers. The only magazine with a general readership that gave sympathetic attention to quantavolution was *Frontiers of Science,* edited by Elizabeth Philips. It failed after several years because it was part of a conglomerate operation that used the bottom line to weed out unprofitable properties. The very small journals, playing to between 300 and 1500 subscribers, were fully unprofitable. Yet without them, there would have been nothing to put forward a viewpoint attractive to millions.

By the rationale of laissez-faire economists, this should not have occurred; in fact, it is normal in the world of education and science. The contradiction between a society's need for creativity and the resources allocated to creativity is stark. It is further exaggerated in the inner organization of education and science, where the more creative the work, the less the outlets for it. New journals in the sciences often form out of failures of the reception system. *Theoretical Physics* was founded because some scholars could not get enough of their material into *Physical Review*. Deg founded P.R.O.D. *(Political Research: Organization and Design)*, to advance new ideas in political science and sociology; it later became the *American Behavioral Scientist,* which was markedly altered in format, approach, and contents when he gave up its editorship in 1965. One of Deg's students, Howard Smuckler, became editor of magazines of Ancient Astronauts and ESP; from the beginning, they were given newsstand circulations of 200,000 copies, with the proviso that wild nonsense be given free rein. The most fortunately situated scholar in the country for communicating occasionally his ideas of quantavolution, sometimes subtly, at times explicitly, was paleontology Professor Stephen Jay Gould of Harvard University, who wrote a regular feature for the magazine *Natural History,* published by the New York Museum of Natural History with a popular circulation reaching a million readers.

Various publicists, such as Sprague de Camp and Theodore Gordon, gave chapters over to mocking or explaining Velikovsky, but their books were not greatly affected by these chapters. One of the best of the publicists was Fred Warshawsky, who wrote *Doomsday: The Science of Catastrophe*. Picking up René Thom's mathematical topological theory of catastrophes, presumably applicable in any field, he applied it non-mathematically, heuristically, in discussing the many works trending toward the quantavolutionary outlook. He undertook with V. a couple of long sessions that curled his hair and set him straight on what to say of V.'s achievements in an article for the *Reader's Digest*. Having escaped perdition, he went on to write a full book on catastrophes, ancient and modern, which was published by the *Reader's Digest Press*. This company made a distribution agreement with Harper and Row, which performed so poorly with his book that Warshawsky complained bitterly to everyone and achieved some promotional effort. The company then closed down, and Harper and Row stopped selling the book, returning its very large remaining stock. Then McGraw Hill bought rights to the book for its back list, to no effect. Over 8000 copies were sold, but 17,000 copies were "remaindered" at a pittance. The *New York Times* ignored the book. Some favorable reviewing occurred. It went out of print after only several years. Note the way in which an author's "property" is kicked around.

The situation, as I last surveyed it, is this: not one major publisher has in print a book on quantavolution, excepting Doubleday, Morrow, and Dell, all with Velikovsky, and excepting, too, the New American Library with a reprint of Francis Hitching's *The Neck of the Giraffe,* in which the head of the giraffe is quantavolution, the neck is the long disdainful connecting link, and the body is conventional biology. (For those who might think otherwise, I should say that Erich von Daniken is an "ancient astronauts" buff, not a catastrophist, except in mood. I say this because I am often asked what I think of von Daniken and I respond that he is not a quantavolutionary, he blithely propounds mysteries without worthwhile solutions, but he is, alas, a cosmic heretic.)

On October 31, 1982 (Halloween) the 15 Paperback Bestsellers (trade) which were listed in the *New York Times* around the U.S.A. carried six (6) titles dealing with the cat, Garfield. The number one bestseller was "Garfield Takes the Cake," then, number 4 was "Here Comes Garfield," number 10 "Garfield Weighs In," number 13 "Garfield at Large," number 14 "Garfield Bigger than Life," and number 15 "Garfield Gains Weight." If Garfield were missing, Rubik's Cube would occupy several of its places, vying with books on diet. The *NYT* defines this class of paper backs as "softcover books usually sold in bookstores and priced at average higher than mass market."

One cannot read Deg's notes and hear him talk without deriving an apocalyptic view of the publishing industry. "It is a doubly sick industry. It is economically sick and it is functionally sick. By 'functionally' I mean physically, ideologically, and morally. It is dominated by cheap non-publishing money, coming from extravagant swashbucklers and conglomerates of merged and paralyzed units. Ownership is alienated from editors, editors from producers, editors from authors. It is characterized by some of the worst labor practices, witness to the shadiest deals, and engages in the thoroughgoing degradation of writers."

This is the way he often spoke. He wouldn't say much and sometimes in a group or committee be quiet, abstracted, even appearing bored. Then suddenly he would be seized, and as if to make up for lost time and to persuade others that he was only speaking because what he was saying was being torn from his lips, he would hammer out the words, scalding rather than sweetening the atmosphere, so that when he finished, there was neither applause nor babble of dissent, but a pause, until someone evasively spoke around him, and when that happened he didn't insist upon his point but subsided for a good while.

Deg could recite a long list of great writers who had put out their own books; he even claimed that most great writers did so. First of all, up until the late Eighteenth Century — Franklin, Voltaire, the Encyclopedists — every writer put out his own books, unless, after burying him, friends or relatives printed his work. In a marginal note to one of his late anatomical sketches, Leonardo de Vinci implored his "neighbors" to see to it that his works would be printed.

The publishing racket (Deg's word, not mine) developed sweetly out of bookstores and printing shops where it belonged and should have stayed, but, by the latter part of the nineteenth century, Balzac was excoriating the thieves and profiteers of the business in an excellent novel, *Illusions Perdues*. Dickens, Dostoevsky and Flaubert sweated to carry their novels first as serials in magazines. But where are the magazines, bad as they were, today — they carry a single chapter but usually the pain of editing a chapter for a magazine is damaging to both the author and his book.

Is it names you wish? (And he would begin.) Walt Whitman, Friedrich Nietzsche, Stendhal, Beatrix Potter — yes, *Peter Rabbit*— James Joyce (an angel helped), Marcel Proust, Rainer Maria Rilke, Virginia Wolfe, André Gide *(The Immoralist* issued in 300 copies), Sigmund Freud and, if you will, Velikovsky himself published his early pamphlets. Colette was published by her husband Willy who even stole her name as author. America's best autobiography, *The Education of Henry Adams,* was put out by the author.

The myth of Thomas Wolfe is used continuously by publishers to show the unknown young writer discovered by the great fatherly editor of a conventional publishing company and led carefully to reveal and convey his beautiful achievements to the world of readers. Even this case is mythical, as the editor involved, Maxwell Perkins, tried to explain in a recent edition of Wolfe's *Look Homeward, Angel.* But the truth will never catch up with the lie until publishing circles come upon a similar myth to serve them.

If Charles Darwin's *Origins of Species* sold out through a bookstore in 1859 it was because writing and printing were still for gentlemanly use and the book was not deposited behind a mass of artificial best-sellers. He belonged to a circle of people that bought books of their friends. Dammit — nowadays you can't even sell a book to a friend! Besides there was a prurient and agnostic public alerted to the sensationalism of the book. Surely you must know, too, that Darwin's thesis was already well-worn and agreed upon; he was selling evolution even though he didn't use the word and the book's *raison d'être* was the silly mechanism of natural selection, which was nothing more than a watered-down Lamarckianism, a slogan for bird-watchers and garden clubs. It was an easy sale.

Deg had one more arrow in his quiver to fire at the now pathetically wounded publishers. They are frauds, announced he.

They pretend to publish the books of the country. Ninety per cent of the serious writing, and I include even novels and poetry here, is put out by government presses of several types, by subsidized university presses, subsidized independent and university institutes, scientific associations, and self-help amateurs like myself. Further, much of the serious writing put out by so-called independent publishing houses is subsidized, by insider deals, involving mutual back-scratching, agreements to arrange publication of one's editors, promotional devices such that no established book reviewer need fear his shit will go down the drain when there are people who will eat it, [I am sorry, but that is what he said], by quiet subsidies, by guarantees of sales, by tricky deals with film-makers, press agents, television companies, and corporations, and you name it.

At this point I intend to escape Deg's diatribes by telling how he came to enter upon his writing campaign and then to publish his own works. Lest you think that such violent opinions as his come out of intense suffering and exploitation, let me once again remind you of Deg's character, acquired in earliest childhood: he could be and was often indignant about a person or an institution or a system, without being hurt by them and even while being helped. In a way, he was rather like his children's generation and the hippies, except that he had the forcefulness and discipline that produce alternatives; he seemed always to have ready a proposal for another way of doing things. In this way, he was more sprung from the nineteenth century utopians: Fourier, Brook Farm, St. Simon, Marx, Henry George. As you will see here, he didn't expect much, he didn't suffer greatly, he didn't mind sacrificing, and he did not dance a jig when he finished the job. I assure you once more of that great difference between Deg and V. Deg did not see himself as a victim; V. saw himself as a victim.

Deg moved into the field of quantavolution slowly and then ever faster. This I would attribute to his heavy involvements between 1962 and 1966 with the *American Behavioral Scientist* and the design and production of the "Universal Reference System," for the computerized searching and retrieval of bibliographic annotations in the behavioral sciences. During the same time, he was writing heavily in political science, especially on the reform of relations between Congress and the Presidency. After he turned from these in the period 1967 to 1972, he wrote *Kalos: What is to be Done with Our World?* Hired by Simulmatics Corporation, and given the assimilated rank of a general with "Top Secret" access by the Department of Defense, he spent a few weeks in and out of Vietnam, tendering advice on psychological operations that were intended to win over the Vietnamese people and to bolster the morale of the Vietnamese Army. (The American generals were jealous of the morale of their own troops.) The job led him quickly into urging measures that were too radical and diversionary for the forces, civilian and military, that were moving in an irresistible death-dance toward the ignominious withdrawal of the United States presence in Indochina.

He was writing poetry and before flying to Vietnam in 1967 he collected his poems and put them to press as the *Passage of the Year;* some of them he framed in what he called an "eccentric," "super-sprung" rhythm.

He gave a copy of the book to Harold Lasswell who said, Yes, he had written poetry when young, at which Deg commented that poetry was more accessible to the senile than the juvenile. He gave a copy to Velikovsky who, it appeared, had published a small book of poems under the pseudonym of Immanuel Ram, in Russian, in 1934. V. read Deg's poems and used a quotation from them on one occasion to persuade Deg of a point. Suddenly it seemed that mankind was a secret crowd of poets.

He then joined with a University instructor who had not studied directly with him, and had met in an annual Department reception, Nina Mavridis, a tough, emotional, polyglot petite blonde, smartly turned out, whom he later married. They went in search of a Greek island house, and he bought a parcel of land on Naxos, which was then a quiet backward island, and there build a stone cottage facing across the straits to Paros.

He turned to several of his former students, graduates, and "drop-outs" from the system, and together they organized an experimental college, l'Universite du Nouveau-Monde, and settled in for a hectic year upon the Alps of Valais, Switzerland. All the while, he visited Princeton, coming and going, keeping in touch with the Velikovsky circle there, and with whoever of his immediate family happened to be home from schools and wanderings around the world.

With the University of Switzerland closed down, the United States withdrawing from Indochina, his work on a new world order totally ignored, his family disassembled, efforts at reforms within New York University ending only in cosmetic changes, and resettled efficiently with Nina in an apartment of Washington Square Village, just across from one of his classrooms, and a block from his office, Deg drove through the resulting energy gap into the field of quantavolution. He completed two books of political science during this period, neither requiring heavy research, but both of which, *Politics for Better or Worse* and the "lectures to the Chinese," *Eight Bads, Eight Goods,* he considered as "state of the art" philosophically, and innovative in format and perspective. Both were "successes," he thought: neither earned much money, $18,000 in the first case, $3,500 in the second.

His University teaching had never in his career cut very deeply into his time for study and writing, partly because he did not "pal around" with students and varnish their wasting time. Too, he avoided committee assignments that seemed useless, and had little need for generalized social encounter. During nine months of the year, he gave an average of twenty hours per week to straight pedagogical work; the rest went into his projects — editorial, political, pedagogical, consultative — and writing. Wherever he had taught, including New York University, he was expected to be a "producer," to do research and writing in return usually for a lighter teaching and committee load. He was usually expected "to bring money into the University," which sometimes he did, and to find funds for his research and activities, which sometimes he did. He used his time fully and completely for these latter purposes, working year-round, seven days week, for three to twelve hours. (Obviously, everything did not "come easy to him," as so many acquaintances believed.) His journal slackened off, through the sixties and seventies, entries occurring only every several days on the average, and even then deprived of events recited in their fullness.

He rarely spent more than ten minutes on the day's newspapers; he watched television several hours a week; he listened little to music and rarely played his trumpet any more, but often was humming and whistling to himself. Except when reading a novel or a poem, he did not read in the conventional way. Reading was an instrument of research and writing. He would pounce upon a book or article and seek directly the point that he was addressing, which had made him pick up the work in the first place. If it wasn't helpful, he would put the work aside. He could rarely be trapped for instance, by some lurid description of a disaster. At the rate of 100 pages an hour, he could tell whether there was anything useful to him in a succession of books or articles. An issue of *Science,* though it might contain 100 pages, would ordinarily occupy 10 minutes, just enough time to see whether there was something of interest in it. He would however spend hours on a relevant two-page article in a strange field — a paleontological article using explicit chronometry, for instance, learning the method used, looking for the expected illogical turn or twist, the weak point in a piece which after all had been fashioned with extreme care, was the darling of the authors' eyes, and had been rigorously criticized by conventional readers.

At first, both current materials and ancient materials on quantavolution were not so easy to find. Stecchini was alone as supplier of references outside of V.'s works. As the network of scholars like Mullen, Juergens, Milton, Crew, Sizemore, Moore, Lowery, James and several dozen others came into the field, the supply of references grew exponentially. *Pensee Kronos, The S.I.S. Review and Workshop* and Corliss' *Sourcebooks and Newsletter* brought hundreds of citations to light. I cannot do less than say that the names of the hundred authors of the articles and notes in these magazines is the measure of 90% of the field. If screened for relevance and translated into quantavolutionary terms, several hundred more names would be added — not that they would gladly accept being added — from the conventional output of scientific books and journals.

In a combination of disgust, impractical judgement and worthy motive, he decided in 1977 to resign all obligations to teach and supervise dissertations and to be at hand for the various faculty meetings; he found the University ready to pay him a third of his salary to engage solely in research until he would arrive at the age of 63, after which he would be considered as fully retired. The agreement was soon followed by a considerable general inflation of the economy, and a reduction in foundation activities, so that he was constrained to stringent personal economy, not so evident on the surface, but oppressive in reality. He had no illusions about the interests of foundations and government research agencies in quantavolution and in fact received no help. He earned a little money here and there, whatever could be done rapidly without taking his mind off of his quantavolutionary studies. He sold a piece of land on Naxos. He sold, too, a small house he had invested in for retirement near Brown University, in Providence, where he had once taught and close friends still lived. These funds and more went into research costs — typing, xeroxing, travel — and to the occasional support of his mother and other family members. Nina, although she finally earned her doctorate, and was a most effective teacher, could not get into and hold onto a position in one of the college systems of the New York area. Whatever money she had, she spent fully and equitably. This is no place to speak of her at length; she was everywhere in those years, but when Deg comes to tell of Naxos, it will be up to him to tell of Nina. By the middle seventies, she and Deg had split, and came finally to see one another as friends only there, on the island, where she bought and remodelled a medieval Venetian house and lived with her husband Peter whenever possible.

Deg's first book in the Quantavolution Series, *The Disastrous Love Affair of Moon and Mars* was written in the early seventies. He had thought for several years that he should write a textbook on what he was then calling revolutionary primevalogy, but before he had settled among several outlines of the work and written a few passages, he reached back for a journal entry written while staying at Pythagoreion on the Island of Samos, and decided to try out the new field with a case study.

Pythagoreion, Island of Samos, July 12, 1968

I have come across and read for the first time closely and consciously the song of Demodocus at the house or Alcinous. How wonderfully it describes what Velikovsky said was the actual set of cosmic events of the Seventh Century before this era, of how bright-crowned Aphrodite loved the god of battle Mars-Ares, and how they repeatedly fucked "in the house of fire," whose master, Hephaistos, finally entrapped them in a net and put them upon a more pious course. The passage must be analyzed Word for Word: the parallelism is beyond coincidence; either Velikovsky wrote the myths of the Greeks, or something like the physical events he describes historically took place.

The story referred to is a brief lyric of a hundred lines, sung in Book VIII of the *Odyssey,* the epic poem of Homer. It tells of a much longer opera ballet sung and danced for Ulysses.

Deg showed his manuscript to Juergens who was surprised at its coincidence with his own electrical theory of the events, which was to appear ultimately as two articles in the magazine *Pensee.* V. would not read it. Deg wished to dedicate it to him. V. said let Bill Mullen read it and if he likes it, go ahead. Mullen did, very much. Cyrus Gordon liked it but could not respond to the astrophysical scenario. Further he suspected Aphrodite to be Venus, not Moon. The English acquaintances of Deg got onto the manuscript when he submitted it to the publisher, Sidgwick and Jackson, who had published *The Velikovsky Affair* in England, and showed it to them. They liked it, but in all conscience could not accept the identification of Aphrodite with the Moon, for they identified her instead with Athene-Ishtar, and the morning and evening star, Venus.

This disagreement meant that the English group was ready to dispute an important point of Velikovsky for, in his application of the *Iliad* to the Martian disturbances of the seventh century, he had found Aphrodite joining with Ares in the Trojan War to fight against Athene. Whereupon, and for other reasons, Aphrodite was assigned to the Moon. Desertions were numerous on this score. When James published a critique of Deg's identification of the goddess, it stood without rebuttal, and Cardona, Rix and others were convinced of James's case.

American publishers were not turned on by the *Love Affair*. W.W. Norton, through Brockway, said it was well written but not to their tastes So it went with one publisher after another, Simon and Schuster, Dodd and Mead, Doubleday, Random House, Harcourt Brace, Stein and Day, Princeton University Press, Harper and Row, Atheneum, Sidgwick and Jackson, Free Press, and even the New York University Press (unless a subsidy were paid). Deg thought he should "toot his horn" perhaps, as his mother used to tell her boys, so he prepared a blurb about it.

He made the *Love Affair* sound as if it might attract the masses, but publishers were quick to point out that the book was serious, learned, of dubious validity, and sophisticated; in a word, forget the masses; indeed, betake yourself to a university press. But Deg knew already that university presses were eager for wide publics, undercapitalized, dominated by editorial committees of the more conventional members of their faculties, and slow and painstaking to a fault. He visited Jerry Sherwood of the Princeton University Press. She returned the manuscript in time with the expected advice. Deg stopped peddling the book. He was too busy with the general work, *Chaos and Creation,* to carry on the sometimes interminable ping-pong of serious publishing.

Time after time over the next decade, he would pause in his work to recalculate the options of his predicament. Naive friends counselled him: "Any press would be happy to consider your books." A publisher encountered would say, cordially: "Let us see it by all means." A "smart operator," a "successful author," would advise, "Get it down to 160 pages — less. No footnotes. One only, not really new, idea." The emerging rule seemed to be: "Never underrate the unfitness of readers, media, and publishers."

Yet it was like a drug, this pushing one into the marketplace, or like television. One succumbed from time to time, had a bad trip, and came away cursing himself for not having avoided the encounter. The condition of the publishing industry in America was unbelievably bad; would that it were terminal. All that could be said of it was that it was freer than publishing in Nazi Germany or Soviet Russia, or for that matter in most other countries. It was as bad or worse than the political system of the United States in meeting its obligations, much worse than the educational system with all its weakness.

But unhappy thoughts of this kind did not obsess Deg; they occurred often for a moment (as when he examined the book review section of the *New York Times,* or looked at a publisher's list). Long before, in the days when his work seemed ordinary, when his means of rewarding and insulting were conspicuously in readiness, publishing his books and articles was no problem.

The society, however, was enveloped in the myth that the publishing process was a logical affair, constrained tightly by the quality of the message between the covers. A writer's fortunes were thought to vary with the quality of his message. So many useless and dangerous myths rule society! Like the myth among scientists of a great many readers perusing their article in a reputable scientific journal — 10,000? 5000? 500? yes, 50, and feel lucky.

Now all of this jeremiad is preliminary to announcing that at a certain point in time, probably it was in 1978, just after he began his final race against dwindling finances, Deg decided that he would, unless intercepted by an angel, proceed to complete his work and then by one means or another publish it himself. Somehow the money would be found, and he thought to publish it in Bombay, where he had connections with friends and a publisher, the Popular Book Depot, which had produced *Kalos* and K*alotics.*

One premise he maintained firmly: he would not be finally frustrated and incapacitated by the publishing system. Another premise was his delusionary Paternoster: that what he attempted might be of great importance to mankind. It was the best work he could set himself to — and who else could do it — noone whom he knew of— and his other great object in life, a new political order of the world, offered at this time no opportunity nor chance of success.

The decision was not easy, hardly definite in fact, because like many decisions he made, it was long foreseen and warmed up on a little burner in a recess of the mind. It was not an optimal solution, by any means. The myth, social binding, and conventions of publishing are so pervasive that noone of his acquaintance thought this procedure wise, prudent, or even possible. All too poignant was his awareness that the controversial matter that he was writing would combine with its unorthodox publication into a hard prejudice against the books. Under such circumstances, more than a touch of megalomania is needed.

He pushed ahead imprudently, erratically, and stubbornly, or so it seemed to others, and they were correct, but they could not see how such failings of character might add up to an achievement. He wrote everywhere and under all conditions, on all sizes and kinds of paper, with pencils and pens of any type, and now and then on typewriters, electrical, or a portable, mechanical one. He read in several libraries, bought very few books, was sent xerox copies of many pieces by Sizemore, Milton, and others, corresponded, and ultimately had made notes on some hundreds of books and articles. These were often caught on the wing, and he was often exasperated upon completing a book to have lost a citation, forgotten the spelling of a name, left relevant pieces now in Greece, now again in New York.

There is nothing special to recommend in his research and writing procedures, except what one cannot anyhow imitate: a wide-cast unerring eye for the salient, the strong background of methodological / especially epistemological — thought and theory, a modest skill at writing, a great skill for synthesizing material, an inborn will to let nothing stand in one's way, a lifetime practice in doing much with little. Once in the while he got help. Donna Welensky, whom sometimes he paid for her typing and sometimes not, whom he came to love for her energy, efficiency, and ineffable kindness to the world, never mind her brawny blonde beauty.

The latter half of the dozen strenuous years were dominated, physically speaking, by the presence of a quiet, deep-voiced, dark-haired, brown-eyed, French novelist whom he encountered first at Naxos, where she was joyfully spending a few francs that her publisher had let her have as a consolation for not publishing her latest book, *The Paladin*. With great difficulty, for her assets were almost literally on her back, she obtained a visa to come to America, and thenceforth Deg took care of her, and she took care of him. In 1982, they married. They lived in New York City, at Princeton, in Washington, on Naxos, and in Paris, appearing more affluent than they were or pretended to be.

They visited her ancestral village, Habsheim, between Basel and Mulhouse, they travelled to England, Italy, Hungary, and Canada. She loved the journeys and loved Deg and adapted quietly, imposingly, to the net of human ties and implausible projects of Deg with a broad, engaging and ever-ready smile. When Elisheva, sculptress forever, met her for the first time, she was awestruck at bones that made her strong hands ache for a chisel and hammer. "How did you find such beauty?" she asked Deg. She could be happier than anybody whom Deg had ever met, under the poorest conditions of life — but then, as he often said to her, and she fully agreed, we are much better off than humanity is or has ever been or will be.

In more than a decade from 1972 to 1983, Deg gave over perhaps no more than eight months to work outside of quantavolution. Almost all of these few months was spent consulting directly and indirectly with the *National Endowment for the Arts* with Carl Stover, a friend of thirty years standing. Given a general directive and promoted by Carl before Nancy Hanks and Livingston Biddle, directors of the Endowment, Deg wrote a number of sketches of what might be done to stimulate a broad range of cultural areas, but principally he committed a trenchant irony called "1001 Questions on Culture Policy" in which, using the format of a book of interrogations, he was able to say all that he wanted to say. The work was an implication that nothing intelligent and basic was being said about public policy on the arts and humanities. Stover even managed to obtain from the Ford Foundation a subsidy with which to send copies of the work to most prominent leaders of the organization and direction of cultural affairs of the United States. Copies were also distributed in Western Europe. The effects, so far as might be perceived, and disregarding the encomia that are easily aroused by techniques of publicity, were nil.

Otherwise, the quantavolution investigations progressed and enlarged grossly. By 1975 the basic *Chaos and Creation* was calving. The theory of homo schizo emerged and went one way, ultimately two ways, in two volumes, one on the origins, one on human nature today. A great fragment fell out of *Chaos and Creation* and became a treatise on exoterrestrial aspects of geology, *The Lately Tortured Earth.* On a sojourn in Naxos, there occurred an idea for an article explaining why the Pharaoh should have pursued the Jews in Exodus; quickly, stimulated by conversations with Anne-Marie, it transformed into a book of exhilarating discoveries and, in the end, *God's Fire: Moses and the Management of Exodus.*

He had already devised a theory of how the solar system might have enacted the set of quantavolutionary dramas which he had been uncovering and classifying. He wrote of it to Ralph Juergens. He found agreement there, and then he achieved the support of Earl Milton. Earl opted to come in on the enterprise of a book; Ralph became engaged, too, but hardly had Earl gone down to Flagstaff, Arizona, to go over their preliminary notes with him, than Juergens died suddenly, of a heart attack. Over several years, in Princeton, Washington, Manhattan, London, and Naxos, and by telephone and correspondance, Milton and Deg worked to complete the book. Its Index, in a unique format, was finished at the Cosmos Club in Washington, D.C., on February 16, 1984.

The *Moon and Mars* book was standing by for revision. *The Burning of Troy,* its title taken from its first essay on the calcinology of Troy IIg, was organized to contain studies, reprints, essays, and notes. *The Divine Succession* was taken up; its central theory, that all gods are of the same family, was put forward; an anthropological and psychological discussion of the major aspects of religion followed. Then, as Deg stood back, gazing anxiously and unproud into the manuscript, there came to him the idea of adding two new proofs of the existence of gods, and also the scheme of a catechism for whosoever might wish to contemplate a possible new religion alongside the old.

There was left only *The Cosmic Heretics,* which I undertook to write.

Its origins lay in Deg's intention, growing over some years, to write an autobiography in half-a-dozen volumes. He still nourishes the thought, cowering over the prospect of its passage through the gauntlet of fast-gathering spiked-leather-fisted knights of time. But perhaps I can also do this job for him.

* * *

In 1980 he sent off *Chaos and Creation* to India for production. Delays were many. Stephanie Neuman lent him $3000 to defray some of its costs. He paid her back two years later. Funds came in from the sale of the book through the mails to lists of friends and of purchasers of William Corliss' *Sourcebooks*. Corliss himself sold copies. But larger sums were needed. They came from an advance of Ben Gingold, a friendly architect who intended to purchase land in Naxos from Deg, from cashing in 10% of the annuities that were to take care of his retirement, from yet another property sale, and from a personal bank loan. Household economies were the rule. The logic was simple: a small saving enabled thirty letters to be sent out, thirty letters might elicit a couple of orders. Deg and Ami moved into a dingy little brick house on an old street of Trenton, in a neighborhood that sociologists call by the menacing term "marginal."

Publishing in India was becoming costly. The Indian rupee which should have lost its international value, maintained itself steadily against the dollar, letting India pay its debts at a loss of export, but then it exported little anyhow. Nevertheless, Deg let himself in for a third round with Indian printers, sending off in early 1982 the bulky manuscript of *The Lately Tortured Earth*.

He rationalized his private publishing company in a memo to readers, but then decided not to print it in his book. Here is a better place for it, so I am carrying it:

A Note on this Edition

This Edition is intended to bring the materials of *Lately Tortured Earth* to the attention of the small number of scholars and students who are directly involved in research into quantavolution and catastrophe. It has not undergone the ideal processing of several expert readers, critics, and editors. It has been published for the very purpose of arousing comment and criticism.

Four major reasons occur for this procedure:

There are inordinate delays and difficulties in publishing through the natural channels of the trade book and textbook publishers and university presses. This book and others in the quantavolution series have already been in manuscript form for some time. It may be better, therefore, to publish the work promptly in this manner than to let more years slip by until finally some convinced entrepreneur will be bold enough to undertake its publication.

Since the work enters upon numerous fields of sciences and humanities, expert readers would be required, a veritable conference of critics, and, logically in each case, a possibly unfavorable critic and a possibly favorable one. Many copies, much time, and thousands of dollars in fees would be needed. Based upon the author's experience with the editorial services of some prestigious publishers, the cost is too high to pay. Publishing the book on the author's responsibility alone will enable hundreds, instead of a score, of experts and students to weigh the validity and utility of the work.

Third, authors of unusual theories and controversial types of evidence are strangers to specialists of most relevant fields. Foundation support, university backing, and publishers' advances are practically impossible to obtain, all of which might otherwise be used to avoid editorial, factual and linguistic peccadillos and to comb more efficiently the library stacks for materials on "non-fields."

Fourth, new high technology has come to publishing, but there is a shameful disparity between the high-level technology abundantly available for the most useless kind of publications, and that which trickles down to assist publications on the broader and deeper problems of human culture and natural history, most of which necessarily occupy the attention of only a few persons. While university presses, never an ideal solution, deteriorate and while commercial publishers vie for scrapulous material, and while publication technology vies for faster addressing and delivery of junk mail and selling computers for games and word processors to enchant the bored secretary, those to whom is consigned the progressive evolution of culture are hard put to survive, assemble, and operate the tools of their trade.

We hope, in sum, that our readers will be fully critical, yet tolerant of our not so sleek editorial packaging.

Delays loomed up in India with *Lately Tortured Earth* so he turned to domestic production. Once again he had to review all of the possibilities for cheap book production in America. His initial constraints were several. He needed a secure conventional binding, preferably cloth or sewn. He could not publish in a large format, say 8 1/2 x 11 inches, because he wanted to put the book before the reader in a familiar form. He needed a bookish type font, an even right margin, running heads and other "luxuries" that American readers had come to expect and demand. He wished to insert many illustrations; this would be costly if they required redrawing or screening.

He observed the rush of new technical systems, computer memory word processing equipment, "perfect" glue binding machines, automatic cameras, small presses of various kinds and alternative xeroxing machines. None of the products and suppliers with whom he treated had a clear perception of what his needs were and he found himself lecturing them about the greediness and unresponsiveness of industry set up to treat deferentially the unconscionable matter of junk mail and the industrial wordage of the culture — and he would sound off sometimes on the gamut of the intellectual pariahs, the serious writers, artists, and scientists.

From time to time, he would play with the design of an ideal system of personal and small-group publishing at a cost the humble creators of culture would afford. He put aside consideration of systems of microform production and distribution, because the fast culture was still too slow to accept them. He foresaw in the meanwhile a word processor with soft-ware for book-setting; a memory capable of handling a book as a whole; soft-ware for intelligent spelling and indexing and storing and addressing networks of acquaintances and potential customers; a big readable screen means of composing tightly and finely; a tape that could be stored and would feed a composer that could be slow but must print out a handsome book font and a generally useful caption font. Then the output, automatically paginated, would be pasted up on cards, the cards then printed in multiple copies on a reliable copying machine that could handle from one to a hundred copies of four pages (11" x 17") at a time, after which a collating machine could fold and merge the pages into a book that would then be placed into a thermal, glue-binding machine, capable of handling up to 500-page text with its covers, be they cloth or card. Next the book would be trimmed, then, if cloth-bound, jacketed with a paper that had been produced by the same system. The small edition, by which Deg meant from fifty to five hundred copies, would be shelved until sold and shipped. Meanwhile the announcements would be coming out through the same system and would be addressed by the automatic print-out of the stored customer and complimentary lists. Small gadgets and work routines would be devised for the interfaces of the system components. The whole publishing company would fit in a garage or basement comfortably. It should not cost more than $20,000, including initial supplies, and a year's maintenance contract. It should be affordable with a $2000 down payment with the balance plus interest in extended payments over a 36-months period Facilities for the bookmaking announcements, billing, accounting required by a dozen books per annum program or its equivalent in magazine and pamphlet production would be provided; actually a much larger output would be possible.

The system he envisioned is quite feasible technically. Beginning in 1981, Deg could set forth the named components and locate their suppliers to provide a complete system in the range of $30,000, but the system would have uneconomic, inefficient, superfluous, and flawed elements. The field was moving rapidly. At some moment, it could be brought together and a revolution in publishing accomplished. Or rather, what would happen is that the great majority of thousands of creative groups of the nation would cut themselves off effectively from the commercial and university press publishers, building firmly and at a cost they might afford the printed communication network which they need to survive. When a company called the Who's *Who of Contemporary Authors* circularized him, asking the usual information and adding a request for "words from the wise," he wrote (May 18, 1981):

> SIDELIGHTS: "Two futurisms for the debased and desperate intelligentsia: A) With the decadence and collapse of the publishing business, creative writers should discover how to publish themselves and reach their own special audiences; commercial publishing is 95% an exploitative delusional myth. B) With the decline and collapse of the existing world system, the free intelligentsia should cut back on writing just anything for money or prestige and begin to assume responsibility for picturing and propagandizing a revolutionary new world order."

He never got around to seeing whether they printed it.

Nothing approaching a new full mini-publishing system was achieved by Deg with the Quantavolution Series. The name "Metron" meaning "Measure" was revived from a personal reporting, consulting, and publishing company he had employed mostly in the 1950's and 1960's to put out the *American Behavioral Scientist*, the *Universal Reference System*, and books and reports. Now it was to be the name of the first quantavolutionary publisher. The means of publication were only half-new, a melange of all ordinary systems. Word-processing with photo-composition by large machines, Compugraphic composition, and old hot-type linotype systems were variously utilized; printing was done by xerox, by ancient letterpress, and by already old-style small offset presses. Bindings ranged from Smyth-sewn cloth-covered board binding to new compact "perfect" thermal binding. Deg designed all the covers and the format, under heavy constraints of format, color, and costs.

The printing and publishing industry was in a technological and marketing revolution and it was annihilating the old breeds of manuscript-evaluator, copy-editor, proof-readers, and designers. All of these operations now were more expensive and provided less reliable and competent services. Deg arranged much of the composition, printing, and binding with Rick Bender of the Princeton University Computing Center and with Skip Plank of the Princeton University Printing Services. They became adept at running small editions in the interstices of time that occur with a large computer and photocompositor.

In all, the labor of his wife and himself as designers, editors, typist, clerks and managers of production and distribution, would have cost $65,000 to purchase as services on the open market. Direct research and overhead costs (actually paid out or otherwise absorbed) came to about $60,000 over the whole time; direct production costs amounted to $41,500; early mailings and advertising cost $6,000. Without any allowances for the author's time or advances against royalties (he being the author), the total real cost amounted to $172,500. The total number of books produced was only about 6,000, and many of these were not intended for sale. The editions were numbered. The average real (but not cash) cost per book, then, not including any compensation for the author, amounted to $28.80 per copy.

When I spoke to him before turning this page over to the printer (taking care not to be seen laughing) the returns had totalled $7,500. He expected receipts to reach $30,000 in a year's time and finish off the balance of immediate direct costs, $17,500, during the second year. This would also exhaust the first edition copies. The main chance of compensating for the $125,000 of other non-monetary but poignantly real costs, would be to sell rights for new editions to other publishers. As for the royalties of the author in our simulated account here, these would have to wait until further new editions were issued, and were ticketed for archival expenses. Apparently, the *avant-garde* or heretical author is frustrated whether by the publishing business or in his own efforts to reach out and communicate.

Deg was continually irritated by the ignorance of the intelligentsia concerning the engine rooms of the ships carrying them. They are brainwashed by the language of Hollywood, in the markets of best-sellers, and in the display qualities of ads of rich corporations. The intellectuals, with few exceptions, inflict upon their creative brethren the oppressive standards of the rotten rich — fame, money, connections. Dick Cornuelle and Deg enjoyed examining some of the exquisite typography, color-drenched illustrations, and perfect printing that went into annual reports of companies which had bought dearly Cornuelle's more than ample writing talents. No expense, no technology, no skill was spared to convey to some thousands of barely interested shareholders and stockbrokers how well or badly the managers had run their affairs during the year. The annual report, no matter how expensively published, was but a trifle in their operating costs of the year. Yet it would have covered the costs of publishing beautifully fifty creative works.

Where are all these creative works? Is that the objection? Most of them are abortions of a culture of intellectual and science prostitution. They do not appear because they cannot be carried to full term. They do not appear because they expire, too, in their creator's archives. And this is why Deg, as he came to the end of the Quantavolution Series and I near the end of telling its history, began to harangue his family and intimates to set up an Institute for Creative Archives. A billion dollars a year, he claimed, is the cultural loss to the American nation of the death of the archives of its creative workers. This was a real loss, not registered in the unselective National Economy's Accounting System. He wanted to do something about it.

CHAPTER SIXTEEN

PRECURSORS OF QUANTAVOLUTION

"Life is like an endless procession, long since begun, which we join as it passes by." So comes down to us a saying of Pythagoras. V. didn't mind joining the procession, but he wanted to be seen carrying the largest idol of science. This sentiment led him to understate the height of the people walking before him, as well of those walking alongside.

The recounting of one's precursors has in it an element of snobbery, like the genealogical research that discovers barons but not brigands, big shots rather than bums. V. was especially careful to admit no disgraceful ancestors and came near to the point of acknowledging no one; *pari passu* he would not recognize any contemporary descendents of non-existent ancestors. This led him into an awkward position where, on the one hand, he was extolling the observations of ancient catastrophists of religion and natural history, but disdaining the multitude of their descendents who were equally impressed by ancient catastrophism; he lost sight of most of the world's people when accusing mankind of a collective amnesia of ancient catastrophes, focusing his mind upon the uniformitarian intelligentsia of modern times.

He was loath to draw sustenance from, and give thanks to, the long line of Christian defenders of the historical and catastrophic accuracy of the Bible, whose works on subjects such as evolution and geology was, for their times, as good as his own in *Earth in Upheaval*. He was unfriendly to religiously committed writers who pursued parallel paths and sought to ignore them. When Donald Patten, who had published an extensive and substantial scientific work on the Biblical Flood in 1966, was introduced to him at a home reception in Portland around 1972, V.'s first words were spoken angrily: You are trying to destroy me, but you will fail in the end! So relates Patten and there is no reason to doubt him, especially when he adds that a while later V. returned to him and apologized. Says Patten:

While I view Ron Hatch as both an associate and protege, as we have developed our model of the dynamics of ancient cosmic upheavals, Velikovsky viewed me as an unwanted protege, not to be encouraged. He seems to have resented the fact that I disagreed with his conclusions in part, and he did not acknowledge or consider that I agreed with him in many ways. Often criticized as he was (and many times unfairly), Velikovsky regarded me as yet another critic trying to destroy his work. He was uncomfortable with my evangelical, Christian faith; I was comfortable with his Zionist bias; many evangelical Christians support Israel strongly, and I am one of them.

Patten was a geographer, hailing originally from Montana. In 1973, he published a second book, *The Long Day of Joshua and Six Other Catastrophes,* all of which events Deg found acceptable in the history of the millennium after -1450 B.C. Deg purchased them in London in 1976 through a member of the Society for Interdisciplinary Studies. In them, he found stimulus and information. Before then, he had heard only a few derogatory remarks about the books.

Patten and his collaborators, of whom the most prominent were Ronald Hatch and Loren Steinhauer, were fully committed to astral catastrophism and built a complete succession of scenarios around orbital intersections of Mars and Earth, beginning with the deluge of Noah. (At first, Mars was exculpated for the Deluge but now Patten would implicate it there as well.) Patten's admiration of V.'s work, which he expressed most strongly in an article of 1982, did not extend to accepting the participation of planet Venus. He presented the Deluge in an unusual structural form; generally his work has this geometrical structure of thought. Like Deg, he was prone to set up categories and lists. He developed also a short-term calendar of the ages.

His brief but friendly criticisms of V. were threefold: that V. was over-influenced by Freud and prone to accept too many evolutionary and uniformitarian doctrines; that he was unquantitative and unsystematic in his geology; and that V. was overconcerned with his critics. I cannot dispute Patten, because these same several views emerge from our own pages as well.

Patten's books, which he himself published, circulated widely and well over the years, and hundreds of thousands in due course watched a 60-minute filmstrip of his ideas presented in English and other languages. He could not be said, however, to have conformed to the ruling formula in Christian Evangelism, which was determined by Henry M. Morris and the leaders of the Creation Research Society, who have held to an age of 10,000 years for the world, therefore constraining creationist science greatly. Deg was next in line of constraints, with his 14,000 years for a holocene period full of quantavolutions, including lunar fission, nor could he believe that the Judaeo-Christian God had laid down this constraint; it was miserably self-imposed, with full blame unto himself. Still, he was grateful for the works tendered him by the creationists and, unlike V., felt no need to disavow them.

V. cited with relish ancient predecessors, but when it came to citing modern scientific ones such as Georges Cuvier, Brasseur de Bourbourg, Donnelly, Hoerbiger, and Bellamy, his lines were niggardly, rather derogatory, and somewhat aside from the point of their predecession. When accused in a letter to the *New York Times* (May 7, 1950) of having taken wholesale from Hans Hoerbiger, an older contemporary, V. rightly answered with details of their divergences and Hoerbiger's failings. But here, as elsewhere, V. held to a narrow view of what constituted the procession of life and science, and precession.

V. had come upon Donnelly's *Ragnarok* in 1940 at the New York Public Library and was depressed by the discovery, according to his own words. Thomas Ferte in 1981 published an account of the numerous foreshadowings in Donnelly's widely known work of less than a century before. But then, V. unsportingly downgraded Donnelly. I have earlier discussed the remarkable case of Beaumont, whose claims were so similar but whose method so differed from V.'s. I mentioned that V. noted to himself that Beaumont must have gotten his ideas from V. by telepathy (though the reverse should be more true, if any credence were to be given telepathy). Discovery of V.'s belief in "telepathy" amused Deg. He was reminded of Hans Kloosterman, the catastrophic geologist leader, whom Deg had joshed for decrying V. as fanciful while himself espousing telepathy. V. might well have agreed with Kloosterman's explanation of the uses of telepathy to Deg, in a letter of May 5, 1976 from Rio de Janeiro:

> Telepathy is not irrelevant to my main line of investigation, because:
> a) Telepathy is possibly important in evolution (see p.e. "The Living Stream" of Alister Hardy);
> b) The biosphere interacts with the lithosphere. And what holds for telepathy holds even more for dowsing, which involves rocks and ground water and ore bodies.

When Greenberg published in 1981 a posthumous note of 1948 by V. on precursors, he reacted too strongly "to put the lie to the idiotic and petty criticism of certain people (e.g. James Oberg) who have accused Velikovsky of failing to mention 'his antecedents' — particularly Whiston, Donnelly, Hoerbiger, and Bellamy — as recently as the Fall issue of *The Skeptical Inquirer,* a trivial publication with debunking pretensions." Then Greenberg advanced three other works that V. might have mentioned, provided he had come upon them, Godfrey Higgins, *Anacalypsis (1*833-60), Comyns Beaumont *The Mysterious Comet* (1932), Harold T. Wilkins, *Mysteries of Ancient South America* (1945). Neither Greenberg nor V. mentioned Nicolas-Antoine Boulanger, a most important predecessor, as I think V. would have granted. Deg carries this story in his journal:

Deg's Journal, November 4, 1972

...I then spoke to Livio [Stecchini]. Did Velikovsky know about Boulanger when you brought his name forward? No, he replied. When I gave him my draft paper to read, he said afterwards that that was the one thing he learned from it, because he didn't like the paper.

This was in the spring of 1963. I asked L. where he found Boulanger. In the Princeton Library. I probably picked up his name as an Enlightenment scientist.

I am relieved. I have been pursuing an unpleasant task. V. does not cite Boulanger, who is a predecessor in that he ascribed a variety of religious beliefs to actual human catastrophes. Yet V. cites an immense number of sources and combed the literature thoroughly.

I recollect V. telling me not long ago that Boulanger was a predecessor, the most important one — not a cause, note well, he didn't say he had read Boulanger. I wondered why he bothered to tell me this. When one is suspicious, of course, one looks hard at any clue. No matter that I admire V. greatly and like him as a friend; one has to chase down a suspicion that he might pull the "silent-footnote" technique on a causal, as against a merely chronological, predecessor.

Another precursor of V. (and of course Deg) was Howard Baker, a geologist who first mentioned Venus as a possible intruder into Earth's space sheath, but had much to say concerning the Moon. Again I resort to Deg's journal:

Washington, February 19, 1979

Yesterday Ami and I spent the day at the Library of Congress to clean up the last of the bibliography and footnotes of *Chaos and Creation*. It is tedious and often unrewarding. Yet I located a copy of Howard Baker's mimeographed book of 1932, another copy of which had been stolen from the Princeton University Library, *The Atlantic Rift*, and 2 articles by Marcel Baudouin from 1916 on paleolithic astronomical symbols, especially the Pleiades. As a bonus, there was a pamphlet from Baker's hand, of 1954.

So far as I know, only the one sentence, by Walter Sullivan in his 1975 book of *Continents in Motion* has ever been addressed to Baker's work, and that [was] a breezy reference in passing, obviously intended to show that anybody could be a predecessor of Velikovsky. V. himself said that he had heard of the book, probably from Sullivan, but when he searched for it, it was gone. I must ask Sullivan some day what assistant dug it up for him.

Baker's work is professional and brilliant, he says that he was working in the field from 1909 to 1954. I shall try to discover more about him. Apparently only 106 copies of the book were mimeographed, and perhaps less were distributed. He argues that Pangea was an all-land Earth, that the moon was pulled in the Mesozoic from the Pacific by a planet now missing, that prior to this, Venus may have interacted violently with Earth, and that the ocean basins were once empty and are now filled with waters from a late disintegration of the same planet (now probably the asteroid belt between Mars and Jupiter) that had earlier caused the Earth's crust to erupt the moon.

There is, in other words, a marvelous correspondence between Baker's ideas and my own, and his method of reasoning, his very mentality, is close to my own. He *sees* the same things on the globe. And he saw all of this before the flood of information of the past 50 years from oceanography and when continental drift theory was held in contempt by American geologists. He does not use legendary material but says reasonably and in measured tones that it can be applied and may support his theories; perhaps had he set a more recent date for the eruption and fissioning of the continents, he would have been able to use the legendary material about which he may have known.

V. had found in legend brief evidence that the Moon was young in the sky. He published it in 1973, claiming that the Moon had been captured, a Hoerbiger idea, and showing no awareness of the large quantity of legendary and geophysical evidence that U.S. Bellamy had brought to bear on the capture theory in several books, especially in *Moon, Myths, and Man* (1936). The main reason why V. dismissed the fission-eruption hypothesis that Fisher, G. Darwin, and now Deg espoused was unusual; for once, V. was saying that such a catastrophe would have been too destructive: "since human beings already peopled the Earth, it is improbable that the moon sprang from it; there must have existed a solid lithosphere, not a liquid earth. Thus it is more probable that the moon was captured by the earth."

On several occasions, Deg would say to V. that he was pursuing affirmatively the theory that the moon was wrenched from the earth in the time of man. V. had no interest in discussing the question. He offered no objection. He would grunt some vague expression like "You are working much, I see..." when Deg would say "Just look at the Pacific Basin..." and then move on the another topic. That he didn't object seemed to Deg a kind of *nihil obstat.*

The mystery of the purloined book of Baker was unsolved. Deg wrote Walter Sullivan one time asking where he had obtained the reference to Baker's work, but received no reply. Deg made a last-minute change in his manuscript to credit Baker's work, not that he believed in credit *per se,* but that he was happy to find like-minded company in the Pythagorean procession of life.

The idea of "precursors," believed Deg, was about as slippery, nonsensical, and morally disturbing as the idea of prior claims in science. In this I certainly agree with him. We know little about how a fruitful hypothesis is achieved and developed. Merely applying words will not help; what are the operations? And he goes on to explain:

> Synonyms for "precursor" might be forerunner, pioneer, predecessor, ancestor, scout, forebears, progenitors, inventor, creator, leader, conductor, pacesetter, guide, steersman, pointer, mercury, bellwether, and precentor. Let us keep "precursor" which is an empty enough vessel to fill with what we want. What do we want to say? The relation between writer B. at T1, to writer V. at T2 is such that V. has heard — forgetfully heard — did not hear of B. V. has set forth a Proposition "M" that is 90% identical (as it operationally describes a set of defined events) with a Proposition "N" of B. V. has arrived at Proposition "M" by employing the same method as B., or did not employ the same method, or did not use any method, or employed a method to arrive at Proposition "M" whereas B reveals no method for arriving at "N."
>
> Suppose V. takes "M" from B's "N." Does he get no credit for perceiving it? Yes, some, you say. But who gets the credit as precursor to V. who was the cause of V.'s perceiving "N" or of reading B? His parents, teachers, colleagues; his type of mind, preparation, briefing, search discipline? His wife for driving him to the library, for cooking food that stimulates the imagination? The librarians over the years?
>
> And what of the precursors of B, who may have directly or indirectly provided him with "N"? We cite Aristotle, knowing he stands for many sources, a few of which he actually names.
>
> And what of all the people who knew and conveyed "N" between B. at T1, and V. at T2, but whom V. did not know about?
>
> Would not V have thought of "M" anyway, and is not the decision to cite "B" as a precursor a socially acceptable choice? Horse thieves are unlikely to appear in genealogies and discredited writers are unlikely to be cited as predecessors. Whether "B" here is Boulanger or Beaumont will make a difference. Deg can testify to this statement; he felt better, and he knew his critics would be more accepting, if he acknowledged Boulanger and did not acknowledge Beaumont as a precursor on one or another point. Boulanger is farther back in time, and more conventional than Beaumont, who seized upon certain quite incredible ideas.
>
> I have scarcely begun to discuss the ramifications, doubts, dilemmas, tricks of the mind, and tactics of the writing scholar. We have been talking of a single, skimpy proposition "M" and "N". Suppose "M" and "N" represent averages of many propositions, then the way in which they are combined, the theory behind their selection, and the style with which they are conveyed are only several of the numerous conditions that may render even a close correspondence between "M" and "N" whether single or an average of a multiple nearly meaningless.
>
> So V. was accident-prone with precursors. It was quite unnecessary.

The absurd attempts of critics to pretend that what he said was not only false, and anyhow not new, could be taken seriously only by fools. But as I have shown here time and time again, he seemed to think that knowledge came in gobs, and he had produced some gobs, and had to defend them against theft by others.

Who were V.'s precursors, I asked Deg, the truth now, and nothing but the truth. Precursors were many, he replied.

> All the ancients were precursors. Beginning with Renaissance times, some score of major precursors have worked. Of these, directly, V. took from Whiston, Donnelly, Bellamy, Brasseur de Bourbourg, and perhaps innocently or amnesiacally, from Beaumont and Hoerbiger. After 1962 he probably took from many people of his circle, both directly and from their references, like Stecchini with Boulanger and Juergens with Bruce, or Schorr on the Dark Ages and Mullen on the Pyramid Texts, but he was writing little after 1962.
>
> On the matter of human psychic origins, he took from Freud directly and from others probably as currents of thought, the psychoanalysis especially. And of course, he was getting a great deal of material from his opponents; we must never forget that. He was a sad man when the Apollo Moon program was cut back, he used Sagan's material on the Venus greenhouse effect to dispute the matter. But I tell you it doesn't matter — not to science, not to the truth of what he is saying, not to me — only to the question of how big a hero was V. — how many scalps on his belt are really his own prizes.

Did V. ever use anything of yours, I asked Deg.

> Perhaps, but I couldn't say. Yes, definitely, he used me to figure out what was happening sociologically to his interests. He soft-pedaled certain of his views on collective amnesia, on anti-semitism, on the wrongness of others like the English heretics, on the inheritance of acquired traits, and such kinds of matters when I was around, though this cannot be perceived in his writings. I am not speaking of tactical advice in his self-defense, of course. All in all, practically nothing.

And you, I asked, what did you take from him? Everything I could, Deg answered.

> I got very little out of conversations, but a great deal from his writings. But I wish to make one point clear. Although V. was my precursor, predecessor, forerunner, etc. I did not accept V. on anything, except for a time his reconstruction of Egyptian history after, say, 800 B.C., and this because it seemed irrelevant to most of my interests. Not until I realized that V. was destroying his own 8th century catastrophic history by moving

kings too far into modern times did I become worried and stop accepting that set of events.

What I mean by "accepting," he continued, is taking for granted, and not reconstructing the same structure alongside his structure. "Accepting" is what, say, a paleontologist does who has a fossil ape and gets it dated at 12 million years by a laboratory on potassium-argon dating and accepts this as his date.

"Accepting" is taking a cloth made by someone else, before going on to embroider it. Everything I took from V. I examined and took apart and put together again. I guess you could call it factory rebuilt." I did not deny him, underrate him, or even disagree with him seriously and often. However, I was building a much larger, more systematic, broader, more scientoid model. I tell you frankly, I had in mind to supersede him.

Did you succeed? Yes, Deg said. How?

Like I told you — putting all that I could of his machine into a larger, more systematic, and broader model. I swung the whole mass of ideas and evidence into a hypothetical model — nothing was true; it simply could well be true. Everything is swung into position for testing: logically, empirically, comparatively. V. worked like a detective who is looking for a culprit, there has to be a murderer somewhere; I work always thinking maybe there was no crime! And if there were, who is the culprit becomes a sociological question, always plural. And I am always suspicious of the detective, too; maybe he staged the crime!

Well, I said, dubiously, how does it happen that your writing often races along breezily and confidently?

It's a matter of style, he said, and of necessity. I am confident of what I am saying, believing that I have put proper limits on it. There is a characterological element in it; I've always written that way, hammering along like a thumping heart, or the old diesel motor of a caique. There's something else, though, purely for the sake of the reader. There is a limit to how many times you can use the word "tends to" or "may" or "on the average" or "holding all other factors constant" in place of "is" or "does". That's one kind of problem; a writer shouldn't carry his miasma of doubts to the extent that he is never clear; actually, every sentence you utter distorts the reality of which it speaks.

Also, when, after having defined Yahweh and Moses and the nature of their "communications," I may be saying "Yahweh then speaks to Moses," I hope that it is understood that this statement of mine is subject to the prior definition of all three keywords, "Yahweh," "speaks," and

"Moses." But the total posture of my work is different. V. accomplished marvels of detection in myth and legends. Also in history. He sets up a contradiction or confusion, then puts forward his resolution. Yet ordinarily he is not self-conscious about his logic, method, and epistemology. He was a practitioner and an empiricist. By contrast, there must be hundreds of pages on the method of myth analysis and anthropological culture analysis in my writings.

Onetime, V., in an unusually frank conversation with Wolfe, Milton, and Rose — at the same set of meetings in fact that produced the euphoric letter that I described in the chapter on "Holocaust and Amnesia" — denounced the coining of words as the tactic of crackpots, and then confessed that he had coined a word; it was "introgenesis." It meant that "everything wishes to make everything else to its own fashion." Existence, whether animal, plant, or even celestial and inorganic bodies, operates by this imperative, to take whatever it encounters, digest it, and reconstitute it with oneself. Introgenesis was marked by him to become the key word in his philosophy. It would have become my philosophical system, he said, if I had not come upon *Worlds in Collision*. Everything wants to swallow up every other thing.

When this burst of philosophical confidences was conveyed to Deg, he wondered at it — it seemed so meaningless — and only years later, when he heard a full statement of it, did he appreciate that V., without realizing it, was simply coining a word (typically he credited words with substance) which referred to his own immense narcissism, the same narcissism that he urged all psychiatrists to fish up from their patients at the beginning of analysis.

The sole coinage of the realm was to be one's own. This wish seems to go hand in glove with the wish for unassailable proof of the purest assay of gold in the coin. V., as he grew old, appeared to be ever more hopeful that some one critical test would occur, some grand fact, that would prove him *right*. The attitude became at times an obsession, in that he would disregard problems or proof that lacked this *capability*. This explains why he became barely interested in myth, while hanging upon every new discovery in space. A fully professional intellectual such as he should have known that there is a) no proof of *right*, b) no *single* right, c) little chance that *right* on a single test would erase wrongs on others, but, too, sociologically, d) one's opponents are not likely to define *right* in one's own terms, e) they are not inclined to come to grips at one's strongest point (even though ideally this would seem proper), f) they will seek to recognize someone else as the originator or predecessor of the chosen point (creating a new issue and argument of an undefined kind). V. was not alone in this regard; he had supporters who worked hard to establish him as champion predictor of the one right critical test result. Still it didn't work.

It seems that all three behaviors join together in an authoritarian character: the ultra-sensitivity to "priorities of claims" to which I referred before, the anxiety over precursors, and the hope for the single critical test. In all of them we discover the intolerance of ambiguity which is a strong trait of the well-researched "authoritarian character" in psychology, and Deg alludes to the research in several of his early writings. There is, too, in all of them, an aversion to the close proximity of others, to a trespass upon one's possessions, a need to define exclusive boundaries.

Dislike of ambiguity is not only "authoritarian" but also "scientific" by the way, for which the antidote is pragmatic operationism, a subject for another essay. Perhaps it is time to venture a clearer statement. How did Deg and V. diverge from their basic narcissism, so that V. fiercely defended his claims whereas Deg untypically and diffidently recollected his claims after dispensing them like the money of a drunken sailor?

Both men, encouraged by their early models, commanded unusually strong energies that they used to conquer their existential fears by creating an independent self, a self not dependent upon others, that would take in the world and refuse to let the world include them. But then V., to enhance his primary ego clutched, contained, and possessed his aberrant egos, his poly-ego, whereas Deg dispersed his poly-ego hoping and expecting dividends to return.

The result was the formation in V.'s case of an authoritarian character, in Deg's case an anti-authoritarian character. (I trust that you will not be put off by the fact that V. had to attack the scientific establishment and that Deg sometimes liked authoritarian causes ("universal national service") and people (such as V.). The authoritarian character led to predispositions to monolatrous, monarchical, and presidential forms, on V's part, while the anti-authoritarian character led to polytheistic and republican forms on Deg's part. On V.'s side, the same character ran continuously the risk of enhanced paranoia; on Deg's part the risk was hypercritical reformism.

I shall not elaborate upon the distinctions farther here, but a rough example may suggest the effect. I selected six well-known historical figures (there is no use in comparing the two men with the cop on the beat, their local congressman, or others whom you have not known): Noah, Moses, Stalin, Trotsky, Theodore Roosevelt, and Charles de Gaulle. I asked a couple of persons who knew both V. and Deg to assign each famous character to one or the other, on grounds of relative nearness. V. ended up with Moses, Stalin, and de Gaulle; Deg was assigned Noah, Trotsky, and "Teddy" Roosevelt. I had, of course, predicted those assignments. The test works out even better by using a scale of "nearness" from 1 to 10.

"Hypercritical "is relative to the standard of evaluation. Deg was uncomfortably aware that by normal practice he was hypercritical, but that by logical and rationally instrumental measures he may have been no more than properly critical. He was elated the first time he saw a sign in a printing shop saying "If things look confused around here, that's because they are." Not only were matters everywhere in worse shape than were admissible, but the only intelligent comment one could make all too often had to begin at least with a negative, and he felt, which I think was true, that he rarely failed to come up with a subsequent constructive resolution. Moreover, the line between critical analysis and hyper-criticalness was often too indefinite to bother with. Furthermore, was he not equally critical of himself, whom he liked exceedingly well?

Now the same kind of self-justification was possible for V. Was it not true that most conventional scholars and scientists were out to get him? Were they not making of him a target for the release of all too many hostilities toward what he represented, an independent, unprotected proud figure of opposition? Didn't the humanists turn him over to the scientific crowd, and the scientific crowd kick him back among the humanist crowd, each proclaiming that he had no place among them? So he was then, a heretic, stimulated continually along the dimension of paranoia. And a goodly number of his supporters, several of whom were close to him but the majority of whom were out in the public, were also exercised in their paranoid dimension and felt better to be able to attach their paranoias like tentacles to such a strong defensible stone.

A great difference between Deg and V. was that whereas V. took the greatest pride in being unbending, determined and assured, Deg was continually seeking knowledge through self-examination and the admission of sins and weaknesses. Thus it came about that V. was a kind of Captain Dreyfuss, every inch of him the reflection of his assailants, whereas Deg was an Emile Zola, vehemently led by the inner necessity to espouse liberty, equality, fraternity and justice. And I have a feeling that V., had he been restored to his commission under the colors of science, would, like Dreyfuss and his family, have begged his supporters to retire from the scene.

When he was writing *Homo Schizo,* Deg came upon the essays of the psychologist Morton Prince, edited by Nathan G. Hale, Jr., where material on multiple personality is contained. What Deg marked in the margin of the Introduction as "terrible" are the following lines:

> [Morton Prince could not] stand aloof from the Sacco-Vanzetti case [anarchists convicted of robbery and murder and later executed], although his opinion at first flouted that of proper Bostonians. On October 30, 1926, Prince wrote to the *Boston Herald,* protesting the prejudice of the trial judge and the incompetence of the government's major witness. The judge, like most lawyers, was lamentably ignorant of the "science of modern dynamic psychology" and had glibly interpreted the defendent's motives in a way which discredited the impartiality of the courts. The witness had purported to describe sixteen different details about Sacco, whom she had seen at a distance of sixty feet, for from one and one-half to three seconds from a car going about fifteen to eighteen miles per hour. Only if Sacco later had been deliberately picked out for her to identify, could she have recalled

such details, Prince insisted. Her "memory" of him was produced solely by "suggestion" and was nothing more than an "unconscious falsification." Later Prince agreed with a committee of review, appointed by the Governor of Massachusetts and dominated by A. Lawrence Lowell, that the conviction had been obtained after a fair trial.

Prince's protest and change of mind had come with the authority of his appointment to a new chair of abnormal and dynamic psychology at Harvard College. Lowell, Harvard's president and [an] old friend, had accepted Prince's offer of $150,000 from an anonymous donor, as well as Prince's services as professor and director of a new psychological clinic that opened in 1927. Prince had insisted that it be attached to the College's Department of Psychology, perhaps as tangible fulfillment of his hope to include psychopathology within that discipline. The clinic was to convey a knowledge of the subject, to conduct fresh research and to treat selected patients. Prince held the chair and headed the clinic for the last two years of his life, with Henry A. Murray as his assistant. He once remarked, "La Salpêtrière is a monument to Charcot. I want no other monument than the Psychological Clinic."

The sacrifice of principles for prestige and self is an everyday affair in science and academia and the victims of misconduct are legion, nor do they receive the glory of execution or the stake.

* * *

When on a snow-enveloped January morning in 1965, Deg's father died, V. projected from the depths of his own character and experience and advised Deg that he would enter now upon a highly creative period. The consoling remark was more revealing of V.'s paternal relationship than of Deg's. Not since he was twelve had Deg noticed his father weighing upon him. Aside from an oration for a Junior High School convocation that he considered too important to let the boy write by himself, and letters that were merely informative and invariably encouraging, Deg's father committed little or nothing of his beliefs to paper. He read and worked upon reams of music as a scholar upon books and papers. Perhaps only a character, not a philosophy, was needed in copying and orchestrating his musical scores — now a soulful surge of Wagnerian triumph, then again a sweet and lively Mozart Overture, and another time he would prepare a Verdi chorus for brass instruments. The only expression Deg came upon when he disposed of the music archive to the New Jersey State Prison System was this: "A rebellion is terribly hard to repress when it is born in men's mind. How can intellectual resistance be killed?" It is not known what occasioned the remark, neatly written on a small note pad.

The heretics, or rebels if you will, carried on with the procession. Deg is now writing Brian Moore in Hartlepool, England:

Princeton, November 17, 1979

Dear Brian:

I regret to report to you and to your colleagues and members of the Society for Interdisciplinary Studies the deaths, within a month of each other, of our friends and colleagues, Livio Catullus Stecchini and Ralph B. Juergens. Besides the personal grief that their passing has brought to us who might count them as dear friends, the loss to pioneering scholarship and science in their demise is great.

Both men left off in the middle of important books and articles, Livio Stecchini on pyramids, on the origin of the gospels, and on ancient measuring systems, and Ralph Juergens on the electrical theory of the cosmos. Professor Earl Milton of Lethbridge University (Canada) has undertaken to review Juergens' manuscripts and I Stecchini's, with a mind towards their eventual publication. Other colleagues are concerned as well.

Both men were models of honest scholars, of personal modesty, and of helpfulness to all who asked something of them. I know that the thousands of women and men who have become related to them through a common interest in the reconstruction of knowledge about ancient history and nature will wish to think of them in companionship and gratitude.

We may hope that the remembrance of their achievements, like a freshly trodden path, will be enlarged now by the usage of the young and bold.

Deg was both disturbed and amused when, in the last years of their lives, Stecchini and Velikovsky disputed the attitude of Plato toward catastrophe, the first stressing that Plato would have catastrophists put to death, the latter regarding Plato as the last direct heir of the catastrophist tradition. They did not communicate for some time before Stecchini's death. The issue is germane to political science because it reveals the conditions under which the elitist political philosopher such as Plato will choose *raison d'état* over truth.

The argument was not resolved, although to Deg it seemed clear enough that Plato was wearing the two caps of scientist and political ruler. When he played the one, he had to recognize the catastrophe of Atlantis and other disasters, and exhibited little confidence in the stability of the heavens. When he played the role of custodian of public morals, he recognized, as few did afterwards, that men behave in imitation of the sky gods. When the gods misbehave, so do men. Hence Plato would severely chastise those who rendered the gods a disorderly mob or perceived disorder as the rule of the heavens.

On November 19, Deg writes to Brian Moore again:

Dear Brian:
Hardly had I posted my letter than the word came that Immanuel

Velikovsky was dead. He died on November 17, at 0800 hours. After a restless night, occasioned by a rapid pulse and feelings of weakness, he arose at first light on the Sabbath and showered. He returned to his bed and Elisheva his wife sat beside him. He murmured several indistinguishable words and took her hand. He became quiet and she saw that he had passed away as if to sleep. He was buried in a private ceremony the next day at a small cemetery not far from Princeton.

He was in charge of himself until the last hour, working daily on his unpublished manuscripts, discussing proposals to film *Worlds in Collision*, and worrying over an article that was half-promised to *Harper's Magazine*. On Monday I had an extended visit with him. We talked of my memorials to Stecchini and Juergens and about the book on Moses that I am completing, and also concerning a brief paper which I proposed to write for *Nature* magazine, setting forth six challenging hypotheses on the worldwide catastrophe of the mid-second millennium.

He urged me to write the article "for tomorrow." I wrote it and talked with him about it on Wednesday. He liked the phrasing of the propositions but disputed my selection of examples and said that he would not become co-author because he had no time to do the necessary research. His powers were fully engaged; he was concerned with the shortness of time, with editorial priorities, with the campaign to advance and defend his ideas.

When I left him as darkness fell, he remained seated. He would usually walk with me to the big door and step out for a moment to breathe the season's air. I telephoned on Thursday and he was working. I still sense that he is palpably at work and will continue working for a long time.

Then, after several years of laboring over Immanuel's archive, his widow, Elisheva, died. Deg wrote a eulogy of her during her last hours.

Sheva

Whiffs of air, a shot of drug, a tube of soup,
a white-breasted meter-maid intruding now and then —
intensive care — to confirm her readings of your organs.

Their prognosis for you is poor you must know.
You don't speak at all well, though you may perceive,
while your intakes and outputs are disordered.
Your heart stands brave above it all,
like a proud cock refusing the falling night.

How I wish you might know of our plan for you:
That you shall be forthwith removed herefrom,
and placed upon your porch above the greening bushes,
overseen by a nervous flitting finch in the beams,

there to sit and listen while Immanuel speaks
of claims and confirmations in words so deep drawn out
that in between them you plan how you will shape
a bust in stone, and next time play that passage *piu adagio*.

Fingering the fiddleneck and banging the chisel,
just and nice your big hands were
that shook my big hands roughly.
Your pot of tea is pouring
interminably into our china cups and, yes,
there was something else — cold white wine of Canaan —
to fetch from the kitchen, but you said "Wait,
one moment, I want to hear this, what did you say?"

I blush to think of injustices done you,
munching buttered cakes and crackers with cheese,
boasting of stalking and snaring man's mind
as the very quarry was serving the hunter's breakfast.
Stroking celestial harmonies from your varnished box
and chipping life into becoming, feeding the animals,
then taking up the phone protectively, "One moment,
one moment, Immanuel is on the line,"
if he wanted to be on the line.
But I did kiss you, did I not, and hugged you, too,
whenever arose the chance in coming or going.

Don't get up; sip your own, your own cup of tea.
Why should it be yours to close the doors, draw the blinds,
bury the dead, argue the law, pay the taxes,
comb the archives, fight the battle, placate friends,
watch Hector's body being dragged around the Trojan walls?
Did you not earn your porch of peace even before the 1950 War began?
Sacrifices so many that never to utter the word was your greatest sacrifice.

Your modest scoffing will not avail
as we burn down the skyscraper for your pyre,
each floor a blazing bargain for your first good, next good, and thereafter.
We have taken the matter out of your hands.
The last chord is not yours to sound.
When the guests set down their cups and leave,
you are to be held close by your loved one
while your ghost rises lightly through the thick dusk air of summer.

* * *

I've told of the three heretics, heroes of V., who were burned at the stake. Do cosmic heretics live long? Plato voluntarily denounced his own catastrophic views; he lived to 80. Whiston was black-balled from the Royal Academy of Science and fired from Cambridge but lived to 85 Boulanger died in his thirties. Carli-Rubbi ended his career as an economist in good style, as far as my inadequate sources reveal. Vico died at 76 but his friends got to fighting over their relationship with him and left his coffin standing on the street. Bourbourg was ridiculed at the end of his life. Ameghino was dismissed finally and posthumously honored; he believed in Atlantis. Donnelly landed on his feet, a versatile populist-utopian, writer and lecturer, and died at 70. Beaumont's papers were destroyed by bomb and fire; he was still writing when he died in his eighties, and Stephanos was still peddling his manuscript when last heard of. Hans Bellamy passed away old and with him most interest in Hans Hoerbiger's catastrophism, which occurred from the Earth's capture of satellites. Claude Schaeffer died in his eighties full of public honors, but not from his great work on *Stratigraphie Comparée.* Frank Dachille died quietly aboard a PanAm airplane to Rome; he was beginning to move back strongly into catastrophism.

Of the fate of certain others, I've spoken elsewhere among these pages. The remainder are too many to census. I don't mean to imply anything. No curse attends to the practice of heresy; most heretics seem to live to old ages. Their only problem is that they cannot fulfill the romantic ideal of dying knowing that their ideas have been accepted. But no one does so, or he is fooling himself if he thinks so. It is easier to found an empire — and much more common — than to found a new model of scientific philosophy, and empire of thought. Christ and his early Christians did so. The Galileo-Newton axis powers did so. John Dewey and his pragmatists did so.

I would compare the cosmic heretics with the story of Leonard Woolf's life. His biography reads like a brilliant, long, and useful career, on the margins of heresy, for he was always a reformer, beginning as a Cambridge student follower of the delightful new philosophy which answered every question by another question: "What do you mean by that?"; proceeding to Ceylon as so efficient a civil servant that he logically arrived at the next step, which was to de-colonize the British Empire; then he became a novelist and a publicist, edited several magazines including especially the *Political Quarterly,* set up his own publishing company, the Hogarth Press, to put out his books and those of his wife, Virginia, and other friends; helped to organize and bring to ultimate triumph the Labour Party; pushed for international government through the League of Nations; supported pacifist causes and creative writers; and best of all kept Virginia Woolf reasonably happy and at work on her novels and also kept her from committing suicide over many years, until she managed in her sixties to end her career by walking to her death in the sea.

Still, when Leonard came to conclude the fifth volume of his autobiography a few years ago, he had decided that the process of life was more important than its imprint upon the world. For in their effects upon the world, most of what he had attempted had failed. Both Ceylon and England had grown more hideous. Peace efforts had failed. International government had failed. Justice had failed. The Labour Party had failed. The publishing industry was much worsened. He had studied hard for twelve years and then labored hard for sixty-four years. So he named his last work, "The Journey Not The Arrival Matters," the reason being that one never arrives.

> All these excuses and explanations of why I have performed 200,000 hours of useless work are no doubt merely another way of confessing that the magnetic field of my own occupations produced the usual self-deception, the belief that they were important...in a wider context, though all that I have tried to do politically was completely futile and ineffective and unimportant, for me personally it was right and important that I should do it, even though at the back of my mind I was well aware that it was ineffective and unimportant. To say this is to say that I agree with what Montaigne, the first civilized modern man, says somewhere: "It is not the arrival, it is the journey that matters."

Of course, if Woolf had believed this in the beginning of his life he would have undertaken few, if any, of his numerous enterprises. It is absolutely essential to society that the young be such fools. And that some of them remain fools forever.

At the end of the third and last volume of his autobiography, Bertrand Russell states what as a boy he wanted to achieve in life and what he discovered in the end. He "wanted, on the one hand, to find out whether anything can be known; and, on the other hand, to do whatever might be possible toward creating a happier world. From an early age I thought of myself as dedicated to great and arduous tasks." Deg had felt precisely the same. It is the narcissistic heroic vision of oneself.

In the end Russell could appreciate that both his works on knowledge and his books on social realities were partially achieved. But he confessed that he could not crown them with a synthesis. He had succeeded in that many people were affected by his works and these were acclaimed. So far, so good, but the failures rankled.

The external world had refused to cooperate with his efforts and was worse, more evil, if anything. The internal world had failed him, too. "I set out with a more or less religious belief in a Platonic eternal world, in which mathematics shone with a beauty like that of the last Cantos of the *Paradiso*. I came to the conclusion that the eternal world is trivial, and that mathematics is only the art of saying the same thing in different words."

Yet Russell was a tough old optimist and "beneath all this load of

Failure, I am still conscious of something that I feel to be victory." The victory consists of still believing, first that a "theoretical truth" must still exist and "that it deserves our allegiance." Second, "I may have thought the road to a world of free and happy human beings shorter than it is proving to be, but I was not wrong in thinking that it is worthwhile to live with a view to bringing it nearer."

Although having some miles still to go and a passel of things to do, Deg might be compared. He never believed in absolute Platonic truth from his first reading of Plato at 15, nor before, nor afterwards, and, being poor at mathematics, he decided early to project the blame upon mathematics, asserting that mathematics were a neat way of speaking and necessarily could not be speaking some basic new truth that sprang *ex machina linguae;* furthermore, there would have to be new mathematics for every important perspective upon the True, requiring therefore many mathematics, whereas mythical and ordinary language could by its indefiniteness suggest all of these perspectives. In either case, language and mathematics were largely dependent functions of thought, though they might, interacting with thought, also determine it somewhat. It can be seen, then, that Deg was a pragmatist, functionalist, and social psychologist. "The truth" remained for him just what it was to the child, a guiding myth which, by much rationalization, was later fashioned into a politics and then a philosophy. Truth functioned existentially, as a hypothesis that worked better than any alternative hypothesis.

Turning to the external world, the same philosophical instrumentalism led him to believe, not that the world would be ultimately better, although this would take longer to achieve, but rather that the world might become either better or worse (in its concurrent configurations with future times), and one should not expect more than that while moving pragmatically and existentially through the process of life.

It begins to appear to me that Deg's moods were externally fairly even, with a frequent enthusiasm and hedonism balancing his hyper-criticality. Privately, as with many people, his moods were more grim and irascible. His journal is not a perfectly true barometer, since he seems to express his critical and negative feelings often and his happiness (a word he detested) less.

Deg's Journal, 6AM Sunday, Jan. 21, 1979

> I derive pleasure from planning the future — my personal future — and thousands of pleasant interludes of 5 minutes to hours of large plans are usually interspersed among the other life operations and taken up euphorically as the whim or impulse seizes me. It is partly this childish pleasure, for I have done it from earliest memory, which leads finally to the drive to shape a world future.

It is written because I have caught myself escaping from some painstaking work on footnotes of *Unsettled Skies* into penciling the best possible calendar I can hope for in the year ahead.

Connected to this impulse is the listing of "things to do." When oppressed by the many little and large obligations, self-imposed and encountered through our hopelessly complex society, I make a list of all that should be done in the next week, month, 3 or 6 months, and so on. Whereupon I feel relaxed and confident, as if it were all done."

When Deg became anxious enough to draw up one of his lists, he unknowingly let us have a way of guessing the ratio of concerns to total time available. Here is his list of stresses, dated late in the quantavoluntionary period; it reveals that the question of chronometry is still plaguing him, as well it might, and that the production of his books and the maintenance of a heretical circle are pressing him too.

Deg's journal, January 15, 1982

Especially worrisome problems (stresses)
1. Inexcusable delay of National State Bank in exchanging a German check for 19,000 DM into $. Am broke.
2. Mom's critical illness and need for continuous surveillance.
3. Whereabouts of 1250 copies of *Chaos and Creation* and their bill of lading.
4. Decrepit and dirty condition of the house on Centre Street.
5. Seemingly impossible contradiction in short-term dating of natural history and the huge defensive effort accumulated *pro* long-term dating.
6. Difficulty starting car.
7. Blocked hot water pipe (frozen).
8. Bad weather— snow, ice, cold.
9. No money.
10. Conflict over debts and title of Clearview house with Sebastian and Edward.
11. Carl's loss of job and pennilessness.
12. Bad domestic and international policies and actions of U.S. Government.

Plus normally worrisome problems e.g. abcessed tooth and dental work needed; Cath's miserable behavior toward me; delays in Anne-Marie's book and her preoccupation with her work; laundry and sewing needs; growing phobia *vs.* long-distance driving; inability to visit or be visited by men with the same interests, especially those expert on what occupies my writing; lack of intellectual and social circles in the area and inability to take time, money, effort to construct (reconstruct) same, in which I might

participate (this has to do with my present life style, and scattered domiciles
_ N.Y., Princeton, Trenton, Naxos).

As a final favor to me, who was much impressed by Woolf's life accounts, Deg prepared a list to end all lists, an accounting of his time over the period covered by this book. He skimmed it across my table to me.

"I did what you asked," Deg said, "but I forgot the four hours it took me to do so. So the Q series took 29,904 hours instead of 29,900."

I didn't believe the figures anyway. Here they are as he gave them to me:

Time Accounting
Hours (Lapsed Time: 21 years, 1963-1983, total hours: 183,960)

1) 53,655   a) Meals, visiting with family and friends (including telephoning), general correspondence, radio-TV-newspapers; b) Housework and shopping, paying bills and taxes, personal hygiene, car maintenance.
2) 54,487   Sleep
3) 29,900   Research and writing, including production and promotion, of Quantavolution Series.
[Note: add 800 hours for early 1984
4) 10,307   Other research and writing.
5) 8,936   Politiking, consulting, and business affairs.
6) 9,651   Teaching, Committee work, doctoral supervision, NYU, 12 years.
7) 2,400   National Endowment for the Arts (excepting book "1001 Questions.")
8) 4,000   New World University at Valais, Switzerland.
9) 500   Kalotic movement for World Government (except in Switzerland, 8 above).
10) 2,000   1 year of hard labor (Naxos).
11) 900   En route somewhere (less project time achieved en route).
12) 1,900   Spent with V. on "the Cause" a) personal: 1190 b) on the telephone: 750
13) 204   Spent with V. on the substance of Quantavolution (not in 3 above).
14) 400   Spent with V. on personal and general socio-political discussions.
15) 2,800   Spent with other heretics (except with Milton, included under 3 above and does not include group time with V., see 12 above)
         a) on the "Cause": 1550 b) on the substance of Q: 1250

184,080   Total hours accounted for
183,960   Total hours to be accounted for 365 x 24 x 21

- 120	Discrepancy
120	Add 5 days for leap years
0	Total Discrepancy

"Do you have any questions?" he said and I said yes, I do: "Why do you include 'personal hygiene' under '1 b)' instead of 1 a)'?" His answer was not nice and I see no need to convey it. He went on to explain other matters that he believed to be beyond my comprehension. He begged me to note that at $40 an hour (he certainly had a modest idea of his worth) he had spent $1,200,000 on the Quantavolution Series. On the heretical movement as such, he had spent the equivalent of $192,000. How did you arrive at the hourly rate, I asked him. It's near to what the University was paying me and about the average for when I operated as a consultant. You see, he said, after you become a tenured professor you can retire on the job, and many do, letting research and writing go by the board. However, such equivalencies don't make sense. If I had gone into business I would have made a great deal more, or a great deal less, because I am a speculator; smooth flows of money do not amuse me.

Earlier were mentioned gross disparities in compensation and resources between the conventional established scholars and the heretics. Here another of Deg's computations presents a shocking state of affairs. The typical prominent professor, at a university of the first or second grade of excellence, may be said to receive these emoluments:

$43,000	salary and fringe benefits
30,000	grants (directly applicable for personal support)
60,000	indirect support (government grants for projects foundation support)
40,000 at 2,000	students who can be put on projects (value of their work) 20 (screened applicants — admissions, scholarships, fellowships)
15,000	use of University facilities (labs, astronomical, machinery, conveyances, University grants)
22,000	assistants (2)
20,000	overhead
7,000	access by influence to periodicals (7 articles $1,000)
20,000	consultation
2,000	personal support to attend conventions
10,000	use of institutional name (mass media, publicity, influence, public relations, legislatures)
1,000	life tenure (worth $200,000 or more)
$270,000	Real income applicable (except for personal taxes) to carrying one's prestige and influence into the arena of scientific controversy

carrying one's prestige and influence into the arena of
scientific controversy

A total of $270,000 annually in emoluments is estimated for a single professor. His tenure is certainly worth thousands per year additionally. Nor have we considered that there must be a cash equivalent for the right to impose upon from 10 to 1000 students a year one's viewpoints, applying sanctions to apparent disbelievers. Because the professor is not selling soap does not mean what he does sell has no cash equivalency. This large sum is some measure, perhaps the best that we can arrive at by speculation, of the annual economic impact of an establishment professor upon his fields of activity. The American public, politicians, and business leaders have only a slight awareness of how great is the influence of professors in society. (Sample surveys, however, show that the population does rank professors in the highest echelons of respect.)

As for the time Deg had given over to the movement, it was little, as you can see, no more than, say, a chairman of the board of a closely-held company would spend on its affairs, much more than, say, V. spent with Einstein, which V. turned into a book (yet unpublished), infinitely more than a day in the life of Leopold Bloom, according to James Joyce, which contained all of the wandering years of Ulysses, ten years in coming home from the Trojan Wars.

Then he said something worth repeating, that the time he spent with other heretics on the cause, and with V., the whole "schmeer" he called it with fine vulgarity, was essential to the Q project. They would all have run around lost if they hadn't been held by their crazyquilt network. The network was essential for morale and V. was the primary reference point; the game worked so that one had to touch base with him in some way, or utter the password, make some symbolic gesture.

Furthermore, working with others on V.'s cause was not like work with a political party or an evangelical sect, where you know what you want and have to believe in it, and there are few surprises, and the question is simply how to achieve them; for V.'s cause excited continually new issues of substantive science — the argon concentration discovered on Mars, the moonquakes, a radiocarbon date, the examination of King Tut's skull, the excavation of Ebla, the finding of ash levels below the sea bottom, and in these and scores of other cases, the heretics had to figure out their possible significance. As it developed, certain people gave themselves over to agitation and publicity, like Robert Stephanos, who accepted answers, while others like Mullen and Schorr were best at evaluating truth and significance, and then there were others, like Lewis Greenberg of *Kronos*, who operated both as agitator and evaluator.

Take the discovery of ash levels below the sea bottoms, a set of discoveries beginning with the oceanographer Worzel, which V.,

Kloosterman, and Deg, among others, were quick to seize upon for their catastrophic significance. What was their extent, their composition, and their age? Did any pertinent facts remain concealed or unsought because of the conventional attitude of the oceanographers? V.'s cause, or let us say, since Kloosterman disavowed V., the quantavolutionary cause was to discover and prove a catastrophe, possibly exoterrestrial. Until they understood the studies, the heretics could not use them. Until they rewrote and extended the logic of the studies, they could not achieve the full use of them.

When the Quantavolution Series was completed, Deg could be asked what portions of this systematic and complete model of cosmogony might he confidently expect to be useful to science, and what might come apart soonest. I give here his answers:

That the basic principles of quantavolution would hold, he was fairly sure: the world has changed largely by sudden, large-scale, intensely forceful events.

Also that the solar system is a broken-down binary and functioned once within a huge sac and plenum of dense gases.

Also that the solar system was born electrically, changed and changes electrically, and only emulates a "gravitational" system when there is too little change to take note of, or build a model upon.

Also that the Earth exploded the Moon one time, and then it was that the continents began their rafting about the globe.

That the morphology of the Earth is almost entirely due to exoterrestrial interventions, including aftermath effects extending for long periods of time.

That biosphere evolution (and extinction) has occurred in generalized quantum leaps.

That the human is genetically and experientially poly-ego and schizoid, and rationality is a pragmatic form of schizoid behavior.

That liturgy, language, history, and literature are schizotypical compensations and sublimations for fear.

That quantavolution as a heuristic model of natural and human history is useful for many scientific and human needs involving past time, and environmental and self-controls.

That historical religions had a crude reality base. Also that Moses behaved as he is described in *God's Fire*.

Deg was not sure of other parts of the model:

That his radical compression of time can stand against the full array of opposing chronometries.

That his microchronic calendar manages to name and divide properly the actual ages of natural and human history.

That gods must exist and that at some point in time they must come to affect the world. (But he insisted upon the axiom that what they are like and when they will operate must stand as open questions.)

That the planets were as fully responsible for quantavolutionary events as he has made them be.

Also he was confident that on many points of detail he would be proven to be in error.

Nor did Deg feel at all certain that the quantavolutionary movement would succeed now, although, if human civilization survived, some model much like it would occur again. Furthermore, he thought it unlikely that quantavolution, if it succeeded in the next century in winning over science, would recognize or acknowledge the heretics of today, but would probably, unless otherwise decreed by a political revolution and for then largely irrelevant reasons, be adopted as a great many bits that would form statistical trends that would quantitatively change the existing gradualist and incremental model until it would appear that the scientific revolution was accomplished by a great many people working independently and empirically until driven together by the facts.

"How would you feel about that?" I asked him.

"It's OK with me," he said, "I'd be so surprised at being right, that I wouldn't think of asking more. Even though it cost me a million dollars."

CHAPTER SEVENTEEN

THE ADVANCEMENT OF SCIENCE

Actors in the dramas of science might learn certain precepts such as:

There is nothing more difficult to plan, more doubtful of success, nor more dangerous to manage, than the creation of a new system. For the initiator has the enmity of all who would profit by the preservation of the old institutions and merely lukewarm defenders in those who would gain by the new ones.

So writes Machiavelli in *The Prince,* which was posthumously published in 1532. He was speaking about politics but the generalization might be enlarged. Probably all who have had anything to do with creating a new science, or trying to do so, would agree with him. Included, even, would be those who could recognize tangible victories in their lifetimes — Galileo, Newton, Hume, Darwin, Pasteur, Freud, Einstein, Planck, and Heisenberg.

The development of science, that is, sustains a branch of sociology: of historical psycho-politico-anthropo-sociology. When this is applied to science, as the science of science, a partial truth such as V.'s concept of collective fear being inherited from the trauma of ancient catastrophes takes its place as a modest useful contribution to the science of science. The more general truth is contained in Deg's mode! Of the gestalt of creation where homo schizo emerges out of a catastrophized ambiance as the true and normal human, who invents science as a typically schizoid set of operations for inducing psychic control and uniting the psychic with control of the external world.

The science of science discloses in the history of the cosmic heretics the "inadequacies" of the American social system in dealing with the challenges of new science. There are three extensions, unhappily, of this remark. One is that the same types of "inadequacies" are characteristic of all areas of American science. The same kinds of "inadequacies" furthermore characterize all other branches of the American social system — political, religious, economic, recreational, and educational, third, the same kinds of "inadequacies" characterize all ethnic or national societies — whether Western European or communist or "Third World."

I shall leave my readers to hunt by themselves for confirmation in the non-scientific areas of American life, whether by means of Deg's other works or the works of better teachers. I abandon them also to their own devices and explorations to discover what happens to new science in other nations. And I do little here to arrest their attention upon instances of nonfeasance and malfeasance in American science, other than by a few examples cited here and there, as by Burgstahler and Barber. I am tempted into one more example, this from a letter which Deg received from the most noted investigator of supersensory phenomena, Dr. J.B. Rhine.

The Parapsychology Laboratory
Duke University
December 16, 1963

Dear Dr. DeGrazia:

It is very good to see the systematic study you have been making of the reception of scientific developments. I am reading with great interest and satisfaction your September number of *The American Behavioral Scientist,* and I hope this number will become widely known in American science.

I have long been convinced that reception is the weakest link in the chain of scientific development in this country, and that the situation has been progressively worsening.

I have, in connection with my own studies, been testing the S.R.S., but I became interested in the problem as part of my study and teaching of the history of science, in partial preparation for the work I have been doing in para-psychology. It has seemed to me that what we are up against in the education of the individual, the growth of the university, or the development of a culture is a perfecting of a fixed conceptual ideal which reduces the possibility of *free* adaptation to new ideas.

I am more heartened by seeing this problem of S.R.S. being made the target of a special study than by anything I have seen since the problem first appeared to my mind...

I have just finished reading a book that, more than any other I have ever read, cuts across a large section of the struggle of ideas with the reception problem in the area of medical psychology. It is Frank Podmore's FROM MESMER TO CHRISTIAN SCIENCE, published by University Books in New York. It is a reprinting. The book itself was published in 1909. Such books as this and John Davies' account of phrenology in America have led me to feel more kindly toward *earlier* periods with regard to their tolerance.

I think I would say I am frightened about the small chance of a true revolution occurring in a major scientific field in America *today*. Western Europe I think is moving in that direction.

But this contrast is not a reflection from my own frustrations. It is true we are having plenty of difficulties, but we are progressing, and we are winning our case, slow though the progress is. But how many explorers die every year in the freshmen classes of our universities! Yes, this is a subject of primary importance. My hat is off to you, Sir!

In the late 70's, Deg began using the term "quantavolution." Not only the increasing number of cosmic heretics, but also restless and probing scientists of the several large fields of geology, astronomy, biology, and the historical sciences had been publishing new materials in which global disasters figured, sometimes mentioning possible exoterrestrial causes, at other times remarking on the shortening of time scales implied in the new discoveries. In paleontology, Stephen Jay Gould, collaborating with Niles Eldredge, was promoting catastrophism in evolution and paleontology as processes of "punctuated equilibria," thus keeping to the fore the gradualist and incremental aspects of natural history and offending as few people as possible.

New York University
September 26, 1980
De Grazia to the Editor, *Discover* Magazine (unpublished):

In reporting the work of Eldredge and Gould, among others, towards rehabilitating some of the constructive aspects of scientific catastrophism, your author, James Gorman, was suffering understandably from verbophobia. Hardly anyone, and for good reason, wishes to advance to the study of sharp breaks and movements in natural and cultural history under the flag of Cuvier. Not only does the term "catastrophism" suggest a long discredited science, but it ignores the "constructive" and "acceptable" features of the "catastrophic" events. (Our world and ourselves were, willy-nilly, catastrophized over time.)

"Punctuated equilibrium" (Gould's term) is admittedly awkward. "Macroevolution" is getting a little closer. I have tried a number of designations in lectures here and abroad, and for awhile "revolutionary primevalogy" seemed the most appropriate. I also tried "saltatory (leaps) theory." Then I began to use "quantavolution" — the study of large-scale change by quantum jumps and found it the most satisfactory and reasonable. I administered a little preference test to students and friends, and "quantavolution" came out ahead of all these other words. Hence I suggest that we stick to "quantavolution" when we refer to intensive, large-scale, temporally-compressed events or periods in nature.

Deg knew he was on a right track with "quantavolution" when he read in Otto Schindewolf the new term "anastrophe" as opposed to "catastrophe" and found in it exactly what he meant, for as Schindewolf had stated in 1961, "faunal discontinuities, as understood by us, involve not just the dying out of the old, but also the more or less sudden emergence of new phyla."

Later, Chicago's Field Museum of Natural History hosted a conclave of biologists called by Eldredge, an officer of the Museum, and Gould. Well-reported in *Science,* it did hot precipitate an organized movement, even in the single field of paleontology. A different kind of advancement of science is occurring — could it be the "partial incorporation of revolutions" that I spoke of earlier? In March of 1983, M.J. Benton of Oxford University wrote in *Nature* magazine on "large-scale replacements in the history of life," whereupon we must add "large-scale replacements" to our list of euphemisms.

Nearly two centuries after Cuvier, thirty-three years (one Jeffersonian generation) after Schindewolf, 23 years after V. and even a couple of years after the laggard Deg, it is written that "there is increasing evidence that major physical changes caused more large-scale evolutionary changes than has competition," and that competition or natural selection "will rarely be the sole cause, whereas it could be postulated that a catastrophic change in the physical environment is sufficient on its own."

When President Richard Nixon and his henchmen were accused of covering up the Watergate Affair, their slogan was "stonewall it"; after a while the message was "we've got to bite the bullet."

Warner Sizemore was keen for influences from many fields and was aware of Deg's embracing the term "quantavolution." Deg writes to him:

Naxos, January 12, 1981

Dear Warner,

After spending Christmas with the relatives congregated in Florence opportunely, Ami and I drove off and were ferried in our Renault 4 across the Adriatic and drove again from Patras to Athens for the New Year celebrations with the relatives there. After we arrived in Naxos, a weeklong storm closed the shipping lanes. There at the Postoffice I found the batch of material from you. Many thanks. The experiments on imitating the rampages of nature upon dead animals and the studies of what happens to them are long overdue, bound to be feasible, enlightening and supportive. I read, too, the article — effusive and popular though it was — in *Brain and Mind,* about Ilya Prigogine's work. It's impossible to tell what may be in it for us, but a search into his books is called for. Certainly they are talking of quantavolutionary changes of system-states. But since the mechanism is entirely abstract, i.e. non-existent so far as they say, I presume that a mathematical model is involved, in which statistical states snap into a new alignment by some set of convergences arising at a juncture.

Crystallization can perform this transformation under environmental stresses. Perhaps half the plant species are instances of proportional structural explosions - New, bigger Boeings are planned, to double the B-747 capacity with little inventiveness. Like catastrophist topological math, there may be mostly wordage here, from our point of view.

The many new ideas that occur to me in my writings appear to emerge from flaws and oversights of science. The philosophy that propagates the point of view that observes these opportunities is largely the pragmatism of James Dewey, Pierce, Mead, and Whitehead, with heavy depth psychology elements out of Freud and Lasswell, these all only being a few. And others like Mannheim on ideological behavior (subtending from Marx) certainly are there as influences. So, I guess I'm in the recycling and recomposing business.

One has to use new images, like the hologram, of course, and devise new images. But I have not yet felt frustrated by an absent "new kind of reality." I hope that I will applaud its discovery, should it come — whether signals from outer space or a kind of intra-organismic communication that is materially effective upon all elements of the organism at once, or whatever.

I detect in the article on Prigogine the eternal hope that a scientific breakthrough will carry a new insistent and moral order. This sort of hope for a Second Coming always puts me on alert. People who can't receive the right kind of vibrations any longer from Jesus, or Buddha, or communism, yearn often for an authoritarian voice speaking out of science like the Burning Bush. That's asking too much of the scientific enterprise. We can probably achieve a better answer by a sober and complete understanding of what we have already learned about the world and ourselves, call it theology, philosophy, no matter.

The universe, including its divinity, will always be an open question, and we shall go on forever, so long as allowed, advancing, defiling, infiltrating, undermining and hovering about the grounds of the question. If there were an answer to the question, we should have to negate all that we think we know about ourselves, the universe, for then we would have to be something other than what we are even in our most megalomanic states. We are already asking too much of ourselves just in order to survive as a species. Again, it is exalting (and arrogant) to play with answers to the question. Anyone for tennis?...

* * *

Chesley Baity was trying to extend her great bibliographic labor in paleo-astronomy by incorporating catastrophism, working through conditional channels that she had persuaded to accept her so long as she did not push quantavolution.

Deg, I said, I can't use your letter from Dr. Chesley Baity, she won't let me. He said why did you ask her, *dummkopf;* you're talking about vital public issues; you're not titillating the crowd with private obscenities. It's a great letter: how she's been trying to get a seminar going on catastrophism at a school where ordinarily you're welcome to sell a course on every other known folly. She's forever asking my advice and then sweetly adding: you don't mind if I don't mention your name? How many more years is she going to waste on this gambit?

I don't know, I said; she's afraid she'll lose the ground she's gained. A few more years and the ground she's gained will be six feet under, he said; and if she has to go, as we all do, at least there'll be her letter on record showing her as a heroine, a wily heretic who knows what she's after, and who knows how she's been led up the garden path by these deans, and university presses, and intolerant astronomers. It'll make sense out of all these years of running around telling people I'm not a heretic, you know, but then oughtn't we consider this and that cosmic disaster. Meanwhile they are laughing at her because she seems a befuddled southern lady, but they wouldn't if they really knew her as I do. The trouble with her is that her husband dominated her for so many years that she still hasn't recaptured the feisty womanhood she inherited from her old Texas stock. I must suggest she read that biography by Sayre of Rosalind Franklin and the British DNA caper.

Now this book of hers dealing with aspects of quantavolution; it's a good collection; good authors. Why is she wasting her years looking for a publisher for it. She can put it out; she's not broke. Did you tell her that, I asked. Yes, I did, and of course she said she wouldn't do any such thing. Another victim of the publishing myth. I said give a couple of thousands to a university press then; they'll publish it. Oh, no I won't do that! Well, then, bury yourself and your authors. The publishers will shed no tears; they'll puff with pride for having kept a bad book off the market. After he said this, I went and checked the list of contributors to Chesley's anthology of *Civilization and Catastrophe*. Of the thirty-six, approximately half have not been mentioned by me in this book and about a fourth have escaped mention in Deg's Quantavolution Series. As you can see, a lot of "reaching out" occurs among the heretics, each in his own style, Chesley-Baity or, as here, Brian Moore, is telling Deg of a new pair of cosmic heretics:

Hartlepool, Cleveland England
9 July 1982,

Dear Alfred:

Thanks for yours of 22 June and I'm glad to hear that the Grecian sunshine is ripening your researches. Great pity you couldn't make our meeting, particularly as I had managed to persuade Victor Clube to come and speak to us about his forthcoming book *The Cosmic Serpent*. I mentioned the book very briefly in the last review as "a catastrophist view of earth history" but had not then seen a copy. Having now read a review copy and met the author I consider it to be a highly significant contribution to the catastrophic cause. Though Clube (astronomer, Royal Observatory, Edinburgh) is conventional enough not to accept orbital changes amongst the planets, what he *does* propose — particularly as it comes from within the establishment — should be enough to lift the level of debate considerably. To summarize briefly: most of Clube's published work deals with the possibility of extra-terrestrial catastrophes in geological times; the book proposes them continuing into historical times at dates very close to those of Velikovsky. His mechanism (though we might not agree with it) is sufficiently well supported by known astronomical data to make the critics consider the implications for mythology/religion/history. He proposes that as the solar system passes through the galactic arms it collects vast quantities of cosmic debris which, in the form of comets, interact with the solar system for thousands of years until by collision/interaction/integration they are thrown out of the system altogether or turn into asteroids. His statistical calculations show that the last series of interactions should have been dying away throughout the 3rd, 2nd and 1st millennia BC. The present Encke's comet is the remains of a giant comet which was on an earth crossing orbit in those times and was responsible for devastation on the earth at periodic intervals. He has an ingenious (though I think inadequate) suggestion as to why the agents of destruction were later remembered as Venus and Mars. He also agrees that Ipuwer/Exodus/end of Middle Kingdom were synchronous and that Egyptian history needs to be shortened by 400 years! The book is defective in many respects, but for a respectable member of the establishment who had not had the benefit of contact with our circles it is an intellectual supernova (well, nova, anyway). Clube wanted to meet you. If you let me know precise dates for your U.K. visit maybe we can still arrange this...

Professor Frank Dachille of Pennsylvania State University had long been a catastrophist in geology; he also was a reader of ancient literature; he piloted airplanes and had been building an airplane in his house at the time of his death in 1983. An acquaintanceship with Deg's work — they met only by phone and letter — led him into the reassessment of his own noteworthy work on meteoritics. A letter of July 29, 1979, shows Dachille engaging in the common quantavoluntary tasks of extending the logic of existing science and rereading ancient documents:

> Dear Dr. de Grazia,
> (...) I meant to mention in my previous letter that at the American Geophysical Union Convention in Washington a paper detailed the possibility

existing in Jupiter of nuclear detonation. This is not new, the idea that Jupiter is in fact a mini-sun, sub-critical, having been about for some time. However, on reviewing the presentation after having read your work and *Worlds in Collision*, I can understand the probabilities of electromagnetic ejecta, and even massive emissions from that planet, and Saturn. You might want to look for a work by P.M. Kolor and L.E. Wharton on this subject. Both are at P.O. Box 142, Greenbelt, Md 20770.

References to Plato in *Worlds in Collision* have led me to an interesting finding, something you must be quite familiar with from you extensive research. The Jowett translation is far from that of Bury, at least with regard to the astronomical descriptions. Jowett does convey some of the information as to sky reversals etc., but I believe his translation more modified by his own notions. Bury was more direct.

My head still swims from my reading of the S.I.S. issue you gave me. The discussions of the Senmut sky maps are captivating but whether from my lack of knowledge or ability, the presentations are most difficult for me to follow. (Is it a British style of writing or is it me?) The electricity paper by Eric Crew is good; I intend to look up his other papers.

Some months after Dachille died, Deg suggested to the State University of Pennsylvania that a memorial meeting be held for him that would treat of subjects upon which he worked and that interested him: meteorites, explosion dynamics, catastrophism in ancient translations, etc. The suggestion caused surprise: Dachille was isolated among the some forty professors of geosciences; he was alone in his heresy, which the Chairman referred to charmingly as "extracurricular"; the Department of Astronomy seemed to be likewise uninterested; the name of V. foreshadowed unwelcome controversy; the campus was not near any large metropolitan center where an outside public would be attracted; besides, all the professors were remarkable people, said the Chairman. Yes, Deg agreed, and they were dying all the time.

* * *

In reviewing the debate over quantavolution and catastrophe over 30 years (for I see no reason to confine this statement to the twenty years of our scope here), I am impressed by the flaccidity and ignorance of the opponents of the heretics more than by any other single phenomenon. Should full-fashioned quantavolution fall before the "truth," it would not be the effect of the opposition but rather of inadvertent blows and self-examination. The opposition has continually pressed the attack with ill-prepared *Volksturm* publicists parroting what scientists say, and then with infantry of the sciences who could only press buttons. The proud creative element of science, the Harrison Browns, Ureys, Neugebauers, Sagans and another score of top-notch scientists and humanists might be court-martialed for their failures, along with those who thought the U. S. Marines in Lebanon had such heavy firepower and such sophisticated gear that they were impregnable to assault and then were penetrated by the simplest of terrorist mechanisms and tactics. This was the "Vietnam Complex," too. Constantly underestimating the opposition; refusing to come to the conference table; seeking allies to help put down the guerrillas among publishing, foundations, universities; laying claim to working for the good of all — are these actions not patent and repetitious on the record?

The opponents of quantavolution — by focusing upon the person of Velikovsky; trying to convert a wide spectrum of interests on the part of hundreds of skilled, intelligent, and creative people into a comic strip; raising the spurious cry of "anti-science" just like the government raises the cry of "reds" and "enemies of democracy;" — ended up heightening the public misunderstanding of science, aroused suspicion against themselves, attracted and promoted the most narrow and bigoted scientists and propagandists to the rank of spokesmen for science. Meanwhile, the humanists and social scientists let themselves be denounced for fools, anti-scientists, and mystics, and be accused of blocking flights to the Moon and wanting to steal jobs from the natural scientists.

The anti-heretics have paid no attention to the scores of heretics who have been building a case for quantavolution all these years. They have spoken of them contemptuously as a mad following that showed up to defend V. or to attack them, failing in every case that has come to my knowledge to read the literature of their opposition. Insofar as V. found it inconvenient to advance his own colleagues, he played directly into the hands of the opposition engaged in making of his work and mission a caricature. Allowing the issues that have emerged in the past decades of this controversy to be centered upon a caricature of Velikovsky is a way of continuously dampening the fires in the hope that they will die. The issues are much larger and important for the advancement of science.

Quite apart from Deg's voluminous work (and even if he had never written a line) there are available millions of words, at least thirty volumes of studies on aspects of quantavolution — and I say nothing of the many distinguished predecessors of V., nor of the hundreds of studies passed as conventional science, that are gems of quantavolution. Nor have I mentioned the mutual teaching and learning going on among hundreds and thousands of students — many of ripened age — that cost their government and school systems and foundations nothing, and risked nobody's capital. Paying for itself, the movement practically registers as zero in the absurd artifice called the Gross National Product.

Files of correspondence and numerous tapes that I hold could be used to demonstrate the level of interaction among the heretics. As they exchange honorary degrees, the eagles of science invariably speak of the need for "interdisciplinary cooperation," of a "melding of the two worlds of science and the humanities." It is mostly pap. They never do it. They cannot do it. But the people they detest and call "anti-scientists" and the "lunatic fringe" do it as a matter of course. They do so because logically their interests and language are unspecialized, because they have slipped their intellectual anchors, and because they must talk to whoever happens to be passing by.

* * *

In Deg's files, I find a brief article about a definition. I mention it to show a kind of particle that floats about unintegrated into a body of science. It is by Walter Federn, an Egyptologist now deceased, who long ago assisted V. in his research. The piece would be almost unretrievable to an outsider for it appears in *Zeitschrift für Aegyptische Sprache und der Altertumskunde* (33 Band 1966, 55-6). There he reproaches those who have retranslated the line "Forsooth, the land turns round as does a potter's wheel," which is from the Ipuwer papyrus, placed now by some scholars to the end of the Middle Kingdom and the Exodus (by those who follow V.'s chronology). Federn says they must not believe the words mean spinning normally in the same direction, but must mean being spun back and forth, as in testing the wheel, as clockwise then counterclockwise. So, Federn declares, the "point of comparison is the reversal of the social order into its very opposite." A great social upheaval is pictured. Or, possibly, I say, it means that the earth itself is gyrating; "The land reverses like a potter's wheel." It is highly probable that it was V.'s employment of Federn that ultimately impelled this dry little piece that drifts unintroduced and unexplained in the slow backwaters of scholarship.

The sociology of science should have field workers auditing conversations at meetings, making tape recordings, too, although Deg, for one, would be annoyed if I spoke of hidden recordings of "goings-on," and would speak of invasions of privacy. But look you where the raw materials of a developing thought-pattern are to be found. I give you an instance where the sociologist of science should be.

Earl Milton was chairman of a symposium on planetary surfaces at McMaster University (Ontario) on June 17, 1974, with astronomer David Morrison, electrician Ralph Juergens and astrophysicist Derek York as speakers. Juergens assigned surface effects to recent transactions between Mars and the Moon. After the chairman called an intermission, the tape recorder was accidentally left spinning, and now a decade later we can eavesdrop upon several people, unknown to us, who spent the intermission by the speaker's table. The tape is not edited. The transcript I give here is partial. The voices are there, but they move so rapidly — and so different are the voices in immediate hasty conversation — and so impromptu the means of transmission and mechanisms employed — and so inadequate the resources here for their study, that the total episode cannot be captured; it is a *soupçon* of the full flavor. At issue is not a "lie" of President Nixon, which is worth millions, and which the nation's media will pay anything to capture, but merely a small truth that an isolated historian, me, is trying feebly to pick up. The balance of the accidental taping only adds to the impression, you have to believe, of an enthusiastic rapid mini-symposium, except that it ends with a new voice, obviously female, arranging to meet one of the voices at "a quarter to eight."

First Voice;... It's an interesting idea and I don't think it has been explored adequately...I was very interested in this discussion...I have done a considerable amount of research in ultra-high current density of discharges, I hope you don't mind my saying that. I think misconceptions, at least as they came out, imply that the conduction went through solid material... *Other Voices* interrupt.

Second Voice: No, no, no, no, you've got to get the charge...[He begins to draw on the blackboard] you see, if we have a surface here assuming of course that we are dealing with spherical surfaces, let's say we have a circle here, and you are going to get a discharge from this point...Now in order to get a discharge from this point I am going to get a small discharge, I am not going to get any arc, I have got to bleed a lot of charge off the surface into this point and then get it off...

Third Voice: I think from, from...I think I can convert the high density discharge phenomena, as Mr. Juergens describes, you initiate a discharge gradient that would allow this to be discharged through the density of the intervening material. At this point the current density which would occur would initiate locally and would spread out as the breakdown progressed and would continue to build up and continue to expand in current magnitude as long as you have more source available and the implication that this could cover the entire Moon if necessary is not at all... *Voices agreeing and protesting... First Voice:* But don't I have a problem here as I start spreading... *Second Voice:* You break that down... *Third Voice:* As long as a discharge is available, and you spread it out and the farther you move out, you are locally vaporizing—as you dissipate energy, you are locally vaporizing solid material which then breaks down and contributes to superconductors, I don't mean superconductive in the terms of superconductivity... *Fourth Voice:* Sure... *Third Voice;* I mean... You are referring to... what you get essentially is a plasma as a result of... *First Voice:* That's right, current density from these discharges can go to the levels of 10^8 amperes per square centimeter and can you maintain... *Second Voice:* As long as there is charge available... As long as it is spreading out it could continue, not over days, but in micro-second discharges...Don't call them sparks...The wire was only the initial source of the plasma. *First Voice:* Yeah. *Second Voice:* During the discharge you have your anode and cathode processes of tremendous pressures on those surfaces due to ion and electron bombardments. Your wire lies between what — between two pieces of metal in these cases — was intended to be a conductor. *First Voice:* But can you do this — explode a wire between two non-conductors. *Second Voice:* Oh, I think you definitely can. Because the metallic nature has nothing to do with it... Only the initial discharge... *Third Voice:* Yes, that's the point...You'll have a discharge when the voltage gradient becomes at a particular level with regard to the density of the atmosphere. *First Voice:* That's the other question...What does the atmosphere have to do with it? *Juergens:* You have to trigger it with electrons dragged out by the field and once they bridge the gap, they ionize the material...[One notes a bit of Juergens' character here, he speaks rarely and in low quiet tones, listens much.] *Second Voice:* If you take a little experiment they perform at the laboratory, if you take a tube here and put on some circuitous track a vacuum tube and come around to here, where the rest of the tube comes around to there, you put a little gap there, say a centimeter across, make the density of the tube at a particular level, you can cause that discharge to come all the way around through there. *First Voice:* Oh, yeah. *Third Voice:* But you will not conduct the material into the center, you will not even conduct the heat into the material except to the manner in which you're vaporizing the surface at a tremendous rate (from the impact), you are vaporizing the material from these discharges... *First Voice:* I agree. *Second Voice:* But the material is not blasting off everywhere at this time, now I am saying that at this time it is not blasting off. It is only to the degree to being charge carriers and to being transmitted inside the arc but the pressure - electron and ion pressure on surface — will prevent a massive expulsion of matter until the discharge is terminated. After it's done, all the material will be vaporized... *First Voice:* Now you are getting to an important point...

This goes on for a minute or two longer. The craters, rilles and mares of the Moon are discussed as if they might have been electromagnetically created. There are quickly disputed points, and then we see a transition occurring from talking about the technology of electrical discharges (from the small crude personal experiment with a piece of wire to catastrophic avalanches of electricity between Moon and Mars). The voices move from the *substance* of science to the *behavior* of science. Let us reproduce this transition, which is important to a science of science.

The voices begin to discuss the "great red spot" of Jupiter, in relation to a newly discovered "red spot" on Venus... A *New Voice* claims the second discovery may be the umbilicus, where Venus spun off... *Others exclaim Objections... Second Voice* says Jupiter's great magnetic field would not let a body escape, nor would a body fly off the Red Spot which is not equatorial. *New Voice* says that there is no reason, only presumption, why Jupiter's field and axis would not have changed at the time of, or after the incident...

Second Voice: But what of Venus' orbit... *New Voice:* That's different, too; Mars is responsible for it in part... *First Voice:* It may be so when we look at it from Velikovsky's perspective...The arguments against, built on the wrong inclinations and so forth, they are held by uniformitarians but they don't explain anything to a Velikovskyite you see... *Third Voice:* Of course, there is a built-in psychological problem. I don't know that it's uniformitarian, but it's built into our Western logic... *Voices of Agreement...New Voice:* If that's nature, we should find out. We should overcome that reaction. We've had our Copernicus, we've had our people who came along and said our world is different from what everyone thinks. We've had ample evidence that this has happened — not frequently — but every five hundred years... And something of this... and may be one of those times... So that's why I say, we ought to drop our resistance to the idea so much and say, well, holy smokes, you know, we've been confused by what we're doing uniformitarian-wise, let's jump over here and play for a while and see what happens, and that isn't the course that's followed, and I don't understand — psychological resistance notwithstanding — the unwillingness of a totally objective person to do that. You see, that's what bothers me. *Third Voice:* I think it's understandable... I think if you consider, if you look at scientists and engineers, they spend years and years in universities buying their education and what you're suggesting is the education I've acquired... is so much garbage... *First Voice:* I don't find it garbage... It's not a waste... The data stand and the objectivity of these measurements stand. It is their *interpretation* of these problems... *New Voice:* You don't sacrifice your education when you change... *First Voice:* No, you don't, that's true...You don't have to throw the baby out with the bath.

All agree. They speak of the strong psychological bent for orderliness in the scientific mind, "neat orderly chambers," dislike of uncertainty. "It's difficult to say I'm wrong!" "It's *easy* to say!" "It's very difficult to say!" "I've had so many years in graduate school. It was all bing, bing, bing, this is it..." Then later the very ideas and outlook changed. *Second Voice:* There are a great many scientists who would never come here to speak or even to listen, they wouldn't even discuss the questions... etc..., etc.

What triggered the transition was a quickly perceived misstep of retrojecting Jupiter's behavior in a uniformitarian way. A second transition then occurs. *First Voice:* people are belongers, I belong to this group, you examine an eccentric hypothesis, then one gets into major trouble, your colleagues branding you a crackpot or idiot. *New Voice:* aren't we suffering from the two-culture problem? *Agreement.* "Velikovsky's cardinal points were in the humanities." *Yes. New Voice:* "Yes, I think so." *New Voice:* They were absolutely unquestioning...

And then *New Voice goes on to argue the factual validity of his proposition,* leaving the discussion of the logic of science and humanities behind and also the straight astrophysics and electromagnetics with which the talk began.

The voices tend to agree in principle: that a consensus of widespread legends is persuasive as to its basic factuality. Now the voices thank each other and disperse, their few moments of exciting discussion ended.

I am afraid that I have lost you, my readers, amidst such a confusion of remarks, but I will regain you if I have merely shown you how the raw materials of this intense human discourse appear. Ultimately we reduce and clarify the process, introducing the logical order on a printed page, but losing some of the intense give and take within the human mind and among different human minds.

* * *

Letters are not so important in scientific discourse as they once were, given the telephone, the xeroxing machines, the airplane, and the comfortable meeting places to be found everywhere in colleges and hotels. They are more important among the heretics than among conventional scholars because they are the cheapest means of communication. Their effect is multiplied too by xeroxing them and passing them around. But even then they are an unsatisfactory record, because they are rendered fragmentary by intervening telephone calls and meetings. Greenberg's and Lowery's correspondence in editing *Kronos* and the *S.I.S.R.* was heavy but would, especially in Greenberg's case, be enormous were it to include transcripts of the phone conversations.

Still, in letters one can follow the kind of internal argumentation that otherwise disappears. Thus Leroy Ellenberger, reconciled with Deg despite his mean attacks upon *Chaos and Creation* (mentioned earlier), began to use Deg as a postal drop, sending him letters, copies of letters and articles, and memoranda. By 1983 Ellenberger was preparing to abandon much of quantavolution and found now that the story of Velikovsky was not without its shady tones, and more important, that Arctic ice cores and bristlecone pine dating technologies were directly contradicting Holocene quantavolutions by their even pattern of annual regression into time; further, that Gentry's studies of the surprising "instant" polonium halos of creation that came from nowhere — parentless — and which threatened the theory of radiochronometry, were probably invalid. You show a total misunderstanding of the Oxygen-18 isotope technique of measuring time in ice varves, he assured Deg, as the *Burning of Troy* with its critique of ice core studies was about to appear.

It seemed that Leroy was on the verge of taking up a macrochronist position in quantavolution, which by 1983 was fast emerging from geophysics and paleontology and which offered respectability to its clientele. One could thereupon dismiss all apparent human experience with catastrophe and get rid of the historical sciences and humanities.

Deg contemplated the prospect sourly. I could, he thought, surrender michrochronism in the event of defeat, but I would rather relabel the total construction as a heuristic exercise machine, good for the circulation of the blood and the sharpening of the critical faculties.

There were always these honest, upsetting or encouraging, epistolary discussions going on among the heretics, many of them — how many? — a score at a time. Here is another one from 1978, going into 1979. The cosmic heretic, Dwardu Cardona of Vancouver, is writing to the cosmic heretic, Irving Wolfe of Montreal:

> Dear Irving,
> If you don't already, you're going to hate me by the time you finish reading this. I'm afraid that, in your cosmic interpretation *of Hamlet*, I do not concur with you at all.
> I should qualify that last statement. I do agree that *Hamlet* has a cosmic connection but not with the Martian close encounters of the 8th/7th centuries B.C....
> The story of Hamlet is, in its skeletal form, identical to that of Horus. To my knowledge, this is the earliest form of the myth we have so far come across. The Egyptian tale was already well developed during the very first dynasties of Egypt. It is that old — and older still. So is *Hamlet*....

This goes on for several pages, one of several letters in the interchange going to show how much of human history and science evolves around the figure of Saturn, the great god of the Neolithic Age and beyond, everywhere in the world.

I will not print Wolfe's reply, equally lengthy, also giving and taking. He has published obscurely (save to cosmic heretics) several articles on the catastrophic imagery of Shakespeare, that when published in book form (he collected a number of rejections) will constitute a formidable body of analysis on Shakespeare, by a new approach.

But then Cardona is also busy with historical astrophysics, and he perceives in Deg's ideas a competitor to his own. Never mind, he has his reasons, and he writes to Earl Milton:

> ...The evidence of myth which points to Saturn having once occupied a position above Earth's north polar regions is voluminous. There is not a race on Earth that has not preserved at least one account which states as much. According to this evidence, Saturn occupied a central position in the north celestial regions, it rotated, and rotated wildly; but, other than that, it was immovable. It did not rise, it did not set. It merely became brighter and more glorious each night as the Sun set. This state of affairs seems to have lasted for ages. *It is the one single dictum of the ancients from which all other beliefs are derived....*

But, of course, there are physical problems, and colossal ones, inherent in the tenet. And that is where I hope you will be able to help the cause.

The problem, stated succinctly, is this: What force, and in what way, could have kept the Earth locked beneath Saturn's south pole?... [one of 3 pages].

And Milton replies:

... As you may know, de Grazia and I are developing a new cosmogony for the planets, one which is consistent with extant mythologies and catastrophic historical events. If Al has spoken to you of Solaria Binaria, then you know something of this cosmogony....

Here is an outline of our speculations about how Saturn and Earth were once locked together. Consider a gigantic dumbbell with the sun at one end and Super Saturn (Saturn was much larger then) at the other. The original planets, Mars, Earth, Apollo, and Mercury, were locked between the sun and Super Saturn, very close to the latter. The new planets, Uranus and Neptune, orbited beyond this inner group. A now distant fragment from an earlier era, the residue of Super Uranus, was receeding from the system. As we see it, the Earth did *not* rotate on its axis such that the Sun was visible daily. The Earth's axis, at that time was aimed along the Sun — Super Saturn line. Earth's "Northern Hemisphere" faced Saturn, the "South," now devastated by the recent tearing away of the Moon, faced the Sun...

And Cardona writes:

I'm glad to see that de Grazia and Wolfe, with whom I corresponded a while back, have not forgotten me. At the time, de Grazia did throw a few crumbs my way concerning his developing new cosmogony and, if I well remember, I cautioned him to be wary of certain mythological identifications. Now I see that de Grazia's Solaria Binaria has been echoed by Tresman and O'Gheoghan. But on all that, a little more later on. (...)

4) De Grazia's super-Uranus needs much evidence. The Uranus of Greek myth seems to be merely an earlier alias of Saturn. This is borne out by Assyro-Babylonian, Sumerian, and Egyptian texts. Annu was the same as Osiris who was the same as Saturn.

5) There seems to be no mythological evidence that the Moon was torn from the Earth. On the contrary, I have come across evidence which points to Saturn as the parent of the Moon. The Moon commenced its celestial career by orbiting Saturn but when Earth itself was torn from Saturn's gravitational embrace, it managed to carry the Moon with it...

(...)

When I wrote to you asking for your help, I did not know that de Grazia had already cornered you. I do not wish to "steal" you away from him. I do believe, however, that we can help each other. For that matter, I thank you for the information you supplied me with concerning the Roche limit. And if it is not too much trouble, I really would appreciate it if you could, if only for a day or so, put your own model aside and weigh the possibility of a Saturn-Jupiter dumbbell formation with Earth locked in between.

And Milton replies, point by point, in an eight-page letter, concluding:

> As with you, I am not out to convert but help. To use only myth is equally as dangerous as to use only a computer to prove Venus' orbit never intersected Earth's. We both know better...
>
> Please keep in touch. I need more data to help you further. Should anything I see in your data be germane to our model, I will credit you and I trust you will do the same re my comments and ideas becoming a part of your cosmogony.

And so on. Cardona has several sympathizers and is seeking to convert Milton and Deg, who in turn are moving rapidly on their own model. Cardona, meanwhile, begins to publish his rich Saturn materials in *Kronos*. Clube and Napier come forth with a cometary model, derived without contact with any of them, in *Cosmic Serpent*, practically simultaneously with *Chaos and Creation*.

A process is here occurring that resembles somewhat the internal competition among the Cambridge, London and California biologists striving to produce the first and most useful model of the structure of DNA, an event of 1953 described by Watson in *The Double Helix*. By 1984 there were in contention the Cardona-Talbott Saturn model, the Clube-Napier galactic cometary model, and the de Grazia-Milton Solaria Binaria model of cosmic quantavolution. All of these were far ahead of, or let us say were distinct from the heavy empirical work beginning to appear concerning meteoritic impacts, clay chemistry, and biological extinctions. Perhaps the tides of particular studies will wash away most of the substance of the models. Such a fate has befallen the model of the victorious biological team, as Stephen Jay Gould tells us:

> It is a credit to the power of Watson and Crick and to the fruitfulness of good science in general that, thirty years later, this Cartesian view of molecular genetics has been superseded, as a second revolution transmutes our view of inheritance and development. The genome, a cell's compendium of genetic information, is not a stationary set of beads on strings, subject to change by substituting one bead for another. The genome is fluid and mobile, changing constantly in quality and quantity, and replete with hierarchical systems of regulation and control... Barbara McClintock is the godparent and instigator of this second revolution. [She published her papers obscurely in her own laboratory newsletter, but, as Gould remarks, she

has lived a blessedly long life.]

And Gould, whom we have come to perceive as a quantavolutionist, can even discover in this movement from the one model to the other, a victory for "rapid and profound rearrangement" over the "implication that evolution proceeds slowly and gradually." Pleased as we may be about this aspect of the change, we are here more directly made aware of the possible short life of even the best of scientific and cosmogonic models.

* * *

Once more I return to the point that almost nothing of the large number of writings in scientific support of, or in modification of quantavolution, particularly as conveyed in V.'s work, has been read by any conventional scholar, including (I stress) those who claimed to have read something by V. prior to attacking him. It is clear that one way of treating with heretics is to go on the principle "Smite the shepherd, and the flock will be scattered." Moreover, anti-heretics lose much of their effectiveness as soon as they discuss work by heretics other than Velikovsky, because they depend so heavily upon a prior inoculation of the public of science with stereotypes against his name.

In this regard, the heretics have suffered by their own behavior. If they must constantly acclaim V. on their first page, like others do Einstein, Marx or Engels, and Freud, it's like prefacing every encounter with a "Heil Hitler" at the worst, or at its mildest, forever snapping salutes between the military, a practice devised to confirm a status system, limit originality, and exclude an outer world.

It must be apparent by now that V. was not without blame. He did not want even one, much less two or a group of martyrs burning alongside him at the stake. He was loath to adopt the ideas or quote or put forward or support anyone who was about to be credited or discredited by a valid contribution that was not *a priori* a confirming footnote to his own work. The idea of a roundtable or true seminar was beyond him. After decades in America he became a citizen but he had always some of the czarism and mosaism of old Russia, that would not let one kick ideas around like soccer balls.

V.'s prominence absorbed all energies penetrating from the outside in addressing him and his claims, diverting attention from all other new work in the field which was in any event dammed up and had to trickle through his notoriety, whether in magazines of general circulation or in the couple of small magazines, which themselves held back most work not directly concerned with his affairs.

Were I to guess the quantity of useful writing appearing as deliberately directed toward quantavolution, I would suggest a statistical figure approaching a Fibonacci series by dodecennial periods, beginning in

1940-1951 at 1000 pages; thus, 2000 pages for 1952-63; 3000 pages for 1964-75; 5000 pages for 1976-87; 8000 pages for 1988-1999: 13,000 pages for 2000-2011; and so on in time, granted there would be no world war or political revolution.

My aim, in quoting heretical correspondence in this chapter at some length (still not one-hundredth of its volume), has been to give evidence of how science proceeds among heretics and non-heretics alike. The published work (which in the case of the heretics has not been read by the non-heretics) is only the tip of the iceberg showing. The same is true in most scientific work. There must be a consensus of sorts between correspondents else they cannot talk: here, with Wolfe, Cardona shares the belief that literature connects with a mainstream of mythology extending to the birth of the human mind; with Milton, (and with Wolfe, too) Cardona shares the premise, arrived at on both sides at the end of years of study, that the planets have moved and changed, even in early human times.

The behavior of the cosmic heretics corresponds closely to that of conventional scholars in regard to their methods of work, and would be practically indistinguishable were it not for the warping of the processes brought on by the heretics' poverty of resources. Back and forth, the shaping form of a new kind of science (like the old) works like a complicated weaving machine, capable of darting up and down and sidewise to pluck its threads, strengthen its seams, and sometimes the machine sticks and threads must be pulled out, sometimes a whole line of thread, as some major patterning element has to be rejected.

In the 1960's the American Psychological Association, through W.D. Garvey and B.C. Griffith, conducted pioneering studies of the communication network of the field with which some 30,000 persons were connected. Of these 30,000, 2000 or less provided almost all the materials that were being circulated as current psychology.

Work published in a psychological journal started on the average 30 to 36 months before publication. Between 18 and 20 months before publication, the work was shaped to a point where it might be reported. Usually, between 15 and 18 months before publication, the reporting process began. Initial communications were highly informal and occurred typically at the writer's institution. After several months a formal report was prepared that in about 30% of the cases came to be delivered at a national or regional meeting. Almost always the audience was below 100, sometimes only a dozen. Copies become available at the Convention, and special papers might be distributed now also by the author(s) through their sponsors such as a government agency. Preprints were usually distributed, between 10 and 200. These were often given to close-in co-workers, acquaintances elsewhere, and persons who had heard about the work and asked for copies. The interval between submissions and publication ordinarily took 9 months or more, but the interval would be doubled if an article were rejected. Few rejected articles failed to gain acceptance somewhere else.

While the publishing proceeded, additional reports were being made to groups and classes. Aside from textbooks, which amount to compulsory subsidizing by students, practically all scientific (and scholarly) publishing is subsidized by scientists as individuals or groups, directly or through tax money whose appropriation and spending they manage to influence.

Exposure of the work by publication is low. The largest journal reaches 30% of the general population of psychologists; specialized psychology journals may reach 1%. The largest journal will expose the title to all; however, one-half of the research "reports will be read by 1% or less of the readership, none by more than 7%, it appears. Half the articles in the largest journals are read by only some 200 readers. Current journal reading amounts to only about one-third of the journal reading of one group of active psychologists studied. Some months later an article becomes retrievable by being indexed in one of the now well-equipped services, such as *Psychological Abstracts,* thus helping people like Deg, who was trying to find out what work was going on regarding "human nature," to find nothing because the term was not indexed.

The Garvey-Griffith study offered proof of what disciplinary leaders know everywhere, that long before the rank and file, and quite long before the public, learns of a new line of research, the leaders know it from personal acquaintanceship, membership on foundation and government boards, and operating at the nodes of communication where manuscripts come in and criss-cross and where money changes hands.

The same process that occurs in psychology occurs on a greatly reduced scale in quantavolution, among the heretical community. The scientific creationists too are loosely organized and operate, also in a small way, like the psychologists. They and the scientific heretics engage in mutual eavesdropping. A somewhat different process occurs among the non-heretical quantavolutionaries, who operate on the fringes of their discipline — psychology, biology, astronomy, anthropology, etc., and are signalled by terms such as "macroevolution," "punctuated equilibria," and so forth. These for the most part are anti-heretical and cling to their disciplinary centers as much as possible. Thus Walter Alvarez, who is himself under fire for a study showing the "iridium layer" marking an end to the dinosaurs in the rock strata is prompt to refer to Deg's work as "anti-scientific". He cannot have read Deg's work or any other considerable literature of the field; otherwise he must be using some narrow and antiquated definition of science, or worse, using the term science for name-calling.

* * *

It is widely believed that all astronomers, all geologists, all physicists, all historians, and all archaeologists have for thirty years been close-minded to the arguments continually brought up by the cosmic heretics.

This is not so. And this stereotype of the resistant and rigid collective mind continually exacerbated feelings on both sides. (As did the opposite stereotype, that all heretics were foolish and anti-scientific.) To illustrate my point, I will turn to Deg again, for he was always concocting hypothetical statistics. (He should have offered a college course on the subject; it is useful for those areas, most areas, where data is trivial or scanty, and the usual resort is to revert to the Aristotelian modes of thought.)

Deg's Notes, Princeton, 1980

The grades of opposition among the probable quarter million of scientists who have formed any opinion on the cosmic heretics should be sorted out. And here I assign estimates in percentages only to illustrate my view. They may be, by my guess, up to 10% off one way or the other:

a) Stereotyped rigid opponents: 19%
b) General dissenters: 35%
c) Specialized dissenters inattentive to major theories: 20%
d) Doubters but interested: 13%
e) Interested and acknowledging truthful elements: 10%
f) Persuaded of the general truth of quantavolution: 3%
g) Persuaded of the general truth and also of some special heretical truths, such as a radical change of planetary motions, or a recent great deluge on Earth: 0.1%

If one were to correlate such figures with the prestige of the opinion aggregates in their own fields, using concepts that I have used in studies of political leadership, we might find that the top elite (1%) would be heavily concentrated in classes a, b, and c; the activist productive scientists (3%) would be spread throughout; the ordinary scientists (80%) would be skewed somewhat higher toward elite opinion but spread throughout; the inert scientists (10%) (recalling that most scientists have hardly heard of quantavolution or Velikovsky as an issue and are therefore not tabulated at all, and that inertness means 'unproductive' ordinary scientists) would be even more skewed toward elite opinion. In consequence of the biases and the gross numbers, we would find the last two categories favoring Quantavolution populated by only a couple of members of the top elite and a few members of the activist productive group. It is understood, of course, that "elite" and "productivity" here may not denote "truth-production" to any great degree; they are terms denoting network and establishment leadership. Thus, if we were placing people, we would shuffle leadership scores like a deck of cards after three aces in a row were drawn.

Also, "forming an opinion" does not denote extensive reading in the field of quantavolution. Furthermore, placement of a person does not suggest his "flip-flopability." For instance, Carl Sagan would probably score as "top elite" and fall under "general dissenters;" but his writings and utterances on occasion signify a suppressed readiness to accept general quantavolution. He would have high "flip-flopability." So would the "activist-productive" *e*-category geologist Derek Ager, who, however, would not have to execute a vigorous flop, just a tilt. Melvin Cook, a geophysicist of the same ranking, would be found in *f*, and would probably move restrainedly into *g*. Robert Jastrow might occur as top elite in the *d* category of interested doubters, perhaps even in the *e* category; he, too, might move up readily.

On the whole, there is much subconscious ambivalence (produced by anomalous and contradictory material) in science plus a goodly concentration of influential near enough to quantavolution theory to accomplish an easy transition. Not one of the top elite of scientists in the country over the past thirty years has read deeply in the literature of quantavolution. That goes without exception for Sagan, although he has been active in the Velikovsky affair.

Deg was here counting as scientists those humanists and social scientists who profess a scientific approach to their fields. He knew of none of these of the top elite who had studied deeply the literature. Probably no more than 1000 persons in the world have been seriously engaged in the discovery and study of quantavolutionary literature over the past thirty years. If Velikovsky's *Worlds in Collision* has been read by a million people, most of the thousand will have read the book, but 99% of the million readers will have read little else of value besides it.

Many a well-known figure of science has had an exoterrestrial skeleton in his closet. Plato would deny the citizenry the right to challenge the divine and natural order of the heavens and proposed severe penalties for such. Yet Plato has for over 2000 years afforded support to quantavolutionists in history (the Atlantis report), astronomy (deviations of the planets) and geology (destruction of early Attica by earthquakes). V. was annoyed when Stecchini stressed the anti-quantavolutionist side of Plato's political writings, and urged upon them a consistency that was not there; at least it seemed to Deg that he could not tolerate a double standard for Plato, that what was true should nevertheless be suppressed for the good of the social order. Here was an example of what was forbidden in principle to a psychoanalyst: V. therefore needed to believe that the truth would free man and wished a social policy that would acknowledge ancient traumas of catastrophe so as psychologically to free him in his behavior today. Given V.'s authoritarian bent, a contradiction of feelings arose which was displaced upon Stecchini's innocent and free-wheeling scepticism and attacked unreasonably. It does appear that Plato was deliberately contradictory. He recognized a chaotic universe while officially forbidding its recognition.

Stecchini performed a similar service with respect to Newton and

Laplace, discovering in both men the inklings of catastrophism. In Newton's case the contradiction between a stable order of the skies of the new science and a biblical literalism ordaining catastrophic belief was explicit, but glossed over by Newtonian science. Stecchini's exposure of the concern of Laplace that destructive cometary visitations were possible, and of his admission that his mathematics, which fixed the modern vision of an impeccable celestial order, simplified reality, was more surprising.

Deg met with additional surprises and came to suspect that when the time came to throw off the uniformitarian guise, scientists would rediscover a general exceptionalism and anomalism in geology, paleontology, evolution, and astronomy. He relocated persons such as Pickering and Wegener. He found that Shapley, who had become the anti-hero of the Velikovskian sociological scenario, had posited exoterrestrial encounters one time, and so, too, Harry Hess, who had filed *amicus curiae* briefs for Velikovsky, and Sagan to whose burst of fame both hypotheses of exoterrestrial communication and rebuttals of Velikovsky contributed.

Some of such characters found a place in the geology of Deg's *Lately Tortured Earth*. Together with the frankly catastrophic writers, such as Melvin Cook and Allan Kelly, they would come to occupy an important substantiating role, like the dissenting minority opinions in U.S. Supreme Court history, when the moment for revising science would occur. Then some of those who had denounced "backward catastrophism" would become forerunners of quantavolution.

But, please note, I have scarcely touched upon the full breadth of the science of science, which would embrace the thousands of cases occurring in the normal operations of conventional science upon conventional offerings to science. Nor can I do so, for I must be done with the case of the cosmic heretics very soon now.

Deg's Journal, en route Washington, October 18, 1966.

Sundry of the quantitatively directed natural scientists have told me and others that they believe Velikovsky to be unimportant and irrelevant because of his qualitative, subjective approach to events in astronomy, physics, and geology. For instance, the work on electromagnetism, radioactivity, interplanetary exploration, and solar system aberrations is learned, studied, and developed in a mathematical setting.

But for what V. is saying, the movements of phenomena are so large and influential as to make quantitative assertions about them unnecessary. What matters to us is that oceans of soil descended from the skies, that numerous eruptions and earthquakes occurred, that gross changes in the sky appeared. These happenings were reported. The reports are ample. Neither the ancients nor we ourselves today would have had the tools, under the circumstances of the events, to describe them and present them in sets of equations.

Deg's Journal, Princeton, January 18, 1968, 10 pm

Every physical law states a proposition that is useful to a culture, with requirements that are relevant to the practical workings of the law, and derives its "eternal truth" from that fact.

The proof, e.g. of Newton's law of inertia, is supposed to lie in the myriad applications of it, in ballistics, industry and transportation. But one need only think of how many enormous discoveries and inventions occurred before Newton's law to see that the law itself does not create the understanding of nature. It only rephrases that understanding in a slightly better and more useful form. It is a mistake to treat each reformulation as more than a useful temporary rendition.

Some natural laws can be made to appear ridiculously simple and indeed they may be such. A body resists changes in its motions. "Nothing changes unless acted upon." Well, why should it? That's the law of inertia. But the opposite of course is true — nothing becomes what it is without having been something else. Etc.

Deg's Journal, October 27, 1972

The revolutionary zeal to refute uniformitarianism and evolution has not considered fully their merits. The doctrine, that the solar system has been stable for millions of years, and that biological evolution and geological changes have occurred almost entirely through small incremental changes over billions of years, seems weak enough, in the light of our reassessment of catastrophic evidences in every area. The recency of catastrophe is plain.

We have had to explain why uniformitarianism triumphed but have done so only cursorily; one does not pause to strip elaborate armor off the fallen foes until the battle is won. When we can return to consider, we shall find that uniformitarianism has, like the Christianity its allies so disturbed, performed functions that we are not yet ready to provide substitutes for, indeed perhaps are not able to discover and recognize for some time.

In Praise of Uniformitarianism

We have said — Stecchini and I, at least — that uniformitarianism was the beautiful philosophy of the Victorian Age and of all those who wished since ancient times to give stability to human affairs. V. has recognized this and says from time to time, cryptically, even in *Worlds of Collision*, that the *Great Fear* remains, and is a cause of war and strife. Uniformitarianism is the culmination of the worldwide amnesia that followed the great catastrophes — (I would call the period ca. 5000 B.C. to 650 B.C. as the *Epoch of Cosmic Catastrophes)* [later extended to 12,000 B.C.] In its triumph, uniformitarianism seemed effectively to reduce to nothingness the catastrophic theories. Great scholars like Eliade breeze over mountains of evidence of the chaos of "the beginnings" without asking whether such chaos occurred; they become a manifestation of primitive minds.

My position is this: that the effects of the Epoch persist; that Uniformitarianism was a successful myth both psychologically and socially, and was in conformity with many scientific discoveries. But far beyond these functions, uniformitarianism is rooted in the provision of the grand assurance that enabled humanity to:

a) Challenge nature
b) Control nature
c) Set up the idea of History as Linear in Time, destroying the popularity of (and essential conservatism of) cyclical theories of history
d) Spawn the idea of *Progress* as the future of man
e) Encourage the faith in stability that promoted the exquisite and productive division of labor in all areas (no rushing to the caves or wombs of overall theology needed)
f) Simplify religion and produce deism, god as mechanic and great designer
g) Give laws immutability
h) Promote the idea of a rational bureaucracy and rationalism generally.

Deg's Journal, New York City, November 18, 1972

Science is protected by a veil of awe and therefore is not usually thought to respond to sociological laws. It does, however, and even to laws about the vulgar sorts of opinion and leadership.

I notice that reforming or revolutionary scientists go back to "discarded," "forgotten," "rejected" sources. (Cf. Velikovsky in "Cosmos without Gravitation" and *Earth in Upheaval.)*

The ordinary supposition is that this is part of the rational system of sciences: viz a) thorough coverage of sources, b) reexamination of misunderstood writings, *etc.*

Actually the explanation of this behavior is *très ordinaire.* Science has only a one-channel mind. It cannot proceed with two theories at the same time.

This may seem ridiculous: "What? The most brilliant intellects among humanity and they cannot hold two thoughts at the same time!"

The absurd becomes acceptable when we realize the deductive and administrative nature of science (Cf. my "Science and Values of Administration"). An enterprise (which science is) seeks one direction, one consistent set of rules of decision, one comfortable theory (if possible), a hierarchy of access and command, and (like an imperial megalomaniac of any world religion) one world-wide code (without culturally and ideologically distinct competitors).

The "old discarded writers" are therefore to be understood as you would view a rabble before it was transformed into an army. Coming early, they did not hear the call, they could not feel the current's strength. Their students, "seeing more clearly, feeling more keenly," rewrote their science to fit the future history of science, that is, to describe the path to be followed. Thus is science administered.

Newton and Darwin are celebrated for unconscious reasons, more than for conscious ones or scientific ones: to cope with increasing anxiety, and yet change from a prescientific to a scientific age:

A) Newton performed a great *theological* role in the transition from geo-centrism to helio-centrism by inventing the c*lockwork universe,* and absolute laws.

B) Darwin's great *theological service* was to give enormous time and minute change (i.e. to reduce Time from quality to quantity) by inventing gradual evolution [by natural selection].

Deg's Journal, New York City, January 1973

It is a formidable block to accusations vs. the reception system of science that "you do not know anyone of great merit who has not been recognized." This is fallacious:
1) One can find such: e.g. Boulanger.
2) Relative ratings are important. Change in rank order from 1 to 30 say, or from "best-seller" to "out of print."
3) People are "infamous" and regarded as "famous" and vice-versa.
4) Famous people now have passed long periods in which they were unattended to : e.g. Aristotle.
5) Famous people are degraded on grounds that, though they *were really great,* they were superseded.
6) Who knows who is *not* known but great.
7) How few scientists on the list are read, and really known, after the first dozen or so.
8) People of great merit may not be able to publish, or they may be without the experimental, research, editorial and critical assistance to make their views plausible or digestible.

e.g. if V. had not been able to hire expert editorial assistance, writing as he did in a language only lately and imperfectly come by, he would not have been able to publish any work of consequence.

e.g. Deg has on occasion recommended student Abner highly and student Boggs modestly, then to discover that Boggs got a scholarship to go on at a first class establishment university while Abner did not go on, went on in a less well-equipped and less influential university and was lost sight of in the production and achievement lists.

Deg's Journal, New York City, 1974

Sidney Willhelm, who has been one of the keenest sociological observers of the Velikovsky Affair, gave two excellent new reasons why V. should have been both accepted and rejected by influential elements of American Society. First, he says, the American democracy has given over to scientists its power and will to regiment ideas: "Reins remain extremely tight upon the creative person through the delegation conferred by the State; by keeping each other in line, scientists avoid direct State censorship." (One thinks, for instance, of how remarkably well the scientific groups have restrained the government from acting forcefully in the extremely volatile area of bioengineering and cloning.) "Thus," says Willhelm, "the forces of resistance find a more difficult time to convince skeptics of the lack of true freedom of inquiry by the absence of an explicit state agency charged with thought control."

Willhelm also points to the psychological compatibility of V's catastrophic theories with the policies of the political elite.

"While it was the longing for peace and tranquility which apparently nourished notions of harmony in nature, today it is the momentum of militaristic destruction which introduces the greater reception toward Velikovsky's controversial interpretations. Modern science owes its growth to wars and the threats of war." The cosmic heretics, with their wars of the gods, and clashes of the planets and comets, are setting an example, unconsciously, for the prospering of militarism and the military-industrial complex.

V. realized these dangers, and coined the idea of 'collective amnesia' with the purpose of exposing this mentality and thus controlling it, while Deg too realized the danger in the association and went further to explicate the original dynamics of *homo schizo,* to build peace institutions, and to devise peace therapies.

Deg's Journal, Washington, D.C., 1979

It may appear shameful that scientists should depend for a new discovery or new perspective upon a lay body of vaguely connected individuals who are interested in an idea. Still, this is not only historically probable; it may be also a logically and sociologically necessary deduction. The triumph of the Renaissance outlook and method in the humanities and sciences was a politico-social-economic-ideological effect. So was the victory of uniformitarian geology and, thereafter, biology in the nineteenth century.

Scientists and specialists, once they receive their kudos, become prideful and seek to shed their origins, retrojecting their present behavior and methods back to their beginnings, taking credit for the recognition and development of their science. The story of Albert Einstein's success, for example, is told almost always as a rational discovery, a steady progress though appraisals and tests, to applications and finally to total acceptance. The full story of his great lifetime success, however, bespeaks a curious figure who caught the popular imagination and was ballyhooed by the press and newsreels under the misunderstood concept of "relativity" until many scientists, no matter how reluctant, had to deal with his idea. Several early opponents of "relativity" (now only a suppressed whisper is heard of this) saw clearly that a "matinee idol" was being foisted upon them. One does not deny Einstein his greatness in pointing out that he might not have wormed his way through the reception system of science and almost certainly would not have received the lion's share of glory if the public and press had not been behind him or, better, dragging him forward.

This is a subject which requires thorough exploration, and has not received such at the hands of science or the history of science. To take up only one point for a moment, few new ideas can penetrate the publications of science; they are pinched capillaries. If they are conveyed, their readership is extremely limited, a few persons, unless they are well-known already, in which event some hundreds read the work. Scientists get little reward from hard reading of anything but items aimed toward their ongoing projects, and they are busy with other affairs. If an idea does penetrate the minds of a very few, the very few must become a group, and must command just enough resources (not so much as to be 'bought off') to become an inescapable pressure against the conventional main front. Then they make a breakthrough, spread out on the flanks, and begin to surround and capture demoralized main body elements.

The winners may not even be correct; they may inspire only one of the many fads that overcome disciplines and the scientific outlook as a whole. If what they espouse is effectively 'true,' a surge of scientific advances occurs and, among other by-products, arouses historians to write (and rewrite) this history. A public, consisting of persons who have time to read seriously, like love letters, the otherwise unreal material, constitutes a heavy factor in assembling, encouraging, calling attention to, and forcing recognition of a new viewpoint or method.

EPILOGUE

Surely, said Deg over the telephone, there must be a better way to write personal histories. He had just read my manuscript. If there is, said I, I don't know it.

It irritated me that he was dissatisfied, perhaps because I am dissatisfied myself. I tried. But there is no easy way of presenting the whole truth about people's lives. The threats of self-censorship and distortion must continuously be warded off, and, if not these, then there may come charging in crying "foul" the police, the torts attorneys, the anti-heretics, and some of the cosmic heretics as well.

I've used many letters of yours, I told Deg, don't you think I should have a piece of paper from you giving me permission, but he said, no, you have them in hand rightfully and it's quite apparent that you are carrying on a public debate in the public interest on a matter of public concern. How can you do your job without reporting what people say, even if they don't like being quoted? If anything, you've been a softy; you haven't used a hundred items I've given to you of myself and others...Wait now, I said, that's just because they would be redundant... O.K., he agreed, but bear in mind how important are the freedom of science and freedom of expression — and truth, and proof of the truth: you couldn't do anything else; ideally you might have printed the whole file and let the documents just march out with fife and drums.

I don't intend to hurt anyone, I said, and he saw I was anxious. Buck up, man, dammit, you're doing a public service. And you've got the First Amendment to the Constitution of the U.S. of A. for a shield. Nowhere else is the letter of the law so close to the spirit of the law.

But weren't you badgering the *Bulletin of the Atomic Scientist* with a suit for slander? Well, he excused himself, yes but I wanted to open up their pages to a discussion, I wanted a chance to reply, and their refusal was damaging to science. It made their scientist readers believe in a phony history and misrepresentations; it was a nasty cover-up. You'd better go back and read what you've said and read the chapter in *The Burning of Troy* on the matter, too. The conduct and progress of science is public business and wrapping it in a cloak of privacy — well, I won't go on, just look at Nixon in the White House, and all that he tried to do in the guise of privacy to make off with his papers and tapes. I didn't file suit; I tried to bulldoze them but they were too smart; it didn't work, nor did an appeal to fair play. Now thanks to you we've had a marriage between Miss Liberty of Expression and the scientists — granted it's a shotgun wedding.

You've gotten me way off the subject, I said. I called to tell you the book is ended. *"La commedia è finita."* All that it needs is a final word from you. Please try to make it positive. I like happy endings.

There was a long pause; then his voice came back on the line, carefully stringing out the words:

> "If quantavolution is untrue, it will stand like a monument to edify all who pass on the road of science... Every one who seeks a new truth in science must become a party to concerns of civil liberty... Science is half psycho-sociology... Of all movements, scientific movements are the most rewarding to their adherents, win or lose, and of all these the most adventurous is cosmic heresy... He who knows how to tell time will decide the fate of the heretics."

"O.K." said I, "that's enough."

"Is it?" he asked. "You have not remarked in your book that Velikovsky wrote his works on catastrophe and quantavolution in the years 1940 to 1960, aged forty-five to sixty-five, which was precisely my experience between 1963 and 1983 when I was of the same age, a curious coincidence—or a signal perhaps that my time is up."

But then, perhaps seized by some worry such as the age of the Greenland ice cores, he exclaimed,

> *"Where are they, Sovereign Virgin,*
> *But where are the snows of yester-year?"*

To which I felt the urge to add

> *" Yes where is the Queen*
> *Who ordered the scholar Buridan*
> *Cast into the Seine in a sack?*
> *But where are the snows of yester-year?"*

INDEX

A

Abell, George O.: 296
Abery, Jill: 95
Adler, Alfred: 201
Aegean Sea: 13, 84 *et passim*
Ager, Derek: 319, 392
Ages in Chaos: 22-23, 67-9, 77, 94, 114, 202, 204, 207, 225, 269
Akhnaton (and see Oedipus): 21, 28, 68, 119, 122, 138, 204, 206, 258
Alvarez, Waller: 211
Amelan, Ralph: 81
American Behavioral Scientist: 18, 26-7, 31, 33, 36-9, 44-5, 66, 215, 217, 265, 272-9, 286, 322, 325, 331, 345, 372
American Political Science Review: 45
Ami: see de Grazia, Anne-Marie
Anderson, John Lynde: 205
Anderson, William: 43
Andriessen, Poul: 62
anti-semitism: 100ff
Arons, D.: 38
Ash: 243
Asimov, Isaac: 283, 285
astronomy: see astrophysics
astrophysics: 178-199 *et passim*
atmospherics: 168-171
authority: 64ff, 356

B

Babb, James: 46
Bailey, V.-A.: 77, 188
Baker, Howard: 350-1
Barber, Bernard: 43, 262, 372
Bargmann, Valentin: 125, 294
Barnes, Thomas G.: 209
Baroody, William: 216, 286, 322
Barzun, Jacques: 44
Bass, Robert: 77, 135, 195, 198, 199
Bauer, Henry: 263-7, 330

Beaumont, Comyns: 95, 150-152, 172, 193, 259, 349, 352-3, 362
Beigbeder, Jean-Yves: 145
Bell, John: 58
Bell, Wendell: 43
Bellamy, Hans: 353, 362 *et passim*
Bender, Rick: 345
Benton, M.J.: 374
Berosus: 203
Bigelow: 52, 77
Billings, Ch.: 145
Bimson, John J.: 81, 85-6, 94, 206
biography: 13-15
Biran, A.: 229
Blake, William: 95
Bockelmann, Peter: 35
Bolsena: 171-172
Bombay: 339
book reviewers: 257-8
Boulanger, Nicolas-Antoine: 349-50, 352-3, 362, 396
Brett, George: 26, 48
Brown, Harrison: 45-6, 224, 324, 378
Bruce, Charles E.R.: 181, 187, 196-7, 353
Bruno, Giordano: 71, 73
Bucaloe: 242
Bullard, Reuben G.: 165
Bulletin of the Atomic Scientist: 249, 268-80, 323, 399
Burgstahler, A.W.: 266, 276, 316-17, 372
Buridan, Jean: 72, 400
Burnard, Rosemary: 96
The Burning of Troy: 116, 192, 214, 318, 341, 384, 400
Buttertield: 226

C

Cadmus: 21
calcinology: 154-168
Calvin, J.: 73
Camvissis, Savvas: 145

Cantril, Hadley: 31, 43
carbondating: 205
Cardona, Dwardu: 77, 86, 191, 301-302, 319, 385-9
Carey, W.: 176
Caskey, James: 165-7
catastrophes: 132, 148ff, 206ff, *et passim*
Catastrophism and Ancient History: 87
Chandler, Craig C.: 155
Chandrasekhar: 224
Chaos and Creation: 67, 85, 95, 115, 175-6, 192, 211, 295, 301, 303-4, 338, 341-2, 365, 384
Charmatz: 159
Chase, Stuart: 43
Chassapis, Spiridon: 178-9
Chesley-Baity, Elizabeth: 168, 375-6
chimpanzee, learning: 128-9, pregnancy: 129, mutation: 130-1
Clapham, Philip: 172
Cleary, Kevin: 145
Clube, Victor: 77, 95, 377, 387
Cohen, Bernard: 69
Colliers Magazine: 23
continental drift: 175-6
Converse, Philip: 45-6
Cook, Melvin: 77, 139, 149-50, 176, 181, 187 195-6, 198-9, 205, 209, 330, 392-3
Corliss, William: 61, 149, 330, 336, 342
Cornuelle, Richard: 145, 161, 176, 322, 346
cosmic heretics, numbers: 213-17, *et passim*
Couch, William T.: 44
Crew, Eric: 77, 85, 149, 194-5, 197, 336

D

Dachille (also d'Achille), Frank: 150, 156, 192-4, 214, 316-17, 362, 377-8
Darwin, Charles: 124-5, 327-9, 333, 371
Davies, James C.: 43
de Camp, Sprague: 295, 331
Deg: *passim*
de Grazia, Alfred: *passim*
de Grazia, Anne-Marie: 62-3, 84, 94, 1453, 213, 298, 340, 342, 350, 365
de Grazia, Edward: 49, 51, 121, 191, 225, 365

de Grazia, Jill: 38-9, 49, 52, 53-5, 143, 226, 241 *et passim*
de Grazia, Joe: 191
de Grazia, Lucia: 251
de Grazia, Nina: 120, 144, 159, 218, 227, 253-4, 316, 335, 337
de Grazia, Sebastian: 17, 19, 141, 144, 250-1, 302, 320, 365
de Grazia, Victoria; 329
debate, scientific quality of: 257ff, 378-80, 388, 391-2
Dell Publishers: 24, 58
desalinization: 52
Dietz, David: 227
discourse, scientific, among heretics: 380ff *et passim*
The Disastrous Love Affair of Moon and Mars: 21, 67, 86, 337-8, 341
Discover: 373
The Divine Succession: 139-40, 341
Dix, Bill: 241
Donnelly, Ignatius: 193, 258, 349, 353, 362
Dotan: 229-30
Doubleday Publishers: 23, 26, 113, 248-9, 330
Douglas, Paul: 292
Drake, Saint Clair: 111
Drioton, Etienne: 203
Dunbar: 208
Dyson, Freeman: 33, 77

E

Earth in Upheaval: 148, 202, 208, 214, 225, 249, 265, 3153, 395
Eddie: (see Schorr)
education: 314-20
Eight Bads, Eight Goods: 335
Einstein, Albert: 65, 69, 71, 76, 77, 112, 188, 196, 266, 271, 368, 371, 388, 398
Eiseley, Loren: 130
El-Arish: 80, 101, 204, 226, 228-30, 232-3
Elisheva: see Velikovsky, Elisheva
electricity: 117, 187ff
Eliade, Mircea: 163, 186
Ellenberger, Leroy: 26, 34, 43, 300-1, 303, 305-6,

319, 384
Encyclopedia of Quantavolution: 88, 90
entropy: 139
epithets of heretics: 267
Ernst: see Wreschner, ernst
Esalen: 121
Eshleman, R.: 222
establishment, 280-1, *el passim*
Euphemeris the Sicilian: 181
Evans, Luther: 31
evolution: 124ff
Exxon: 153-4

F

Fairservis, Walter: 133
Farinholt, L.H.: 324
Farkas, Robin: 153
Feldman, Bronson: 121
Firor, John William: 168-71
Fleming, Donald: 325
fluoridation: 276
Foundation for the Study of Modern Science (FOSMOS): 228-9, 234-41, 320
foundations: 320-9
Fournier, Levi: 145
Francesca, Francie, Franny: 51-2, 54, 228, 241
Franklin Institute: 58, 215
Franklin, Steven: 300-1
Fraser, Michael and Chloe: 80, 85
Freeman, Lawrence Z.: 131
Frelinghuysen, R.: 145
Freud, Sigmund; 14, 21, 64-5, 67-8, 108-9, 113, 118, 122, 127, 137, 186, 237, 333, 371, 375, 388
Friedlander, Michael: 293
Frutkin, A.: 38
Fuhr, Ilse: 204

G

Gammon, Geoffrey: 77, 81, 85, 206-7, 308
Gaposhkin, C. Payne: 324
Geb: 115-16
Gentry, R.V.: 209
Gerhardt: see Roesler

Ginsberg, Allen: 49-50
God's Fire; 138, 341 (and see: Moses)
gods, proof of: 140; succession of: 189-90
Goldman, Ralph M.: 43
Goldschmidt, R: 214, 281-84
Gordon, Cyrus: 82, 86, 266
Gould, Stephen Jay: 319, 331, 373, 387-8
Grace, Joseph: 258-9, 305
Greenberg, Lewis: 57, 77, 81, 83-4, 90, 101-2, 143, 165, 250, 296-7, 299-302, 305-6, 308-9, 319, 349, 368, 384
Gurr, Ted: 36-7, 45

H

Hadas, Moses: 32, 44, 77, 82, 218, 266, 275, 323, 325
Hadrian: 119
Haldane, J.B.S.: 328
Hall, Jay: 145
Hammond, Philip: 226, 228, 229, 230, 232
Hanna, Isnander: 222
Harper's Magazine: 27, 215, 225
Harris, Matthew: 113
Hatch, Ronald: 348
Hebrew U. of Jerusalem: 22
Hecksher, August: 44
Heinberg, Richard: 90, 137
Heinsohn, Gunnar: 216
Helmholtz, Hermann L. F. von: 262
heretic, definition: 284-5, 291; numbers: 319; organization: 213ff: weaknesses: 292ff; *et passim*
heroes: 71ff
Hess, Henry H.: 32-3, 77, 149, 174-5, 227, 229, 264, 266, 268, 272, 393
Hester, James: 286
Hewsen, Robert H.: 305-7, 319
Heycock, Rayburn: 80, 85
Higgins, Godfrey: 349
Hillenkoetter, R. H.: 32-3
Hitching, Francis: 332
Hoerbiger, Hans: 353
Holbrook, John Jr.: 229-30, 232, 239, 241, 244, 249

Holmes, O.W.: 40-1
holocaust: 100ff
Homo Schizo, *Homo Schizo I & II:* 67, 94, 111, 118, 124, 126-7, 140, 163, 210, 252, 310, 357
homosexuality: 119-21
Hoyle, Fred: 319
Huber, Peter: 282-3
Hueber, Anne-Marie: see de Grazia, Anne-Marie

I J

instinct: 126-7
introgenesis: 355
Isaacson, I.: 206 (and see Schorr, Eddie)
Isenberg, Arthur: 70
Jacob, Phil: 145
James, Peter: 77, 81, 84, 86, 89-90, 94, 297, 306-8, 336, 338
Jan: see Sammer, Jan
Jastrow, Robert: 37, 82, 218, 309, 392
Jerusalem: 28
Jews, Jewishness: 24, 112-14, 138, 143 *et passim*
Jill: see de Grazia, Jill
Johnson, Earl S.: 299
Joshua: 68
Joyce, James: 117
Juenemann, F.: 301, 319
Jung, Carl: 121-2, 186, 226, 237
Juergens, Ralph: 27, 33, 52-5, 65, 77, 149, 181, 187-9, 192-7, 206, 214-15, 225, 228, 233-6, 238, 240, 247, 264, 266, 316-17, 323, 336, 341, 353, 359, 380

K

Kallen, Horace: 32, 44, 77, 229
Kalos: 241, 298, 334, 339
Kaplan, Jeremiah: 89
Kaufman: 77
Keller, George: 158
Kelly, Allan: 150, 156, 193, 214, 393
Kennan, G. & A.L.: 251

Kennedy, John F. & Jacqueline: 46, 231-2
King, Ivan: 280, 285
Kloosterman, Hans: 173, 349, 369
Kluger, Richard: 226-7
Knight, Frank: 292
Kogan: 52
Komarek, Edwin U.: 156-8
Kramer, Richard: 225-8
Kronos: 56-7, 60, 81-2, 84, 87, 90, 105, 133, 173, 208, 216, 250, 280, 283, 297-308, 336, 368, 384, 387
Kruskal, Martin: 189, 264-5
Kugler, F.X.: 222-4
Kuhn, Marion: 249
Kuhn, Thomas: 33, 38, 39, 325
Kurtz, Paul: 218, 295-6

L

Lackenbauer, Ilse: 145
Larrabee, E.: 27, 214-15, 268-9
Lasswell, Harold D.: 31, 33, 126,134, 252-5, 266, 320, 325, 335, 375
The Lately Tortured Earth: 154, 176, 341-3, 393
Latham, Harold: 32
law: 29-30, 33-5, 269-270, 273-4
Lazar, Theoodor: 122
Lear, John: 54-5
Leary, Timothy: 48-50
Lee, Charles: 256
lenses, ancient: 179-180
Leonardi, Pietro: 171-2
Lethbridge University (Conference): 75
Lewonton, R.C.: 125
Leys, A.R.: 43
Libby: 204-5
Livio: see Stecchini, Livio Catullus
Lowery, Malcolm: 77, 81-2, 85-6, 94, 116, 195, 307, 336, 384
Luckerman, Marvin: 87, 319
lunar fission:176, 350-1
Lundberg, George: 43-5
Lyell: 130

M

Maccoby, Hyam: 81, 84, 106
Madariaga, Salvador de: 32
MacCrea, W.H.: 195
Macmillan Company: 22, 26, 32, 34, 43, 48, 89, 214, 331
McCarthy, Joseph: 37, 43
McClintock, Barbara: 387
McClintock, Stuart: 54-5
McGraw-Hill Company: 27, 215, 331
McKie, Euan: 81-6
McKinnon, Roy: 209
McLaughlin: 48
McMaster University (Conference): 380-4
McNulty, Ted: 34
Mage, Shane: 282-3
Mailer, Norman: 225
Mainwaring, Bruce: 165, 226, 229, 239, 249, 320
Mandelkehr, M.: 172
Manetho: 203
Margolis, Howard: 221, 265-80, 323-4
Marina di Massa: 36
Marinatos, Spiridon: 164, 178
Marinos, G.: 161
Martinson, Carl: 145
Marx, Christoph: 99-100, 102-6, 109, 114, 116, 216, 249-50
Marx, Karl: 121, 138, 328-9, 388
Matelli, Dante: 220
Mayur, Rashmi: 145
Menzel, D.H.: 220, 271
Merriam, Charles: 124
Merriam, Robert: 292
Michelson: 77, 282
Miller, Alice: 216, 249
Miller, Arthur S.: 43
Milton, Earl R.: 73, 77, 83-4, 96, 102, 136, 138, 149, 176, 181, 187, 188, 190, 1920, 195-7, 211, 247-8, 299, 301, 318-19, 330, 336, 340-1, 355, 359, 366, 380, 385-7, 389
misanthropism: 47
Mitchell, G.W.: 37
Molcho, Schlmo: 71-2
monotheism: 137-8

Moore, Brian: 44-5, 79-81, 85-6, 90, 94, 305-7 310, 336, 358
Morris, Henry: 72, 348
Morrison, David: 283, 380
Moses: 21, 28, 67-9, 74, 84, 106, 137-8, 143, 249, 341,354, 356
Mulholland, J.D.: 282-3
Mullen, William (Bill): 57, 117, 191, 205-6, 242, 247, 316-17, 336, 338, 368

N

Napier, William: 77, 95, 387
narcissism: 63, 74, 118
NASA: 38
Nature: 374
Naxos: 13, 84, 144, 161-2, 190, 337 *et passim*
network, social theory of: 287ff
Neugebauer, Otto: 33, 51, 56, 218, 222-4, 378
Neuman, Stephanie: 24, 145
New Scientist: 249
New York Herald Tribune: 23
New York Times: 37, 46, 51, 70, 243, 249, 290, 332, 339
New York University: 48-50, 58, 313-16
Newell, Norman: 34
Newgrosh, Bernard: 81, 95
Newsweek: 58
Nikmed of Ugarit: 21
Nilsson, Heribert: 208
Nina: see Nina de Grazia
Nut: 115-16

O

Occam, William of: 71
Oedipus: 118-19, 258
*Oedipus and Akhnaton:*19-21, 31, 63, 67-8, 119, 137, 249
Olson, Ken: 134 *et passim*
Onassis, A.: 231-2
O'Neill, John J.: 22
Orbell, George: 45
Origin of Species: 125

P

participation, obstacles to: 263
Passage of the Year: 72, 334
Patten, Donald: 135, 330, 347-8
Pensee: 73, 81, 174, 187, 189, 197, 199, 205, 242, 248, 280, 282-3, 336, 338
Peoples of the Sea: 62, 69, 86, 231, 233, 243, 247, 249
Pes(h)tigo Fire: 156
petroleum exploration: 152-4
Picasso, Pablo: 20, 227
Pfeiffer, Robert H.: 77, 249
Pickering: 393
Pirez, Diego: 71-2
Pirogine, Ilya: 374
Planck, Max: 262, 272, 371
planetary motions: 178ff
Plank, Skip: 345
Polanyi: 218
Politics for Better or Worse: 335
popular science writers: 79
Portillo, Jose Lopez: 252
precursors: 151-2, 347ff
Prescott, Bernard: 81
Price, Don: 39, 198
Price, George McCr.: 258
Prince, Morton: 357
priority in scientific discovery: 125, 293-4
prisoners, V. letter to: 137
psychology: see *Homo Schizo,* anti-semitism, amnesia, jewishness. personal names, *et passim*
publishing: 21-2, 37, 57, 264, 277-9, 282-3, 290, 330-46; heretical: 81ff, 87ff, 102ff, 298ff, 398
puerperal fever: 40-1
Putnam, James: 22

Q

Quanta: 87, 90
quantavolution, defined: 14, 373-4; growth of writing on: 378-9: theses: 369-370; Quantavolution Series, writing and production: 298ff, 339-46

R

Rabinowitch, Eugene: 221, 270-1, 273-5
Ralph, Elizabeth: 222
Ram, Immanuel: 335
Ramses II and His Times: 86, 243, 250, 308
Ransom, C.J : 77, 135, 187, 247, 319, 330
Rapp, George: 166
Reade, Michael: 77, 81, 307
Reader's Digest: 23, 331
reception system of science, chs I, II, pp372 *et passim*
religion: 134ff
Reich, Wilhelm: 116, 237
Ressa, Ch.: 145
Rhine, J.B.: 372, 373
Rilli, Nicola: 155, 373
Rix, Zvi: 60, 82-3, 86, 99, 108, 118
Rix, Melitta; 99-100
Robins, Don: 305
Rockefeller, Nelson R.: 59, 286
Rockefeller, Rodman: 59, 314
Roesler, Gerhardt: 162-3
Rose, Lynn: 73, 77, 102-3, 105, 111, 138, 197, 247, 249, 259, 283, 307-8, 316, 319, 255
Roy, A.E.: 195, 198
Russell, Bertrand: 13, 66, 363

S

Sacco & Vanzetti: 357-8
Sagan, Carl: 64-5, 77, 82, 102, 218, 247, 280, 282-6, 294-5, 307, 309-11, 353, 392
Safire, William: 249
Sammer, Jan: 57, 90, 101, 104, 137, 249, 319, 378
San Francisco, AAAS Panel: 64-5, 249, 319, 378
Santillana, Giorgio de: 77, 179, 186, 320-21
Saturday Review: 54
Saunders, Michael: 311
Schaeffer, Claude: 77, 155, 160, 172, 206, 213-15, 362
Schenkman, Albert: 45
Schiaparelli: 262
Schildkraut, Karl: 130
Schindewolf, Otto: 173, 208, 214, 374

Schliemann: 155, 168
Schmidt, Arno: 117
Schmidt, Barbara: 145
Schorr, Eddie: 21, 164, 166, 241-2, 247, 316, 353, 368
Science: 39, 171, 222, 248, 290, 336
Science and Mechanics: 137
scientific fictions: 173-4
Scientists Confront Velikovsky: 103, 281-3, 285
Scott, Andrew: 168
Scott-Meredith: 102-3
Sebastian: see de Grazia, Sebastian
Stratigraphie Comparee: 213
Serveto, Miguel: 71-2
sexuality: 115ff
Shanklin, Douglas: 40-1
Shapley, Harlow: 32, 34, 37, 42, 46, 48, 65, 243, 271, 274, 288, 307, 311, 328, 393
Shazar, Zaluccan: 221, 231
Sheldon, Eleanor: 326-7
Shelley-Pearce, Derek: 206
Sherwood, Jerry: 167, 338
Sheva: see Velikovsky, Elisheva
sibling rivalry: 250-1
Sieff, Martin: 67-8, 77, 81, 85-6, 89-91, 93-4, 135-6, 206-8
Simon, Herbert: 43, 269-70
Simon & Schuster: 227-8, 231, 338
Simpson, G.G.: 214
SIS, SISR: 44, 80, 86-7, 173, 175, 195, 199, 209, 280, 297, 305-6, 336, 384
Sizemore, Warner: 26, 48, 56-7, 60, 80-4, 102-3, 105-6, 114, 143, 214-15, 229, 241, 250, 197-203, 306, 319, 336, 340, 374-5
Sloan Foundation: 275-7, 322-3
Sloane, William: 44
Smart, W.M.: 195
Smith, Charley: 89
Smith, Peter: 175
Snow, C.P.: 33
Society for Interdisciplinary Studies: see SIS
sociology of science: 26, 380ff *et passim*
Solaria Binaria: 83-5, 184-8, 190-2, 196
Spangler, George: 205, 209

Sperling, Jerry: 165, 167
Stargazers and Gravediggers: 23, 69, 136, 214, 243
Stecchini, Livio Catullus: 17-19, 25-7, 31, 39 45, 48, 57, 82, 111, 113, 117, 122, 148, 179, 184, 186, 192, 214-15, 223, 225, 228, 240, 244, 262, 264, 283, 316-17, 321, 350, 359
Stein, Gertrude: 20
Steinhauer, Loren: 348
Stephanos, Robert (Bob): 123, 151, 172, 193, 215, 322, 226, 229, 240-1, 368
Storer, Norman: 37, 282, 301
Stover, Carl: 340, 145
Sullivan, Walter: 51, 219, 318-9, 350-1
suppression of a book, techniques: 259-62 *et passim*
Switzerland (Swiss Alps): 13, 59, 64, 74, 120-1, 144, 335
Sykes, N.J.G.: 209

T

Tagiuri, Renato: 43
Tagliacozzo, Giorgio: 185-6
Talbott, David and Stephen: 57, 77, 81, 189, 247-8, 319, 330, 387
Temple, Robert: 85
Thom, Rene: 331
Thomsen, Dieterick: 70
1001 Questions on Culture Policy: 341
time, cosmic: 198-9, 201ff *et passim,* personal: 366-7, *et passim*
Tobia, Annette & Peter: 130-5, 316-17
Tompkins, Peter: 38-9, 122
Trainor, Lynn: 77, 318
Tresman, Harold: 81, 85, 249
Tumin, Mel: 69

U

Umschau Verlag: 104
UNESCO: 216, 286
uniformitarianism: 394-5
Urey, Harold C.: 277-8, 378

V

V.: *passim*
Vail, Isaac: 189
Vanderpool, Catherine de Grazia: 165
Vanderpool, Eugene Sr.; 167
Vaughan, Raymond: 197, 319
The Velikovsky Affair: 33, 41, 84-5, 104, 106, 324, 330, 338
Velikovsky, Alexander: 251
Velikovsky and His Critics: 80, 231, 282
Velikovsky, Elisheva: 22-4, 53-5, 75, 84, 90, 101, 104-5, 107, 113-14, 138, 165, 213, 213-22, 235-6, 239, 242, 244, 246, 249, 303, 305-6, 340, 360
Velikovsky, Immanuel: *passim*
Velikovsky, Ruth: 106, 235, 246
Velikovsky, Shulamith: 306
Velikovsky, Simon: 22, 29
Venus: 28, 53, 65, 116, 180, 183-5, 191, 223, 293, 350
Vico, Giambattista: 185-6, 362
Vietnam: 13, 58-9, 144, 286, 334
Villon, Francois: 72, 400-1
Vitaliano, Dorothy: 163-4, 166, 181, 330
von Daniken, E.: 30, 254, 332
von Dechend, H.: 179, 186, 320-1

W

Wallace, Alfred: 125
Warlow, Peter: 85, 299, 306-7
Warner: see Sizemore
Warshawsky, Fred: 331

Washington,: 340-1
Waxman, Spelman: 52-3, 55
Washington Star: 280
Weaver, Warren: 323, 325
Wegener, Alfred: 175, 393
Welensky, Donna: 340
Westcott, Roger: 319
Weyerhauser, Susan: 145
Whelton, Clark: 71, 84, 93-4, 318
White, Lynn Jr.: 45
Whipple, Fred: 26, 38
Wiener, Norbert: 149
Wilkins, Harold T.: 349
Willes, Margaret: 85
Wittenberg, Philip: 32, 269
Wolfe, Irving: 73, 138, 191, 247-8, 301, 319, 355, 385-6, 389
Woolf, Leonard & Virginia: 13, 362-3, 366
Workshop: 95
Worlds in Collision: 22-3, 26-8, 32, 342, 66-7, 69, 114, 138, 180, 193, 206, 214, 272, 304, 315, 317, 355, 360, 378, 392, 394
Wreschner, Ernst: 131-2, 172

XYZ

Yeats, William Butler: 14
York, Derek: 380
youth:64-5
Ziegler, Alfred: 175
Ziegler, Jerry: 117
Zysman, Milton: 216

ABOUT THE AUTHOR

Alfred de Grazia was born in Chicago on December 29th, 1919. His father was a musician and conductor. After receiving his doctorate from the University of Chicago in 1948, he taught social theory, political psychology and behavior, and social invention for some years at University of Minnesota, Brown University, Stanford University, and finally New York University, lecturing widely in establishments of higher learning at home and abroad. His works on politics and theory helped to establish scientific method in the field of political science. He worked at military psychological operations in three wars. In World War II, he was affected to OSS in the first year of its inception and fought in seven campaigns in North Africa, Italy, France and Germany. A lifelong occupation with social invention was reflected in the founding of the *American Behavioral Scientist,* the design of information systems, and in public policy studies that began with the First Hoover Commission on the reorganization of the federal government and continued through the publication of over a dozen books, and participation in numerous commissions and panels. In 1964, he published *The Velikovsky Affair.* From 1963 he began a new series of studies on ancient catastrophes and their effects upon natural and human history, which eventuated in the Quantavolution Series. He also published poetry, plays, and several autobiographical works. He lives presently in France and in Greece, with his wife, French writer Anne-Marie de Grazia. His brother Sebastian was a political scientist, laureate of the Pulitzer Prize. His brother Edward a prominent First Amendment lawyer, his brother Victor was Deputy-Governor of Illinois. His daughter Victoria is a historian and member of the American Academy.

His two million+ visitors/year website: http://www.grazian-archive.com

See also: Q-MAG.org - the magazine of Quantavolution: http://www.q-mag.org

LATEST PUBLICATIONS (all Metron Publications)

America's History Retold (Vol. One): Conquest, Colonialism and Constitutions (2012)

America's History Retold (Vol. Two): Originating American Ways of Living and Working (2012)

America's History Retold (Vol. Three): Shaping Earth's Cultures and Powers (2012)

A Taste of War - Soldiering in World War II (2011)

The American State of Canaan – the peaceful, prosperous juncture of Israel and Palestine as the 51th State of the United States of America (2009)

The Iron Age of Mars – Speculations on a Quantavolution (2009)

***

THE QUANTAVOLUTION SERIES (All Metron Publications, 1980-1984 and 2009)

Chaos and Creation: An Introduction to Quantavolution in Human and Natural History

The Lately Tortured Earth: Exoterrestrial Forces and Quantavolution in the Earth Sciences

Homo Schizo I: Human and Cultural Hologenesis

Homo Schizo II: Human Nature and Behavior

God's Fire: Moses and the Management of Exodus

Solaria Binaria: The Origins and History of the Solar System *(with Prof. Earl R. Milton)*

The Divine Succession: A Science of Gods old and New

The Burning of Troy: Essays and Notes in Quantavolution

The Disastrous Love-Affair of Moon and Mars

Cosmic Heretics

The Iron Age of Mars: Speculations on Quantavolution (2009)

OTHER WORKS OF ALFRED DE GRAZIA

Michels, Roberto, *First Lectures in Political Sociology*. Translated, with an introduction, by Alfred de Grazia. University of Minnesota Press, Minneapolis, 1949. And Harper & Row, New York, 1965.
Public and Republic: Political Representation in America. Alfred A. Knopf, New York, 1951.
The Elements of Political Science. Series: Borzoi Books in Political Science. Alfred A. Knopf, New York, 1952. And second revised edition: *Politics and Government: the Elements of Political Science*. Vol 1: The Elements of Political Science and Vol. 2: Political Organization, Collier, New York, 1962. And new revised edition: Free Press, New York; Collier Macmillan, London, 1965.
The Western Public: 1952 and beyond. A Study of Political Behaviour in the Western United States. Stanford University Press, Stanford, 1954.
The American Way of Government. National Edition. Wiley, New York (1957). There is also a "National, State and Local edition," publ. by Foundation for Voluntary Welfare.
Grass Roots Private Welfare : Winning Essays of the 1956 National Awards Competition of the Foundation for Voluntary Welfare. Alfred de Grazia, editor. New York University Press, New York, 1957.
The American Behavioral Scientist, Magazine, founded by Alfred de Grazia, Metron Publications, Princeton, N.J., 1959; acquired by SAGE Press, 1965.
American Welfare. New York University Press, New York, 1961 (with Ted Gurr).
World politics: a study in international relations. Series: College Outline Series, Barnes & Noble, New York, 1962.
Apportionment and Representative Government. Series: Books That Matter. Praeger, New York, c1963
Essay on Apportionment and Representative Government. American Enterprise Institute, Washington,1963.
Revolution in Teaching: New Theory, Technology, and Curricula. With an introduction by Jerome Bruner. Bantam Books, New York, 1964. (Editor, with David A. Sohn).
Universal Reference System. *Political Science, Government, and Public Policy: an annotated and intensively indexed compilation of significant books, pamphlets, and articles, selected and processed by the Universal Reference System*. Prepared under the direction of Alfred De Grazia, general editor, Carl E. Martinson, managing editor, and John B. Simeone, consultant. Princeton Research Pub. Co., Princeton N.J. 1965–69. *Plus* nine more volumes on the subjects of: *International Affairs; Economic Regulation; Public Policy and the Management of Science; Administrative Management; Comparative Government and Cultures; Legislative Process; Bibliography of Bibliographies in Political Science, Government and Public Policy; Current Events and Problems of Modern Society; Public Opinion, Mass Behavior and Political Psychology; Law, Jurisprudence and Judicial Process.*
Republic in crisis: Congress against the Executive Force. Federal Legal Publications, New York, 1965.
Political Behavior. Series: Elements of political science; New, revised

edition. Free press paperback, New York, 1966.
Congress, The First Branch of Government, editor, Doubleday – Anchor Books, New York, 1967.
Congress and the Presidency: Their Roles in Modern Times, with Arthur M. Schlesinger, American Enterprise Institute for Public Policy Research, Washington, 1967.
Passage of the Year, Poetry, Quiddity Press, Metron Publications, Princeton, N.J., 1967.
The Behavioral Sciences: Essays in honor of George A. Lundberg, editor, Behavioral Research Council, Great Barrington, Mass., 1968.
Old Government, New People: Readings for the New politics, et al., Scott, Foresman, Glenview, Ill., 1971.
Politics for Better or Worse, Scott, Foresman, Glenview, Ill., 1973.
Eight Branches of Government: American Government Today, w. Eric Weise, Collegiate Pub., 1975.
Eight Bads – Eight Goods: The American Contradictions, Doubleday – Anchor Books, New York, 1975.
Supporting Art and Culture: 1001 Questions on Policy, Lieber-Atherton, New York, 1979.
A Cloud Over Bhopal: Causes, Consequences, and Constructive Solutions, Kalos Foundation for the India-America Committee for the Bhopal Victims, Popular Prakashan, Bombay, 1985.
The Babe, Child of Boom and Bust in Old Chicago, umbilicus mundi, Quiddity Press, Metron Publications, Princeton, N.J., 1992.
The Student: at Chicago in Hutchin's Hey-day, Quiddity Press, Metron Publications, Princeton N.J., 1991.
The Taste of War: Soldiering in World War II, Quiddity Press, Metron Publications, Princeton, N.J., 1992 – 2012 (Revised)
Twentieth Century Fire-Sale, Poetry, Quiddity Press, Metron Publications, Princeton, N.J., 1996.

www.ingramcontent.com/pod-product-compliance
Lightning Source LLC
Chambersburg PA
CBHW031611160426
43196CB00006B/94